THE POLITICS OF TERRITORIAL IDENTITY

THE POLITICS OF TERRITORIAL IDENTITY

STUDIES IN EUROPEAN REGIONALISM

EDITED BY
STEIN ROKKAN & DEREK W URWIN

Sponsored by the European Consortium for Political Research/ECPR

SAGE Publications
London · Beverly Hills · New Delhi

To Stein
who inspired it

Copyright © 1982
SAGE Publications, Ltd.

For information address
SAGE Publications Ltd
28 Banner Street, London EC1Y 8QE

SAGE Publications Inc
275 South Beverly Drive
Beverly Hills, California 90212

SAGE Publications India Pvt. Ltd.
C-236 Defence Colony
New Delhi 110 024, India

British Library Cataloguing in Publication Data

The Politics of territorial identity: studies in
    European regionalism.
    1. Ethnic groups — Political activity
    I. Rokkan, Stein   II. Urwin, Derek W.
    306'.2 (expanded)   JF1061

Library of Congress Catalog Card Number   82-80204

    ISBN 0-8039-9788-4

Contents

Acknowledgements

It may be true that 'no man is an island'. But by the same token it is also true that it is almost impossible for an individual scholar to become deeply expert about the politics and social structure of more than a few countries. The table of contents of this volume indicates the extent to which it is the result of cross-national collaboration between scholars from several academic disciplines who have been willing and able to combine their intimate knowledge of specific countries and situations with a comparative framework that would permit the furtherance of cross-national analysis. It is with pleasure that I acknowledge the work of those who have contributed chapters to this volume.

Such extensive collaboration was due not only to the commitment of the various authors. We are deeply grateful to the Volkswagen Stiftung and the Norwegian Research Council for Science and the Humanities for the generous funding which made the whole exercise possible, not only in permitting the participants to meet frequently, but also through providing resources which enabled us to accumulate an extensive body of regional data sets for Western European countries. These files have been recoded to conform to a common format which was adopted in 1980 by the International Federation of Data Organizations as a common standard for all European data sets held by its members. All the data sets held by the Institute of Comparative Politics, University of Bergen, and the related computer-mapping facilities developed during the project are, of course, freely available to all who wish to have them. It is hoped that this body of data will serve future collaborative research which could advance our knowledge of the structure and dynamics of territorial politics.

The European Consortium for Political Research (ECPR), the body sponsoring this book, provided the necessary organizational infrastructure. Recruitment of potential participants was carried out under the auspices of the ECPR, which also, through its on-going programme of workshops, enabled the group not just to meet regularly but also to invite other scholars to participate in its discussions. The group first met during the ECPR Workshops in Berlin in

1977, and appropriately held its last plenary session during the Workshops in Brussels in 1979. In between, the group met twice in 1978, first as part of an ECPR Workshop on Nationalism held at Loch Lomond, Scotland, and later in Bergen.

Several other scholars participated in these meetings, and we are grateful for the advice and criticism that they offered. We would like in particular to thank the following: Erik Allardt, University of Helsinki; Denis Balsom, University College of Wales, Aberystwyth; Jack Brand, University of Strathclyde; Jim Bulpitt, University of Warwick; Jan Ulrich Clauss, European University Institute, Florence; Wanda Dressler-Holohan, University of Grenoble; Jorgen Elklit, University of Aarhus; Tom Garvin, University College, Dublin; Christian Gras, University of Strasbourg; Olafur Grimsson, University of Iceland; Hervé Guillorel, University of Nanterre; Christian Haerpfer, University of Vienna; David Handley, University of Geneva; Martin Heisler, University of Maryland; James Kellas, University of Glasgow; Adrian Lee, Plymouth Polytechnic; Jeanine Lothe, University of Namur; Tom Mackie, University of Strathclyde; Ian McAllister, Australian National University; Peter Madgwick, University College of Wales, Aberystwyth; Karl-Johan Miemois, University of Helsinki; Kerstin Nystrom, University of Lund; Thore Olaussen, University of Bergen; Edward Page, University of Strathclyde; Ricardo Petrella, Commission of the EEC; Colin Rallings, Plymouth Polytechnic; Henry Valen, University of Oslo; Jon Wagtskjold, Norwegian Social Science Data Services, Bergen; Maria Wieken-Mayser, European University Institute, Florence. Many of these contributed detailed and suggestive papers in the project, and we can only regret that exigencies of space have prevented us from including more in this volume.

Several graduate students and research assistants at the University of Bergen have been involved in the project, carrying out much of the necessary tasks of data compilation, cleaning and analysis: Sigvor Bakke, Pål Bakka, Jan Eek, Helge Holbaek-Hanssen, Bjarne Kristiansen, Jofrid Lied, Jon Erik Lindberg, Jens Lorentzen, Johnny Pedersen, Leonhard Vårdal. In particular, we would like to acknowledge the contribution of Pamela Adams, who served as senior and coordinating research assistant for much of the lifetime of the project. Finally, we would like to thank Jorunn Nordanger, Randi Taule and Astrid Blom for their valuable and necessary secretarial assistance.

Last, but certainly not least, I would like, on behalf of my co-authors, to acknowledge the efforts of Stein Rokkan. It was Stein who inspired and initiated the whole project and who, until his untimely death in 1979, firmly steered it in very specific directions. As far as is possible, I have attempted to remain faithful to his conceptions, and I hope that the end result is not too far removed from them: it is in any case gratifying to see how many of the contributions to this volume bear the hallmarks of his ideas. It may perhaps be thought unusual for a volume to be dedicated to one of its editors: but then, Stein Rokkan was an unusual man, and with his death political science has lost its most tireless advocate and practitioner of the value of collaborative cross-national research. It is with both pleasure and sorrow that we dedicate this book to his memory.

Derek W. Urwin
University of Warwick, Coventry

1

Introduction: Centres and Peripheries in Western Europe

Stein Rokkan
University of Bergen
Derek W. Urwin
University of Warwick

For much of the twentieth century the concept of the nation-state has been widely accepted as a norm for territorial organization, and Western Europe as its home. As late as 1960 the likely consensus of opinion upon the violence and irredentism of the German-speaking population of the Alto Adige would probably have been that it was an atavistic intrusion into the usually placid waters of Western politics — the exception, as it were, that proved the rule. Closer to the accepted norm was the situation and behaviour of the Swedish-speaking minority in Finland. The Swedish Finns may have had their own political party, but they accepted not just the rules of the game, but also the legitimacy of the state in which they found themselves in its existing territorial form: their self-image, in other words, was of being Finnish, not Swedish. Organizations and activists that claimed for a regionally based group an identity distinct from that of the 'national' population, and aspired to some form of territorial change and political autonomy, were for long more likely to be condemned as misguided criminal terrorists or ridiculed as quaint and irritating anachronisms.

However, events of the past two decades have turned 'regionalism' and 'ethnonationalism' into something of a political and academic vogue in studies of Western industrial societies. Not without reason, for in these countries there flowered in the 1960s and 1970s a conspicuous 'regional-cum-ethnic' protest against regimes and political centres. The political alliances engendered by the Norwegian EEC referendum of 1972, which brought back to the surface a more historical pattern of geographical politics, is equally as striking an example as those movements that have more explicitly and visibly voiced ethno-national sentiments in, for example, Belgium, Spain or the United Kingdom. Even the two countries that conventionally have been taken to epitomize the extremes of state organization in Western Europe have been disturbed. Federal Switzerland had to live with two decades of heightened agitation in the Jura for disaggregation of the canton of Berne. Nor has France, the 'model' unitary state, remained immune: since the late 1950s there seems to have been an escalation of regionalist political activities, even in areas that do not possess a population or minorities that can lay claim to some kind of ethnic or historical distinctiveness.

Such trends and events have forced political science to look again at some of its constructs, for they cast some doubts upon the validity of prevailing theories of social mobilization, economic development, political integration and the desirability and successful impact of redistributive welfare policies. In many ways the doubts were a result of the fact that political science had displayed little sympathy for groups that were regarded as having lost in the historical game; moreover, since ethnic politics were, in a sense, a recent 'discovery', it was difficult to treat the groups as social givens. This is not the place to elaborate upon the 'traditional' concepts and the dissenting and revisionist additions that have subsequently enriched them. Let it suffice to say, first, that, as the flood of recent publications indicates, the theories and concepts behind the earlier dismissal of regional, peripheral and/or ethnic politics and those seeking to explain the more recent prominence of nationalist parties and groups have been the subject of much discussion. Second, the focus has been primarily upon the relationship between ethnic loyalties and political behaviour, and at least three broad approaches may be identified in the research output of the past decade: an intra-state version of the theory of international economic dependency and marginality (internal colonialism), where the survival of ethnic distinctiveness rests upon a cultural division of labour; a stress upon the maintenance of ethnic differences through the presence of marked boundaries, which are more important than what they enclose, with territorial demarcation as significant as behavioural differences; and a broader approach, which stresses the dichotomy between centre and periphery seeking to place ethnic variations in a general framework of geopolitical location, economic strength and access to loci of decision-making. At any rate, the academic landscape has changed considerably, perhaps more so than the actual political landscape.

That most, if not all, states in Western Europe are multi-ethnic, with several layers of identity, is perhaps now part of the conventional wisdom. All, therefore, should contain the potential for the political expression of regional/ethnic protest. How, when and to what extent there is a metamorphosis from social distinctiveness to political expression are questions of empirical research as well as theoretical speculation.

If we were to seek an explanation in recent developments, we might look no further than three parallel processes of change that have affected Western Europe since 1945. First, there has been a continuing internationalization of territorial economies and a persistent erosion of inter-state boundaries through a growing diffusion of messages, ideologies and styles of organization. Second, demands on and expectations of the resources and manpower of the machineries of each state have increased: the pressures to expand welfare and educational services and to intensify aid to the less productive sectors and regions of the economy have increased the cost of infrastructural requirements. Third, there has been the multiplication of efforts to mobilize peripheries, regions and even localities against the national centres, and an assertion (or reassertion) of minority claims for cultural autonomy and for separate powers of territorial decision-making. The simplest model linking these three processes of change might be:

increased internationalization of transactions → decreased resources for central control → increased regional protest.

But this model would not help us in an effort to account for differences across Western Europe. True, it might be of some value in a study of trends over time, country by country; but it would not contribute much towards an understanding of the sources of variation in the strains between centres and peripheries and across regions in Europe.

Instead, what is required is a broader historical perspective. For example, much of the literature of the 1970s implies a mere revival of the kind of ethnic concerns that were manifest in 'pre-modern' periods, and so, for instance, a less than perfect, or even temporary, incorporation of such interests in European party systems. And it is true that many of the current regionally based movements and parties can trace a pedigree back to the opening phases of mass mobilization. Alternatively, in regions that today are the homes of such organizations, there has almost invariably been some political manifestation of regional discontent: indeed, it might almost be said that the period from 1945 to around 1960 constituted an aberration. However, it is salutory to point out that such heightened mobilization is not general: there are, moreover, areas distinctive in some aspects of identity — for example, Schleswig, Alsace and Bavaria — which display a steady decline in the strength and vitality of region-specific political mobilization. These conflicting trends must be tested against the more traditional view that the previous failure of separatist or ethno-nationalist movements could be attributed to the integrative functions of political parties in the early phases of mass politics; and despite the recent concern with regionalist politics, it is reasonable to state that political parties have been highly successful in mobilizing support cross-regionally: for most parties in most countries, failure to compete successfully in one region has usually reflected a similar inability over the whole state. From the mention of political parties it is but a short step towards stressing the contextual framework within which regional mobilization occurs. It is undoubtedly important to incorporate a regionalist perspective into an analysis. But the centre — simply because it is the centre — still occupies a crucial political position in determining not only the generation of regionalist aspirations, but also their content and likely prospects of success. Moreover, assuming that we know what is ethnic, whatever its basis may be, we cannot automatically proceed to accept that the ethno-national assumption of a linkage between state legitimacy and ethnicity is incompatible with loyalty to an existing state.

In short, centres and peripheries and the groups they contain exist in symbiosis. The first step should be to develop a typology of territorial structures that would combine information on the historical sources of strain in each territory with information on the strategies of unification distinctive of the state-building elites. Historically, we can distinguish two major sources of territorial strain: cultural distances, whether linguistic, religious or, in a diffuse sense, 'ethnic' between core areas and 'less privileged' peripheries; and economic conflicts between regional centres competing for the control of trade and productive resources. Some possible indicators, for example, of cultural distance would be the grammatical structure of the languages, the ease or difficulty of communication, the costs of both elite and mass integration across cultural communities. Similarly, the centralization of economic transactions can to some extent be gauged by indicators of the rank size structure of the city networks.

Centres can be minimally defined as privileged locations within a territory where key military/administrative, economic and cultural resource-holders most frequently meet; with established arenas for deliberations, negotiations and decision-making; where people convene for ritual ceremonies of affirmation of identity; with monuments that symbolize this identity; with the largest proportion of the economically active population engaged in the processing and communication of information and instructions over long distances. Centres, then, are both locations providing services and nodes in a communications network. These two simple concepts of degree of centralization and centrality pull together the key concepts in any consideration of central structures — resource endowments, distances, communications — measured in terms of the three conventional dimensions of differentiation: political control, economic dominance, cultural standardization. Typically, a centre controls the bulk of the transactions among resource-holders across a territory; it is closer than any alternative site (location) to the resource-rich areas within the territory; and it is able to dominate the communication flow through a standard language and a set of institutions for regular consultation and representation. By contrast, a periphery is dependent, controlling at best only its own resources and more exposed to fluctuations in long-distance markets; is isolated from all other regions except the central one; and contributes little to the total flow of communication within the territory, with a marginal culture that is fragmented and parochial, yet not fully dominant across the politically defined territory. In all these arenas a periphery depends upon one or more centres, and its predicament cannot be understood in separation from the latter.

From these conditions of centrality we can derive the simple dichotomy betwen monocephality and polycephality. A monocephalous territorial structure is one where there is a marked primacy of only one area, or even only one city, on both definitions of centre and along all three (political, economic, cultural) dimensions of differentiation. By contrast, polycephality indicates a more even and widespread diffusion of central features across territory, and a possible spatial segmentation of the different types of resource-holders, a chain of functionally distinctive centres each with its own profile of elite groups.

The combinatorics of the two dimensions of historical strain in Western Europe can be set out in a simple fourfold typology of conditions for territorial unification. However, as Figure 1 indicates, a typology based upon cultural distance and economic conflict would still bring together in the same cells countries of distinctly different territorial structure. To achieve some further refinement, conditions for differentiation must be combined with a classification of distinctive styles in the organization of sovereign territories. Again, we may begin with a simple dichotomy. On the one side there are centralizing strategies: all regions and peripheries, whatever their cultural or economic status, are incorporated under one universal system of standardization. On the other side there are strategies of federalizing accommodation, where all regions and peripheries, while being incorporated into one territorial unit and subordinated to one system of collective decision-making in at least matters of defence and foreign policy, nevertheless possess some guarantees for the protection and survival of their basic cultural distinctiveness as well as some autonomy in decision-making. In between these two poles there is a large grey area of mixed strategies: standardization only within the core or over most of the territory; separate and varying arrangements for resistant peripheries that have actively expressed a political will to remain distinctive; standardization of legal rules, but provisions for equal or separate status for several linguistic and/or religious standards.

Strategies of unification have obviously varied with conditions in each territory, but a brief reading of history will indicate that there is no one-to-one fit. Strategies of centralization have been employed even against seemingly heavy cultural odds, while federalizing accommodation has emerged across culturally very similar territorial units. The contours of a three-dimensional typology of centre-periphery structure along these lines would not exhaust the richness of the histories of territory construction in Western Europe, although it could provide a means of beginning to systematize research on sources of variation in regional policy as well as in the strength and intensity of regional and peripheral protest movements. However, before we can do this we need to develop a comparative perspective on current pressures upon the territorial 'nation-state', and for this we clearly have to go back into history, to the crucial phases in centre formation and the attempts to build territory-wide identities.

These points may be phrased rather differently. If we study, for example, the lengthy list of ethnic or linguistic minorities that have been identified in Western Europe, then it is clear that only in a few instances has any associated sentiment upset the prevailing forms of political activity by becoming a significant issue or organizing

FIGURE 1
A typology of conditions for territorial unification

		Economic distinctiveness/competition	
		Stronger	*Weaker*
Cultural distance	*Greater*	*Spain* (Castille v. Catalonia, Viscaya, Andalusia; polycephalic) *Switzerland* (four language communities; religious differences; polycephalic) *Belgium* (two language communities; religious and economic contrasts; monocephalic)	*France* (some linguistic heterogeneity: Alsace, Brittany, Corsica; Paris dominant) *United Kingdom* (distinctive communities in Scotland, Wales and Northern Ireland, but London dominant) *Finland* (a distinctive linguistic minority, but monocephalic)
	Lesser	*Germany* (linguistic homogeneity; some religious differences; polycephalic) *Italy* (largely linguistically homogeneous, with small minorities along the borders; marked economic contrasts; polycephalic) *Netherlands* (largely linguistically homogeneous; religious contrasts; polycephalic) *Portugal* (linguistic homogeneity; marked economic contrasts; monocephalic)	*Austria* (largely linguistically homogeneous; markedly monocephalic) *Sweden* (highly homogeneous and monocephalic) *Denmark* (largely linguistically homogeneous and monocephalic) *Ireland* (highly homogeneous and monocephalic) *Norway* (strong linguistic variations; monocephalic) *Iceland* (highly homogeneous and monocephalic)

force that has intruded upon the state level of politics. But whatever the outcome of attempts to place such identities in perspective, the recent assertion or reassertion of some of them has been a salutory reminder of the tenuous, if not artificial, nature of the nation-state concept: political boundaries are contingent, not permanent. And it is to the state — as a territorial as well as a social and legal concept — that we must return and, within the perspectives of strains and strategies, seek to disentangle the relationship between territorial regions and regionally concentrated groups. What separates political movements based upon the latter from the mass is the nature of their claim upon the state: they identify with, and make claims upon central government on behalf of, territories and groups that are not coincident with state boundaries and national populations. While not all people in a territory may identify with or support such a movement, the latter may demand control of, or adjustments in the control of, the territory, and hence over all those who live there, irrespective of their political opinions.

Since all such movements tend to identify with both a territory and a group, it is clear that we have to deal with two interrelated spatial dimensions, which may be termed 'membership space', that is, membership of a group that possesses, and probably is aware of possessing, some common sociocultural stigmata, and 'territorial space', that is, identification with and occupation of a specific geographical area. These two dimensions are not necessarily coincident. They can, with perhaps not too much licence, be related to two ideal-type constructs of the historical processes of state-building and nation-formation in Western Europe: the generative definitions so derived are, in fact, rather similar to the classical Gesellschaft-Gemeinschaft dichotomy.

A territorial definition of space might, in historical terms, best be related to dynastic expansion, where a single centre and its elite gain, through conquest, the effective control of large territorial areas. Consequent upon the establishment of acceptable and reasonably secure boundaries, there is a more deliberate effort of unification. The centre will seek to ensure its political and economic dominance through an efficacious system of administrative control, and to create unity out of diversity through a constant and conscious policy of cultural standardization. In other words, the objective is to make synonymous identification with the state and occupation of territory: the end result would be a pure nation-state. The boundaries themselves may be defined as the line within which there are friends.

In contrast, a membership definition of space may best be approached through the Weberian concept of 'politische Verbänd', the coming together of local groups, each with a distinctive cultural identity, for specific political or economic purposes. The driving motive is essentially defensive: through agreement these groups establish a covenant of mutual toleration and protection. To preserve the covenant is to preserve territorially varying rights and cultural identities, as well as to preserve the state. For a membership space, the boundaries are externally defined: they constitute the line beyond which there are enemies. In contrast to territorial space, identity here rests in membership of one's own group rather than in the state, which consists of a collection of distinctive identities. In terms of the earlier discussion, territorial space implies monocephality, while polycephality is characteristic of membership space.

The centre-periphery polarity would be easily resolved in these ideal constructs. While economic variations might continue to exist, with the complete supremacy of territorial space political centralization and cultural standardization would be assured: without the existence of an identity other than with the state, there would be no distinctive peripheries. Similarly, with utter membership space there would be no single centre, but rather a chain of economic and political centres that may be linked to distinctive cultural identities.

For these ideal-type constructs, therefore, the centre-periphery distinction would be irrelevant. In order to use them analytically in an examination of territorial structures, we can relate these two opposites with the previous discussion on historical sources of strain and strategies of unification by disaggregating them into nation-building and state-building components. Approximations of the two opposites may be considered as the poles of a continuum of nation-building which may be divided into four broad categories.

1. *Territorial space predominant.* There is an unambiguous central point of control in all three dimensions of differentiation — cultural, economic and political — with only weak or no regional institutions that could serve to shelter a distinctive identity or act as recruitment or mobilization agents for protest. Where political protest does arise, it tends not to have any strong positive regionalist or peripheral perspective, since there is little or no distinctive identity other than that with the state. Protest against the centre, when it emerges, might be said to be almost cathartic in expression; that is,

political mobilization tends to be anti-centre rather than specifically in favour of something. In short, it tends not to constitute a direct challenge to the prevailing territorial structure of the state.

2. *Territorial space dominant, but with strong membership space characteristics.* For most of the territory, identification is synonymous with the state. In limited areas, however, a membership identity distinct from that of the state territory has been preserved, and there may exist some kind of institutional infrastructure and/or vaguer social stigmata which can sustain the identity and, given the appropriate catalysts, serve as focal points of mobilization. Especially if they are geographically remote from the centre and possess an economy that is distinctive, weaker and/or subordinated to that of the rest of the state territory (especially the centre), they may more appropriately be described as 'peripheries', which may have the potential to generate political protest.

3. *Membership space dominant, but with strong territorial space characteristics.* The idea of a covenant, with the acceptance and toleration of diverse identities prevails. However, the membership format is subjected to strains generated by internal pressures. Strain arises through the emergence of one or several competing centres with territorially standardizing ambitions and objectives. Whether there is only one unit seeking to establish itself, through military-administrative or economic strength, as the dominant centre in the covenant, or two or more units seeking a similar objective, the result will be the same. The greater diversity of identities and a wider and more equitable spread of units in terms of resources make a simple characterization of centre versus periphery less applicable: it is more appropriate to describe territorial political mobilization in this context in terms of regional tensions and conflicts.

4. *Membership space predominant.* As in the ideal construct, there is little or no conflict between the several units, each with its own centre. Since there is acceptance of differences, internal variations do not give rise to tension and territorial political mobilization directed against the structure of the state. Where strains do arise, they are caused primarily by external pressures, for example through some kind of identification of the convenant members with different neighbouring states. These relationships can generate tension where two or more neighbouring states pursue distinctive policies that impinge directly upon the covenant, or where the members of the latter are pushed, through their differing identities, to support one side or the other in a broader international conflict

that does not bear directly upon the fundamental nature of the convenant. However, if the state does survive, then the covenant of toleration, with its implication of multiple membership identities, is strengthened.

The characteristics of the institutional-structural maintenance implicit in the ideal constructs relate more clearly to strategies of unification. Both centralizing strategies and those of federalizing accommodation can be dichotomized, giving a similar fourfold classification of state-building.

1. *The unitary state*, built up around one unambiguous political centre which enjoys economic dominance and pursues a more or less undeviating policy of administrative standardization. All areas of the state are treated alike, and all institutions are directly under the control of the centre.

2. *The union state*, not the result of straightforward dynastic conquest. Incorporation of at least parts of its territory has been achieved through personal dynastic union, for example by treaty, marriage or inheritance. Integration is less than perfect. While administrative standardization prevails over most of the territory, the consequences of personal union entail the survival in some areas of pre-union rights and institutional infrastructures which preserve some degree of regional autonomy and serve as agencies of indigenous elite recruitment.

3. *Mechanical federalism*, introduced, as it were, from above by constitutional means. A pattern of territorially diversified structures exists across all areas of the state, accepted or even introduced by the centre. This pattern of diversity is, nevertheless, accommodated within a hierarchical system of control, with a centre that is politically and institutionally stronger than any other constituent part.

4. *Organic federalism*, imposed from below as a result of voluntary association of several distinctive territorial structures. These retain their distinctive institutional outlines with wide discretionary powers. Control by the centre is limited, having to take cognizance of the large degree of institutional autonomy residing in the constituent parts.

These basic ideas are necessarily crude, and it is not difficult to see that there is a large degree of overlap between the two dimensions of state-building and nation-building. This, however, is little more than a reflection of the complexities of reality, and to combine them in a simple typology of Western European state- and nation-building processes provides a useful overview of the con-

FIGURE 2

A simple typology of state-building and nation-building processes in Western Europe

	Space/identity characteristics			
	Territory →→→→→→→→→ Membership			
Strategies of unification	Anti-centre	Peripheral protest	Regional tension	Covenant/ toleration
Centralizing ↑				
Unitary state	France Denmark Norway Italy Iceland Sweden Portugal Austria	Finland (Swedish Finns)	Belgium	(Berne)
Union state	(Brittany) (Alsace) (Occitania) (Corsica) (Carinthia) (Sardinia)	Netherlands United Kingdom (Friesland) (Cornwall)	Spain (Galicia) (Andalusia)	
Mechanical federalism	(Alto Adige) (Val d'Aosta)	(Scotland) (Wales)	Germany (Flanders) (Wallonia)	
Organic federalism	(Faeroes)	(Northern Ireland)	Bavaria (Basques) (Catalonia)	Switzerland (Jura)
Accommodating ↓				

tinental mosaic. Such a typology is produced in Figure 2, which shows a reasonable distribution of countries across the matrix from the cluster of unitary, standardizing states to the federal covenant of Switzerland. The typology also indicates where further research might be pursued into the relationship in any one state between the two dimensions. The north-west/south-west diagonal represents a good fit between state- and nation-building histories. Placement in cells away from this diagonal indicates the degree to which there is incongruity between the two processes, and therefore also the potential for varying kinds of territorial politics. Moreover, regions and groups have been placed in the cells. Figure 2 is not meant to include an exhaustive list of Western European regions and/or peripheries; it merely includes those with some degree of historical and/or contemporary political distinctiveness, and indicates suggestions of possible central solutions that might satisfy the respective peripheral demands. The typology is suggestive only: there are problems with the placement of some countries as well as many of the peripheries. It is, however, a summary of the problems that this research project intended to study.

To develop a comparative perspective on the current pressures on the territorial nation-state and on peripheral political mobilization, participants in the project were asked to go back in time to incorporate in their studies a consideration of the crucial phases of centre formation and the building of territory-wide identities. Some of the participants were asked to present broad-ranging reviews of how strains and strategies have produced varying patterns of and changes in the territorial structure of contemporary Western European states. Others, from a 'worm's-eye' viewpoint, concentrated upon the successes and failures of peripheral mobilization in a number of 'problem areas'. Both approaches are represented in this volume, though the division is artificial. The chapters on the territorial structures of Germany and the United Kingdom, for example, review the status of and political mobilization within several peripheries, while those on the Basques and Galicia begin with a consideration of the territorial structure of Spain. While there was general agreement on a wide-ranging list of key elements of analysis, there was no insistence on a standard list of themes: contributors were asked to present a state- or periphery-specific account of the most characteristic features of each development, but with some attention to possibilities of systematic comparison with other cases in Europe. In addition to the subject matter of the following chapters, papers and research notes were presented on a further range of countries and peripheries: England, Wales, Scotland, Northern Ireland and Cornwall; Ireland; Denmark, Iceland and the Faeroes; Schleswig and Bavaria; Flanders and Wallonia; France, Brittany and Corsica; the Alto Adige; Spain and Catalonia; the Swedish Finns. Some, regrettably, never reached final fruition. Others that did had to be excluded because of the limitations of space. The guidelines that were followed in the final selection were to seek a balance between the two approaches, to achieve the widest possible coverage across countries, and to give priority to those states and regions where the central problem area seems to be most acute and/or which seemed to be more interesting theoretically and in a comparative perspective. However, the collective findings of the group will be incorporated into a forthcoming companion volume, *Economy, Territory, Identity* which will present a comparative overview of the territorial structures of Western Europe and the phenomenon of peripheral mobilization.

The chapters on Norway and Belgium point to the importance of a strong unitary tradition. Frank Aarebrot stresses that there was no room for regionalism in Norway: only peripheral protest could arise, and this, partly because of the relationship between the country's move towards independence and the institutional features of Norwegian history, was directed more to capturing control of, or at the very least strongly influencing, the policy of the centre. Similarly, it might be added, implicit in César Díaz López's analysis of post-Franco Galicia is the notion that the province, by supporting so strongly the leading Spanish party, has thrown in its lot with the centre: this is not a takeover of the centre, but perhaps can be taken as an expression of hope that more benefits might accrue to the province, for example through patronage, than would be the case with outright opposition. By contrast, André Frognier and his colleagues point to the failure of the unitary Belgian state to achieve both an economic and a cultural balance between Wallonia and Flanders. The resulting conflict between the two, particularly because of the changing nature of territorial economic prosperity, might more appropriately be regarded as a case of regional tension.

By contrast, Germany failed to produce a unitary state. The chapter on German developments shows how constitutional conformity and attempted centralization was imposed upon a mosaic of regional interests that to a greater or lesser extent resisted standardization: only with the Federal Republic was there achieved, within the context of German structures and developments, a more accommodating balance between the territorial units concerned. In contrast to German historical polycephality, the United Kingdom more or less successfully achieved monocephality out of disparate territorial structures: Derek Urwin's chapter characterizes the United Kingdom as an old union state, which tolerated distinctiveness in its various peripheries through a particular style of membership accommodation. More recent strains on the British territorial structure are seen as a result of an intensification of standardizing policies and economic change, themselves partially a consequence of monocephality, within what is still a diffuse and variable territorial framework.

Apart from Wales, neither in Germany nor the United Kingdom is language a major source of territorial differentiation. Can we say that language buttresses and aggravates peripheral distinctiveness and political mobilization? All the other cases reviewed here have different linguistic standards as a seemingly major element of territorial tension. A first group comprises the 'success stories'. One such success within a group accommodation that entailed territorial parity and local unilingual status was that of the Swedish Finns. In comparative terms, this accommodation was assisted by the fact that they did not occupy a distinctive piece of Finnish territory, and most particularly by the fact that much of the Swedish-speaking elite were to be found in the Finnish centre. Risto Alapuro, in his study of the successful periphery of Finland, demonstrates how this accommodation was possible through the distinctive geopolitical location of Finland as an interface between Sweden and Russia, the changing nature of the trade of the various Finnish regions and the growth of an integrated cross-local economy, and through the commitment of the original Swedish-speaking elite to the idea of a Finnish state. Federal accommodation in Switzerland, and the possibilities permitted by this structure for further disaggregation *within* the overall federal structure, form the background to David Campbell's analysis of the social bases of support for Jurassian 'separatism', where language is intertwined with religion and economic change. Campbell shows how possession of a distinctive language was not sufficient to preserve the integrity of the Jura as a separate canton in the face of territorially based religious divisions and economic differences. By contrast, Alsace represents a success for the centre. An interface between the German and French linguistic standards, and with dialects of Germanic origin, Alsace was exchanged back and forth between the two neighbouring states. The careful plotting of Alsatian developments since 1918 by Solange Gras shows how the dialect was claimed as the symbol of

Alsatian distinctiveness only to be eroded by persisting French intransigence on the central issue of linguistic recognition as well as by international events. Alsace is a history of resistance and then resignation to cultural incorporation.

These chapters are parallelled by others which take up unsolved territorial problems seemingly linked to linguistic differences. However, all the contributors seem to conclude that the case of language has been overstressed in studies of contemporary regional and peripheral political protest: some quite specifically point more at economic variations and economic change, reiterating the stress placed by David Campbell and Solange Gras upon the importance of integration into a cross-territorial labour market. The chapter on Belgium, for example, discusses how language is only one of a nexus of differentiating factors between Flanders and Wallonia: language may be the bedrock of distinctiveness, but territorial strain is related also to other economic differences between the two provinces. Especially interesting is the argument that Brussels has been only one element of a centre which both culturally and economically extended well beyond it, and that economic change over the past few decades has in a sense created a competing Flemish centre. Perhaps the same point is made even more strongly by Marianne Heiberg, who views language more as a symbol for — or even an excuse by — Basque political activists. The rural heartlands of the language are divorced from the urban-based groups that have provided the primary driving force of peripheral nationalism, a point that also emerges in the study of Galicia by Díaz López: in both these Spanish peripheries language serves more as a guise for economic discontent and problems, and these vary across different groups within the periphery.

Internal divisions within a cultural periphery, in fact, are at the heart of Marianne Heiberg's assessment of the Basque problem. This is but one aspect of a general point that emerges from many of these chapters: there is rarely a straightforward polarity in identification and/or conflict between a national centre and a culturally distinctive periphery. Derek Urwin, for instance, shows how, historically, the German political parties, all reflecting to some extent the imperfect integration and standardization of the Reich, could adopt varying centrist and regionalist stances on different issues; while David Campbell forcefully demonstrates the extent of internal difference within the Jura. The complexities of peripheral politics are excellently illustrated by Solange Gras's disentangling of the numerous groups, all with different identities and different motives, that came together to argue for autonomy for Alsace. It is rather striking that the self-image of regionally based political movements is that they are advocates of a future polity based in the periphery, which would protect and strengthen the distinctive culture and identity of that area, and that their perceived enemies are the agents of a central polity and bearers of a central culture. However, it is not inconceivable, for example, that in the Basque country Spanish-speaking students with Basque sympathies might consider it striking a great blow for peripheral nationalism to assassinate a Basque-speaking policeman — an example that shows how relatively easy it is to turn the ethno-national approach on its head. There is no simple centre-periphery polarity across culture, economics and politics. The predicament of peripheral politicization lies in the incongruity between cultural roles, economic roles and political roles.

2

Territorial Structures
and Political Developments
in the United Kingdom

Derek W. Urwin
University of Warwick

In the world of political science the United Kingdom has occupied a special position, traditionally portrayed as an outstanding example of successful integration — almost, as it were, the first 'melting pot'. Until the 1960s the United Kingdom was accepted as a text-book example of a homogeneous society where influences and characteristics were equally significant and effective throughout the whole territory. Contributory factors often mentioned were:

1. a *common set of values*: Almond and Verba (1963), for example, claimed that 'the whole story of the emergence of the civic culture is told in British history';

2. a relatively unbroken *historical thread of continuity*, producing a unitary form of government with high legitimacy and a legislation designed to be effective everywhere in the state (e.g., cf. Eckstein 1962; Beer 1965);

3. the binding effect of a *nationwide two-party system*, the consequences of which are concisely summarized in Peter Pulzer's epigrammatic statement (1967: 98) that 'class is the basis of British party politics: all else is embellishment and detail'.

The ubiquity and predominance of class became the cornerstone of homogeneity: for many, there was homogeneity *because* of class divisions. The consequent political landscape is simple and pristine, with each class having its own party spokesman. Implicit in this landscape is the prejudgement of failure for, or irrelevance of, all other political movements: the latter could be dismissed as colourful anachronisms or annoying trivialities.

The 1970s witnessed a revisionist trend. Rose (1971) described the United Kingdom as a multinational state, drawing attention to the differences that had and still did exist between the constituent parts of the country. A more radical approach was adopted by some Marxist-inspired scholars: most provocatively, perhaps, Hechter (1975) described the British situation as one of internal colonialism, with an English imperial centre systematically controlling and exploiting a 'Celtic periphery'.[1] What the 'new' studies had in common was a stress upon the *territorial* basis of British politics. The justifications for such an emphasis were several.

1. Recent political events had raised a question-mark against the degree of unity in the United Kingdom. Since 1969 Northern Ireland has rudely reminded everyone that it is part of the British state. At the same time, there was after 1966 a rise in Plaid Cymru support in Wales and the more explosive advance of the Scottish National Party (SNP). In addition, an examination of the historical record would indicate sharp discontinuities in British development with periods of pronounced instability.

2. The territorial distribution of classes has not been constant throughout the state. It has followed the regional patterns of industrialization. Certain regions have had a high working-class component, and where these have been distant from London there has been an inherent territorial dimension in the 'them-us' dichotomy: the bosses traditionally have been identified with the centre.

3. Religion has played a significant role in British history, and fed into the primeval party system of the eighteenth and early nineteenth centuries. The marriage between monarchy and Church, between local squire and parson, led Macauley to describe the Church of England as 'the Tory party at prayer'. While of declining importance, religion has continued to exert some influence upon political behaviour. Religious differences, moreover, have possessed a territorial component. There is a geographical distinctiveness to English Nonconformism, while with reference to Anglicanism Scotland, Wales and north-east Ireland have been strongly 'nonconformist', and the rest of Ireland wholly Catholic.

4. Last, but not least, the United Kingdom was and still is the result of a merging of several units highly distinctive in cultural terms, with each possessing and articulating its own sense of identity. While this persisting national consciousness generated a powerful political nationalism only in Ireland, all have always been recognized, in one way or another, as distinct entities within the United Kingdom; and despite the unitary constitution this distinctiveness has been recognized at governmental level.[2]

The significance of geography

Insularity is the most striking geographical feature of the British state, which at the peak of its territorial expansion on the islands did not share a landward boundary with any other political entity. Even today, Northern Ireland is the only part of the United Kingdom that borders another state. Although small in area, both islands display an elongated shape that historically hindered the efficacy of land communications. Economically and politically, England — and especially its southern regions — has perhaps always been the central unit, for here were to be found extensive and fertile lowland areas capable of sustaining intensive agricultural production and fairly large populations.

It is conventional wisdom that this island position has been of crucial importance in British history, with the encircling seas constituting a formidable barrier that successfully protected the islands from military invasion after 1066, and with the intrusion of the Irish Sea in turn hindering English efforts to subjugate Ireland. However, it is valid to regard the sea as a barrier probably only for modern historical times, since the development of rapid landward means of communication. In the British case, the sea served as an 'invincible' shield against invasion only after the rise of reasonably strong states on the islands and the building of effective naval forces.[3] Before the establishment of national states and marine power, the British Isles were exceptionally prone to penetration from outside. Together, the two islands have over 11,000 kilometres of coastline, and the extremely lengthy seaboard permitted easy access at almost any point. The British Isles were a reception area for waves of ethnic groups from the Continent. Their invasions and settlement at different times (but climaxing in 1066) and at different points of entry produced a very mixed population: the British Isles were 'balkanized' (Coupland 1954: xv). The population of Britain was relatively homogeneous perhaps only in the Celtic era prior to the Roman invasion and domination of England (cf. Raftery 1964; Heslinga 1962); afterwards, homogeneity survived only among the Celtic populations of the more remote upland areas to the west and north of both islands.

Water was a more effective and efficient means of communication than land, and its importance remained long after the growth of territorial states. England's rise to world prominence after the sixteenth century was not least due to sea power. That the English centre did not regard water as a barrier can be indicated simply by pointing out that for centuries the English kings in London ruled large areas of France, and that their claims to this territory persisted long after it had finally been lost with the relinquishing in 1558 of the last continental foothold in the port of Calais.[4] Moreover, the elongated shape with its several indentations has meant that no one area of the islands is too far removed from access by water. The combination of sea and river permitted the most effective means of transport, communication and central penetration until the development of the railways. The early cities were mostly either ports or located on navigable rivers. Roads were simply bad: as late as the mid-eighteenth century the single stagecoach that linked London and Edinburgh took two weeks over a journey that was universally accepted as being extremely hazardous.

The most simple internal topographical distinction is that popularized by Mackinder (1907) between lowland and highland Britain, where contour patterns and climate produced, before industrialization, different styles of human settlement and economic activity. This dichotomy has proved tenacious, and, for example, has been employed recently by Hechter (1975: 58) to differentiate 'English' and 'Celtic' styles of social organization. It is much too simple, however, to equate Celtic Britain with highland Britain. Much of Ireland is not upland, while parts of England are not lowland. In the nineteenth century there were cross-cutting spatial differences within England in terms of agricultural rents and wages that did not clearly relate to such a topographical dichotomy (cf. Chambers and Mingay 1968: 142). Such geographical distinctions may be meaningful in a political context if we refuse to treat England in particular as a single entity. The lowland heart, where population density was historically greatest, lies predominantly in southern and eastern England. The remainder is more upland, and perhaps different. It may be better, in combining geography and political history, to distinguish within both centre and periphery, as well as between them. These general comments may seem trite, but they are an essential part of the British story, and form a background to any review of the territorial structure of the British state.[5]

The uniting of the kingdom

The collapse in the ninth century of Northumbria, a prominent unit in the Anglo-Saxon heptarchy of states, meant that centralization would spread outwards from the south. The history of incorporation begins with the rise of the kingdom of Wessex in 802, culminating with the formal union with Ireland in 1800. Disaggregation occurred with the secession of 26 Irish counties in 1922. Events in Scotland, Wales and Northern Ireland suggest that further disaggregation is not totally impossible.

It is common to assume that the process of integration involved four distinct units. Certainly, Ireland was a separate island, while the borders of Wales and Scotland have been more or less fixed for centuries. Yet there has always been a problem with the English residual. Since the thirteenth century there has never been a distinctively English state. And while English kings may have laid claim to sovereignty over the whole island group, their effective authority was more often limited to central and southern England. The northern English marches were not fully absorbed until the sixteenth century, remaining in association with, rather than being annexed by the south (cf. Bulpitt 1975).[6] For centuries northern England was a kind of no-man's-land between the core areas of England and Scotland. It was to be a laboratory for central policies of integration that later would be pursued in other regions of the British Isles. Similarly, the subjugation and integration of Cornwall extended over several centuries: Cornish rebellions persisted until 1648 (cf. Halliday 1975; Hatcher 1970; Rowse 1969; Coate 1933).

Because of internal English variations, and because of differences in degree of association of the 'Celtic' kingdoms with England, a more refined centre-periphery distinction makes more sense historically: (1) a central core consists of London and the immediate environs of the south-east; (2) an outer centre comprises Wessex, East Anglia and the Midlands; (3) an inner periphery embraces Cornwall, the northern English shires and Wales; while (4) Scotland and Ireland form an outer periphery. These distinctions correspond roughly to phases of territorial expansion. The centre was defined by the tenth century, with the inner periphery coming under central influence during the next 400 years. The establishment in the sixteenth century of the Councils of Wales and the North indicated a need for special treatment of these areas. The outer periphery, despite central claims and efforts, remained more independent of London until later.

The dominant political position of England in the process of integration was aided by three factors: its geo-economic advantages, its more effective central authority, and the political flaws of its rivals. While regional differences and conflicts existed in England, contributing in part to the two internecine struggles of the Wars of the Roses and the Civil War, they were not as significant in their consequences as the deep rifts between north and south in both Scotland and Wales, or as de-enervating as the perpetual feuding of the Irish Gaelic chieftains. Soon after it became the seat of government, the old Roman headquarters and port of London achieved a political and economic dominance that it never relinquished. By contrast, it is doubtful whether the early Welsh and Irish proto-states possessed a centre in any conventional sense: centres were established only later, as control nodes of English influence.[7] In the north, Edinburgh became the chief royal residence in the early thirteenth century, but the undisputed Scottish centre only much later: even then, it was undisputed only in so far as the Highlands maintained a stance of indifference, a state of affairs that lasted until after the union with England.

In comparative terms, English economic developments were rather unique. From the beginning, towns belonged to the royal demesne: while possessing commercial privileges, they had no degree of political autonomy. Moreover, urbanization in medieval England was very limited outside London. The towns were never sufficiently numerous or large either to stake a claim for an independent or autonomous status or to challenge the hegemony of London. In the countryside the involvement of East Anglia in the medieval wool trade gave English agriculture a commercial orientation which it never really lost. England led Europe in eliminating internal obstacles to trade, and at a relatively early time territorial magnates began to view land and labour in economic terms rather than merely as a source of social prestige, political influence or military strength. In addition, much of England was fertile soil for agricultural production, and could more easily sustain a greater population level, while the low-lying territory of the English centre made communications easier. At any rate, England clearly dominated the British Isles by the end of the Middle Ages in both human and material resources, a superiority that was reconfirmed in subsequent centuries (cf. Wallerstein 1974; also Minchinton 1969; Wilson 1965).

The early extent of English central authority and penetration was impressive: linguistic developments, a national customs system and the structure of common law had all given the medieval state a strong stamp of uniformity (e.g. cf. Strayer 1970: 49). Yet the style of central control was significant. From the Magna Carta (1215) onwards, limits were successfully imposed upon arbitrary central authority. In a sense, the English Crown wielded influence because of a compromise with the territorial magnates, in order to advance their separate interests. Anderson (1974) calls it a 'concurrent centralization', of both royal power and aristocratic representation in a territorial legislature. The preferred central stratagem of penetration and standardization was indirect rule, through the co-option and co-operation of local magnates. The resolution of territorial issues was to be achieved through the juxtaposition of a national parliament and a locally based 'bureaucracy'. Legislation and taxation operated at the parliamentary level, while administration and the maintenance of order remained at the local level. While the indispensability of Parliament was not assured until after 1689, it early acquired a strong negative ability to limit royal legislative power. Parliament may not have had a high degree of fiscal control, but after the late thirteenth century it was accepted that the decree of new statutes by a monarch should first receive parliamentary consent. By contrast, the provision of local administrative and judicial services was the task not of a central bureaucracy, but of non-hereditary sheriffs — royal appointees, yet drawn from the local magnates and gentry. Pre-industrial England remained 'a confederation of overlapping communities, politically united only on the unusual occasions when the representatives of these communities were summoned together in Parliament' (Underdown 1971: 24). Parliament increasingly became the arena where issues of territory could be accommodated with central policies and the indirect management of the provinces by London.

Despite (or because of) the lure of territorial extension in France, this English state almost inevitably came into conflict with Wales and Scotland. The subjugation of Wales was successful; that of Scotland failed. Welsh independence came to an end in 1277. The Crown assumed direct ownership of about one-half of the territory. The rest was granted to lords of the marches, who ruled in the king's name, but who like their counterparts in the northern shires were in practice semi-sovereign magnates. The Crown laid down provisions for the territory in the Statute of Wales (1284), but until the sixteenth century Wales was not subject to parliamentary laws, and its citizens did not possess the rights and duties of their English counterparts. The formal absorption of Wales in 1536 (Rees 1967) was in essence the result of a diktat by the English Parliament. The two immediate catalysts were the marriage dispute of Henry VIII and the decline of Wales into a state of lawlessness. The break with

Rome lent a greater urgency to national mobilization and centralization, and the Welsh coastline had to be secured against possible invasion by Catholic enemies (cf. Rees 1967: 19).[8] Acts of 1536 and 1542 conceded some autonomy to Wales within the confines of English centrist philosophy, and gave it representation in the London Parliament. At the same time, Welsh customary law was replaced by English law; the new Anglican church was made the established church of Wales; and there began the process of anglicization of the landowning classes. The Council of Wales (abolished in 1689) was to strive for the legal, administrative, social and ecclesiastical integration of Wales with England. Simultaneously, the combination of political centralization and religious reformation brought the remainder of the inner periphery more firmly under central control, most importantly in the north. Tudor policies had initially provoked a potentially serious uprising in the northern shires, the Pilgrimage of Grace of 1536 (cf. Scarisbricke 1968: 444-52). Once this was crushed, a new Council of the North was established to work for conformity with central directives. Cornwall, too, was locked much more firmly within the central orbit.

There had been English settlement around Dublin since the twelfth century, but its density was never sufficient to ensure predominance. Substantial migration to Ireland did not commence until the seventeenth century and the breaking of the Catholic lords of Ulster. While military attempts to pacify and control Ireland had continued sporadically since the twelfth century, again it was the Tudors who initiated and pursued a more systematic policy of management by oppression. It was Mountjoy, the commander appointed by Elizabeth, who ensured the military annexation of Ireland (cf. Quinn 1966; also Anderson 1974: 130-3).[9] The new strategy involved not only the subjugation of a host of Catholic chieftains, but also the control of an Anglo-Irish aristocracy that often had been as intransigent as the indigenous Gaelic population. The seventeenth-century plantations in Ulster were part of this 'grand design', and by 1688 nearly 80 percent of Irish land was controlled by English and Scottish Protestants (Bottigheimer 1971). Catholics were excluded from the parliament that sat in Dublin, and active discrimination destroyed the vitality of the Catholic upper classes. But English policies were only partially successful, and Catholic enfranchisement in 1793, after pressure from London, only made the situation more radical and unstable: the impact of the oppressive penal laws could not be immediately swept away.

Short of ignoring Ireland entirely, the major remaining option that could secure English jurisdiction and defend English interests was formal incorporation. A formal union was ratified in 1800 by Westminster and by the short-lived Irish legislature in Dublin (cf. Bolton 1966; also the historical overview in Beckett 1966).

Scotland remained independent until 1707. Two centuries of warfare after attempts in the 1290s to conquer the country had done little more than consolidate the boundary between the two states, although considerable English influence in and upon Scottish affairs persisted. By the sixteenth century the Scottish state still did not possess effective control over all its claimed territory (cf. Urwin 1978: ch. 2). Simultaneously, the Scottish Crown could not but be aware of the fact that dynastic marriage had placed its holder only one or two lives away from the greater prize of the English throne. In 1603 the Scottish king inherited the English succession, and moved his residence south. Subsequent attempts to establish a broader union than that of the two crowns more frequently came from Scotland. They eventually succeeded in 1707, with the ratification by the two parliaments of the Act of Union.

There were several different paths, therefore, towards incorporation. The English inner periphery was annexed directly, though originally through a 'trade-off'; Wales was absorbed in the face of limited and unco-ordinated opposition; the merger with Scotland resulted from a treaty between two independent states; while the incorporation of Ireland was more a blunt political takeover that followed upon centuries of abortive military conflict and oppression. But in each instance incorporation was seen by the centre not necessarily as desirable, but as the only possible option after the failure of structures of indirect control.

Despite the lack of a substantial military arm in the Tudor apparatus, it was the sixteenth century that first saw a serious drive towards the political integration of the British Isles (cf. Bindoff 1966; Elton 1972). But control remained indirect. The House of Commons successfully resisted monarchical demands for extraordinary legislative powers and a large standing army. At the same time, the monarchy was financially weakened by military involvement on the Continent. To recoup its losses, much of the ecclesiastical property expropriated by the Crown under the Reformation was sold: in the long run this strengthened the gentry, increasingly the regional backbone of Parliament, and weakened the greater nobles (Dobb 1963: ch. 5). The persistence of indirect control meant that Scotland and Wales — and even Ireland — retained an impressive degree of freedom from positive interference until the nineteenth century. Scotland in particular retained its traditional institutional infrastructure almost intact. It was the improved communications made possible by industrialization and technology, and the new demands upon government to play a more active role in social life, which in the late nineteenth century stimulated an escalation in regional tensions.

Political union did not mean cultural homogeneity, although the Cornish language died fairly rapidly after the seventeenth century.[10] The union with Wales had not meant the active persecution of the Welsh language; Welsh may have been denigrated as uncouth (Coupland 1954: 186-95), but the language survived almost undisturbed: the effects of union were negative rather than positive, being designed primarily to enforce English as the only language of official business. The decline of Gaelic in Ireland was more pronounced after the seventeenth century: by 1851 only 29 percent of the population was registered as Irish-speaking. By contrast, Scotland had never been linguistically homogeneous. Gaelic was in decline from the tenth century, and by the end of the Middle Ages was spoken mainly in the Highlands. In time, lowland Scottish lost much of its distinctiveness as its variety of English moved closer to that spoken south of the border. The disastrous decline of Gaelic after the eighteenth century was due as much, if not more, to lowland Scottish agencies as to policies emanating directly from London.

Centralizing efforts after the sixteenth century owed much to religion. The English Reformation was political, not theological; and conversion to Anglicanism, successfully carried out by the Crown, meant that national and religious allegiances reinforced each other, and contributed later to the consolidation of a legitimate regime. The legal discrimination against Catholics persisted through to 1829 (Machin 1964), confirming and widening the wedge between Ireland and the rest of the British Isles — especially as more practical aspects of anti-Catholic prejudice endured long after formal emancipation (cf. Norman 1968). Religion was also the driving force behind Welsh incorporation. The 1534 Act of Supremacy had placed Wales under the Archbishopric of Canterbury, while it remained outside the direct jurisdiction of king and parliament: it was necessary that the country also be placed under parliamentary jurisdiction (Rees 1967: 18-21).[11]

In Protestant Europe the British Isles, and especially England, are unique in that they fostered a high degree of religious pluralism

(Schoeffler 1960: 322-4). Nonconformist sects experienced their greatest impetus during and after industrialization.[12] Their particular success among the new lower strata in the industrial regions of England (cf. Inglis 1963; Rudé 1964: Gay 1971) added a further territorial dimension to the political picture. Some clear regional variations emerged in the distribution of English Nonconformism: it tended to be stronger in the English inner periphery (Cornwall, the North) and in parts of the outer centre (a belt extending from the eastern Midlands into East Anglia).

In Wales, religion mixed with language to affect later political developments. Welsh benefited from the translation of the Bible and prayerbook, but the Anglican Church in Wales, even though many of the lower clergy were Welsh, became associated with the anglicized gentry (Williams 1950: 89) and synonymous with central authority, for it conducted services only in English. Nonconformism, absent during the seventeenth century, grew out of the religious and educational movements of the eighteenth to dominate the principality one hundred years later: it was its Welsh character that ensured Nonconformism supremacy (Morgan 1963: 8-12). The episcopal Church of Wales, after years of bitter struggle, was finally disestablished in 1920. Scotland too became wholly Protestant, but in contrast to England the Reformation was a popular movement that triumphed over the Crown. Moreover, the country became Presbyterian, not Anglican, and the Church of Scotland remained as the established church after political union: episcopalianism was to find few adherents in Scotland (cf. Donaldson 1960). The Reformation failed conspicuously in Ireland, apart from the Presbyterian plantation in Ulster. Ireland was always regarded as more different than Scotland or Wales (Curtis 1968), and by the nineteenth century Catholicism had become synonymous with Irish nationalism.

By the nineteenth century, therefore, cultural patterns had been consolidated. English was the dominant language everywhere, except in Wales: in Ireland and Scotland the Gaelic language had been driven back to the western and northern fringes (Figure 1). Welsh thereafter entered into a steady decline, with less than 20 percent of the population speaking it by the 1970s.[13] In contrast to increasing monolinguism, Anglicanism dominated only in England. Scottish Presbyterianism ought more appropriately to be linked with Nonconformism. Wales too became largely nonconformist, but these sects, while strong enough eventually to force the disestablishment of the Church of Wales, were so numerous that no

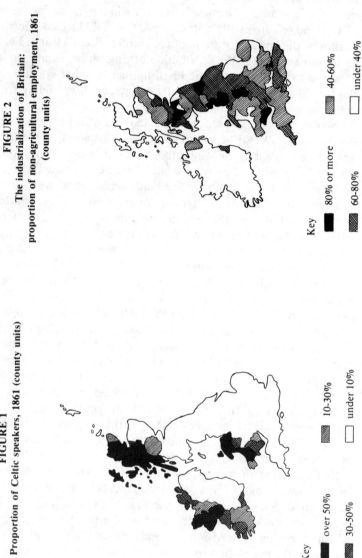

FIGURE 1
Proportion of Celtic speakers, 1861 (county units)

Key

over 50%
30-50%
10-30%
under 10%

FIGURE 2
The industrialization of Britain:
proportion of non-agricultural employment, 1861
(county units)

Key

80% or more
60-80%
40-60%
under 40%

one could predominate. Outside its north-eastern tip, Ireland remained overwhelmingly Catholic. These religious distinctions have persisted ever since (Table 1).

TABLE 1
Religious identification in the British Isles

	Anglican	Presbyterian	Other Protestant	Catholic
	%	%	%	%
England	69	1	10	10
Wales	45	—	45	7
Scotland	3	68	3	16
Northern Ireland	24	29	10	35
Republic of Ireland	4	—	—	95

Source: Rose (1971: 61).

Industrialization and economic structure

The contraction of agriculture was well advanced by the seventeenth century, when it was responsible for no more than around 40 percent of the total national output (Deane and Cole 1962: 78; also Clarkson 1971). By the first census of 1801 only 36 percent of the British labour force was engaged in agriculture: by 1911 it was down to 8 percent. Clearances and enclosures had progressed intermittently since the Middle Ages, and the nascent characteristics of at least English agriculture were crystallized by the technological advances of the eighteenth century (cf. Chambers and Mingay 1968): Ireland and mountainous Scotland lagged far behind.[14] Outside these peripheries the waves of clearances solved the peasant problem. British agriculture, small in size (cf. Table 2), was closely integrated in the broader social structure. The logical conclusion came in 1846 when, by repealing the Corn Laws (e.g., cf. Read 1964: ch. 3), Britain endorsed the decline of agriculture (and by implication the political power of the landed interest) and accepted that the necessary food supply would be obtained through imports paid for by the profits of industrialization. The abolition of the Corn Laws hurt Ireland badly: it no longer enjoyed a privileged position in the English grain market. At any rate, the ability of a rural periphery to mount a serious political challenge was progressively and severely circumscribed.

TABLE 2
The decline of agriculture in Britain, 1831-1911

	1831	1851	1881	1911
	%	%	%	%
England	31	27	10	8
Scotland	30	30	14	10
Wales	40	34	13	10
Ireland	64	53	54	40

In the 1700s British industry was geographically diffuse. One hundred years later this low-level diffusion was in the throes of the Industrial Revolution. By 1821 the number of workers in mining, manufacturing and industry surpassed those in agriculture. The greatest changes occurred around the coalfields in northern England, central Scotland and South Wales. The best indicator for a graphical display of economic change is the proportion not dependent upon agriculture. Figure 2 shows the regional distribution in 1861. The non-development of Ireland (apart from Belfast

TABLE 3
Personal wealth by region, 1803-91

	£ per capita tax assessment			
	1803	1851	1871	1891
England	10.2	11.0	15.2	18.1
Scotland	4.8	7.9	12.0	15.1
Wales	5.7	7.2	9.5	11.7
Ireland	—	3.0*	4.9	8.4

* Income tax was introduced in Ireland only in 1855: the 1851 figure refers to 1855.
Source: Rose (1971: 67).

and Dublin) immediately stands out. On the British mainland, the only comparable regions to Ireland were the Scottish Highlands and central Wales. Economic developments, moreover, were much more extensive in central Scotland and northern England than in the southern core region. The north-south dimension was important in nineteenth-century English history (Briggs 1959: 51). England remained the wealthiest unit, as measured by tax assessments (Table 3). However, Scotland and Wales displayed an equally strong rise in wealth: Scotland's growth was particularly impressive, and by 1900 it had almost achieved parity with England.[15]

Hand in hand with industrialization went population growth and urbanization. The British population trebled between 1801 and 1901. Population growth was concentrated in the new industrial regions, and led steadily to an increase in England's share of the British population at the expense of the other regions (Table 4). Rural areas suffered a relative decline, especially in Wales and Scotland, where by the twentieth century the bulk of the population came to be concentrated in a geographically restricted industrial core (Table 5). Demographic imbalance produced a new twist to territorial differentiation within Scotland and Wales — as did the rise of Belfast in Northern Ireland — and contrasts strongly with English demographic developments.

TABLE 4
Population distribution in Britain, 1801-1971

	1801	1851	1901	1951	1971
	%	%	%	%	%
England	72.0	75.6	79.8	81.9	82.9
Scotland	13.6	13.1	11.8	10.1	9.4
Wales	5.1	5.0	5.2	5.2	4.9
Northern Ireland*	9.3	6.3	3.1	2.8	2.8

* The territory of the Republic of Ireland has been excluded.

Sources: Hammond (1968); McAllister *et al.* (1979).

More spectacular was the *absolute* decline of the Irish population. By the nineteenth century Ireland suffered from the characteristics of an impoverished peasant society: overpopulation and underemployment. Its tenuous link with survival was shattered

TABLE 5
Regional changes in population distribution throughout the British Isles, 1801-1966

	England		Scotland		Wales		N. Ireland	
	1801	1966	1801	1966	1801	1966	1801	1966
	%	%	%	%	%	%	%	%
Central region*	21.9	25.5	38.3	71.1	18.8	63.4	19.6	47.9
Remainder of country	78.1	74.5	61.7	28.9	81.2	36.6	80.4	52.1

* The English centre is defined as London plus the six south-eastern counties of Essex, Hertfordshire, Kent, Middlesex, Surrey and Sussex; the Scottish as the nine central counties of Ayr, Clackmannan, Dunbarton, Fife, Lanark, Midlothian, Renfrew, Stirling, and West Lothian; the Welsh as the two south-eastern counties of Glamorgan and Monmouth; and the Northern Irish as the county of Antrim.

Source: Census data. The Irish figures for 1801 are from the first Irish census of 1821.

in the 1840s by the great famine. Although there had been large scale emigration earlier, it now became endemic. In 1801 Ireland had held 33 percent of the British population: by 1881 it had declined to 14 percent, and this absolute decline has persisted ever since (e.g., cf. Cullen 1972).

Britain was not a particularly urbanized country before the Industrial Revolution (cf. Laslett 1965). London, growing from 35,000 in the mid-fourteenth century to half a million 300 years later, dominated in terms of sheer size. The urban population was about 25 percent of the total British population in 1801. By that year, while London had already passed the million mark, no other city had reached a level of 100,000, and only eight were larger than 50,000.[16] The urban population reached its majority by 1851. Not only was urbanization complete by the turn of the century; suburbanization was already well advanced.[17] The greatest expansion was that of the industrial cities in the North, some of which, between 1801 and 1891, increased in size by between 600 and 1,000 percent. By contrast, the growth of London was 'only' 392 percent. The sheer size of London is one of the most striking features of British urbanization. Its dominance can be expressed by its share of the total population of the 25 largest cities (including London itself). During the nineteenth century this figure declined steadily from 52.9 percent in 1801 to 39.5 percent in 1891: thereafter it rose slowly, and stood at 44.3 percent in 1971.[18] But London now dominated an urban country. Urban areas are officially defined in terms of local government status: on this basis, between 70 and 80 percent of the British population in 1971 were defined as urban residents (except in Northern Ireland, where the figure was 55 percent).

The previous centuries had generated a political structure of territorial control and constitutional government that could, without unduly severe strains, absorb industrialization. While the rise of new industrial cities temporarily counterbalanced London,[19] industrialization in time gave an impetus towards further centralization, as more and more resources, both political and economic, were established in the capital: it confirmed and redefined the United Kingdom as a strongly monocephalic state. The same is true of communications developments, even though the canal era and, after 1830, the railway era originally arose in the industrial districts. The railway grid, almost inexorably, stretched towards London, rather than being diffused outwards from London.[20] Later transportation networks on land and in the air simply reinforced this centripetality, which was further confirmed by developments in the mass media, with their pattern of limited regional coverage within a national framework.

Communications brought everything closer to the centre, as technological innovation improved possibilities for stricter central control. That certain areas were left out of this economic transformation was due perhaps more to their remote location and lack of appropriate natural resources than to the fact that they were the surviving core regions of the cultural minorities. Industrialization in Britain was, for a long time, very much a 'spontaneous' development. There was no central pressure for regions to specialize in a particular form of production. Spontaneity was particularly marked in the case of Scotland, where capital investment and development was largely financed from indigenous sources. Over-specialization in heavy industry by Scotland and Wales, which led to economic triumph in the late nineteenth century and economic problems in the twentieth, was not the result of a centrist conspiracy or exploitation. Indeed, nineteenth-century economic change generated inequalities *within* Scotland, England and Wales that were greater than the overall inequalities between them. Industrialization must be considered in conjunction with Britain's great imperial expansion. The market was greater than the two

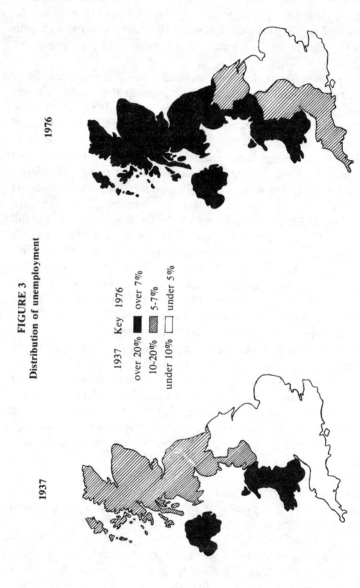

FIGURE 3
Distribution of unemployment

1937   Key   1976

1937
over 20%   over 7%
10-20%   5-7%
under 10%   under 5%

1976

1937

Source: Medlicott (1967: 277); McAllister *et al.* (1979)

islands, and the participation of Wales and Scotland in this vast economic system not only bound them closer to England, but also permitted them to develop and demonstrate a British identity.

Contemporary regional economic problems are the consequence of an economic specialization compounded after 1918 by far-reaching structural industrial change (cf. Pollard 1969). The decline of the staple industries of central Scotland, southern Wales and northern England was parallelled by the rise of new industries, located more towards the British centre.[21] A regional economic contrast between the English core and the remainder of the United Kingdom is therefore much more of a contemporary phenomenon. Unemployment is perhaps the most simple and dramatic indicator of this regional contrast: after 1918 it has always been appreciably lower in southern England than elsewhere. Figure 3 illustrates how, in both 1937 and 1976, unemployment levels increased with distance from London and the English centre. The new realities of the interwar decades ushered in the first attempts at regional economic policy-making and planning: as a consequence of the processes of incorporation, concern previously had been with possible regional political structures. With the primacy of economics and the development of a more clearly class-oriented party system, the political questions disappeared from view, not to arise again until the 1960s.

Economic and political developments over the past two centuries have substantially increased the degree of standardization. With secularization, religious differences no longer play an active political role, except in Northern Ireland where, however, they may be as much a symbolic language as anything else (cf. Rose 1971; Harris 1972). The Celtic languages have no area of strength outside rural Wales and the extreme north-western fringe of Scotland. The United Kingdom has a very small agricultural society: the most agrarian periphery seceded in 1922, and in 1971 the highest propor-tion of agricultural employment — in Northern Ireland — was only 8 percent. The United Kingdom is a strongly urban and industrial country, in terms of population if not of territory.[22] Improved transportation has placed at least 50 percent of the population within easy reach of London, while the mass media have a very strong nationwide impregnation. Government welfare policies, moreover, have brought about a reduction in regional inequalities.

Are there, then, any meaningful social structural differences across the four territories that can both be quantified and possess potential political significance? A recent study has attempted to

compile a list of important social indicators to illustate their territorial dimension (McAllister *et al.* 1979): it would seem unnecessary at this point to do more than comment briefly upon their
findings. The degree of difference between the constituent parts of
the United Kingdom was determined by comparing conditions in
each with those in the state as a whole over 21 conventional
indicators of socioeconomic wellbeing, as well as cultural and
political measures. Table 6 presents a summary of the most central
of these socio-structural indicators: by taking the United Kingdom
base as 100, and the figures for each territory as a proportion of
that, the four countries are contrasted with each other and with the
total British state. Table 6 points strongly towards homogeneity
and standardization. Northern Ireland is the most 'deviant' area,
but even so the degree of social differences is relatively small.
Moreover, a comparison with similar data between 1951 and 1976
suggests that the degree of difference has progressively narrowed.
However, while the four countries may vary in their respective
placements across the indicators, the cumulative effect is that of a
territorial gradient. When the indicators are organized in such a
way that they consistently, albeit vaguely, correspond to something
that may be described as 'the quality of life' or 'advanced industrial
society', England clearly ranks first, and Northern Ireland last.
There is very little difference between Wales and Scotland, but
overall they lie closer to England than to Northern Ireland. These
data confirm the distinctiveness of the Ulster rump in the British
state, while stressing that the degree of difference is restricted.
Social differences across the territories of contemporary Britain
suggest a picture of hierarchy within homogeneity.

Demographic and economic change produced significant variations within each of the four territories: Scotland, Wales and
Northern Ireland each have an industrial core region and a rural (in
the first two, also sparsely populated) periphery. The extent to
which such internal variations are important will depend upon the
definition of region and upon controlling for the fact that in any
modern society the degree of interregional variation will almost
automatically increase with the extent of territorial disaggregation.
In the United Kingdom the problem is compounded by the fact that
it is difficult to detect politically meaningful regional units that
have survived the passage of time. Confusion is especially rife in
England, and has been reflected by the diachronic absence of a
standard regional format by which governments have presented
information. McAllister *et al.* (1979) disaggregated the United

TABLE 6
The degree of regional difference
(selected social indicators)
(UK = 100 in all instances*)

	England	Scotland	Wales	N. Ireland
Urbanization	102	93	92	72
Population born in territory	100	102	91	101
Infrequency of emigration	103	90	81	89
Life expectancy	100	98	99	94
Infant mortality	102	98	107	79
School education above the minimum	101	98	110	64
Further educational qualification	100	117	92	92
Home ownership	104	61	112	96
Housing quality	101	97	96	80
Industrial earnings	100	101	101	93
GDP per capita	102	97	90	78
Car ownership	102	82	105	91
Public expenditure per capita	96	120	111	141
Employment	100	99	98	95
Middle class	107	87	89	101
Non-agricultural employment	100	99	98	95
Average value	101	96	98	91

* Figures above 100 indicate a greater incidence within the territory concerned; those below 100, a lesser incidence.

Source: Derived from McAllister *et al.* (1979: 12).

Kingdom into 28 regions (though the definition of these units varied across the four territories). Their conclusions are what might be expected: that there tend to be greater disparities within each country than between, and that these disparities peak in Scotland and Wales. Again, however, the evidence points not only to a lesser degree of difference than might have been expected, but also to a persistent narrowing of the gap over the past three decades. It is difficult, therefore, to use straightforward quantitative indicators as an explanation of territorial political differences. We are left with the more diffuse world of culture, a world that embodies both the ambiguous concept of identity — be it national, ethnic or communal — and attitudes, based upon territorial perspectives, towards government and central decision-making as the bedrock upon which regional political mobilization may be based.

Territorial structure and politicization

There might have been a design to the extension of central control over the British Isles, but there was little systematization of the process of incorporation other than the basic principles of territorial settlement and compromise, and indirect rule through collaboration with local elites. The tone was one of tolerance and indifference. With Parliament as the central forum, the style of politics was territorial. After the Glorious Revolution of 1688, political stability was strengthened with the extension of clientelist electoral practices that were built upon the basic characteristics of both society and political tradition (cf. Plumb 1967). By the nineteenth century union was a fait accompli, except in Ireland. Yet the style of politics fashioned by the institutional framework was English and medieval, and it did not change much before the 1880s.[23] The persistence of the traditional territorial constitution determined that electoral politics would essentially be local politics. Not every borough or county constituency was rotten or in the pocket of some magnate: indeed, here and there partisan competition could be quite keen. But in the last resort, competition or control rested upon local influence and the aspirations of the squires. The most appropriate description is that of Hanham (1959): elections were not general.

Over and above localism, electoral politics were impregnated with a certain degree of geographical distinctiveness, relating most noteworthily to religious and cultural patterns, and these were as noticeable as the urban-rural basis of party competition between Liberal and Conservative.[24] Liberal strength in England followed quite strongly the distribution of Nonconformism. In the decades before mass mobilization, the most notable distinction was between the constituent parts of the United Kingdom. Nationalist politics had begun to make their mark in Ireland, while Wales and Scotland had moved slowly, but inexorably, into the Liberal camp. The extreme weakness of Conservatism in the latter two countries was perhaps the most striking feature of nineteenth-century politics.

The democratizing electoral reforms of the 1870s and 1880s had two important and somewhat contrary consequences. First, they provided a stimulus to party organization and mobilization, at both national and local levels. Personal influence remained significant in many places, and candidate nominations could still be purchased in various ways. Nevertheless, the net result of the reforms was a more genuine national party system with both remarkably even nationwide turnout patterns and a relatively uniform national swing. Second, they helped to generate a debate upon the territorial constitution. Vastly greater popular electoral opportunities coincided with a massive expansion of central government activities. For some, the prize of government was greatly enhanced; for others, Westminster became something that should be resisted, though perhaps not at whatever the cost. Ideas on local self-government, home rule, federalism — even national self-determination — were widespread. The new debate was to have consequences for the party system, for it sprang out of the historic regional and cultural differences that had survived despite the nationalization inherent in Britain's industrialization and urbanization.

The simplest indicator of nationalization is the degree of partisan competition. Historically, the British electoral system encouraged a large proportion of uncontested seats. The reforms sponsored a higher rate of competition, but not dramatically so: before the First World War the most keenly contested election was that of 1885, the first after franchise reform. During this period Ireland stood out as a region where partisan competition had largely ceased: outside Ulster, with a Conservative-Nationalist cleavage, nationalism was hegemonic, and those contests that occurred were primarily between nationalists of various ilks. Elsewhere, an earlier pattern persisted. In Scotland, Wales and northern England it tended to be Liberals who enjoyed an uncontested election: in the other regions it was the Conservatives and their allies. The regional distribution of uncontested seats reflected the territorial nature of party strength (cf. Table 7). The Conservatives were strongest in the English core and Northern Ireland, the rest of Ireland was a totally nationalist camp, while the Liberals ruled more or less supreme in Wales, Scotland and northern England. In view of future events, it might also be pointed out that Liberals were far more likely to abstain from competition.

The split after 1885 in the Liberal amalgam over Irish Home Rule in a sense opened the way for the Conservatives to take the lead in national politicization; because of their stronger identity with tradition, the less likely were the chances of radical changes in the ter-

TABLE 7
Political party strength in Britain since 1885,
by region (selected elections)

	1885	1910 (Jan.)	1922	1929	1945	1959	1970	1974 (Oct.)	1979
	%	%	%	%	%	%	%	%	%
England									
Conservative	47.5	49.3	41.9	39.2	38.4	50.0	48.3	37.9	47.2
Liberal	51.4	43.0	26.9	23.6	9.6	6.0	7.6	20.2	14.9
Labour	—	6.9	29.0	36.7	50.1	43.9	43.4	39.0	36.7
Scotland									
Conservative	34.3	39.6	23.6	35.9	41.0	47.2	38.0	24.7	31.4
Liberal	53.3	54.2	41.9	18.1	5.6	4.1	5.5	8.3	9.0
Labour	—	5.1	31.2	42.4	47.5	46.7	44.5	36.3	41.4
Nationalist	—	—	—	0.2	1.3	0.8	11.1	30.4	17.3
Wales									
Conservative	38.9	31.9	23.1	22.0	23.9	32.6	27.7	23.9	32.2
Liberal	58.3	52.3	32.5	33.5	15.0	5.5	6.8	15.5	10.6
Labour	—	14.9	42.0	43.9	58.4	56.4	51.6	49.5	48.5
Nationalist	—	—	—	0.0	1.1	4.8	19.7	10.8	8.1
Northern Ireland									
Conservative	53.9	56.4	(55.8)	67.9	52.2	77.2	59.6	—	—
Liberal	16.7	19.5	—	16.9	—	0.6	1.5	—	—
Labour	—	3.8	—	—	16.9	11.2	8.1	1.6	0.6
Nationalist	28.8	20.3	(44.2)	6.6	16.3	11.0	21.4	29.7	28.0
Southern Ireland									
Conservative	11.2	8.6							
Liberal	2.0	—							
Nationalist	86.2	88.3							

ritorial constitution. There was only limited change in the electoral
landscape. Class did exist as a source of conflict, but while issues
may have been about class interests, they were not fundamentally
between class interests. The political temperature was not raised
significantly by the advent of the fledgling Labour Party after
1900, which peaked at only 78 candidates in January 1910. Until
the First World War, Labour remained a junior collaborator of,
and in the shadow of, the Liberals (cf. Bealey and Pelling 1958).

Class issues had perforce to fit into an older style of politics.
Religion and territory persisted as the major determinants of
voting.[25] Wilson (1966: 26) stresses that, right up to 1914, the
Liberals were 'underpinned by the force of nonconformist conviction' (cf. also Koss 1975). But secularization, economic change and
government actions were steadily weakening the dissonance between and influence of church and chapel. Yet paradoxically, the
impact of religion was honed by its association with territory. The
Irish question became the Irish problem, and hung over British
politics like a sword of Damocles.

The impact in Ireland of the revised territorial settlement was
traumatic. It broke the vestiges of the Protestant ascendancy and
stoked the fires of protest against central involvement and intervention. It inspired the formation of a host of organizations, all rooted
in a comprehensive mass mobilization and with independence as
their objective. The large Irish intrusion in Westminster after 1885
and the Liberal advocacy of home rule set the terms of reference
for the next three decades. British parties played little or no role in
Irish electoral politics. The Conservatives offered token opposition
here and there, and consolidated their position in the Protestant
north-east: the Liberals more or less disappeared from the island.
In a sense, Ireland after 1885 passed beyond the orbit of British
politics. There was also a significant impact upon the rest of the
United Kingdom. Most immediately, it led the Liberal Unionist
schism eventually into association with the Conservatives. In
England, this strengthened the religious nature of partisan politics.

Most importantly, neither Scotland nor Wales followed the Irish
nationalist footprints. Both had benefited from British economic
expansion, and both shared in the optimism of the imperial era.
Also they were still separate from England to a considerable extent.
Neither had suffered from the degree of central intervention or
settler ascendancy experienced by Ireland. Moreover, territory and
religion did not run in complete harness, in that the Welsh chapel
and Scottish kirk had perhaps even less in common with

Catholicism than had Anglicanism. Wales seemed to be more preoccupied with the issue of church disestablishment, while Scottish discontent was based more specifically on political mismanagement of, and inadequate attention to, Scottish affairs: these Scottish grievances were largely satisfied in the late 1880s. Both, and especially Wales (cf. Morgan 1963), remained loyal to Liberalism. The major difference with Ireland was that there the inspiration and demand for home rule came from Ireland itself. The possibility of Scottish and Welsh home rule was provided more by the British Liberal leadership, and the the two countries were largely content to wait upon Irish developments.

The 1914-18 hiatus in 'normal' politics represented a decisive turning-point in territorial electoral development. The postwar years were confronted with continued and widespread social and economic change, a dramatic intensification of organizational life — especially of trade unionism — and a radical extension of democratic principles in the electoral-institutional framework. Two political factors were also significant. The centrifugal tendencies within the great Liberal amalgam were intensified by personal antagonisms among its leadership. Above all, the war speeded up the elimination of Ireland from British politics. Since 1885 British politics had to a considerable extent danced to the Irish tune: with the polarization in the island between the forces of Sinn Fein and Ulster Unionism (e.g., cf. Kee 1972; Harbinson 1973), and the eventual secession of the 26 counties in 1922, the way was open for British politics to develop further. One immediate consequence was the freeing of the Irish immigrant vote from the Liberal fold. In addition, the insulation of Britain from the lingering Irish problem was assisted by the devolvement of legislative powers to a Stormont parliament in Northern Ireland. The total domination of political and institutional life in the province by the Ulster Unionists (cf. Budge and O'Leary, 1977), their persistent application of discriminatory practices against the Catholic minority and the limited representation allotted to the Irish rump in Westminster avoided the resolution of the basic territorial problem of Northern Ireland. As long as effective and undeviating Unionist control of the province could be assured, that resolution would be indefinitely postponed. In a sense, therefore, Northern Ireland also moved outside the British political orbit, while remaining part of the state.

Despite a slow organizational response to an even greater electorate, party competition became truly national for the first time: only 75 of the 587 seats in Britain were uncontested in 1918: in 1929 the United Kingdom total of unopposed returns sank to a mere six. Proportionally, the greatest number of uncontested seats were in Northern Ireland. Elsewhere, only in Wales was there a persisting lack of competition, shared between Liberalism's grip on the Welsh cultural heartland and a new Labour ascendancy in the southern coalfields. The intensification of competition was helped by, but cannot be attributed wholly to, the emergence of Labour as a mass party. There was a proliferation of candidates of all kinds. Independents flourished, despite new and stringent financial penalties for poor electoral performances. The well-known Liberal disintegration and internecine conflict was equalled in the 1920s by confusion among Conservatives and socialists. While the left was more or less organizationally consolidated by 1924, the Conservatives persisted with a willingness to collaborate with any locally approved candidate: most importantly, a tacit Liberal-Conservative alliance came to characterize much of rural Scotland until the 1940s. Yet no single party contested all the seats until the 1950s, though Conservative and Labour both first approached saturation point in 1929.

The major partisan change was the triumph of Labour (cf. Cowling 1971) and the eclipse of the Liberals. While social change contributed towards Liberal enfeeblement, the party's decline was hastened by continuing internal conflict and a greater persistence than its rivals in a selective policy of constituency contestation. The party nevertheless retained an extensive vote; in 1929 it still hovered around the break-even point in the penalizing simple-plurality electoral system. The 1931 crisis was the party's nemesis: the number of Liberal candidates slumped from 501 in 1929 to only 112 in 1931. The radical transformation cannot wholly be blamed upon a National Liberal secession, for there were very few (43) National Liberal candidates in 1931. The Liberals never recovered: they disappeared as a major electoral force in the first instance because they did not compete.[26] Where it survived in reasonable force, Liberalism rested upon the nonconformist tradition (e.g., cf. Miller and Raab 1977): within England this religious connection held across industrial areas and the pockets of agricultural concentration. In Scotland and Wales it was driven back to the rural fringes where it continued to reflect, for a while, more traditional peripheral values and social network politics. The party's decline, in a sense, emphasized more pointedly its historic stance as an 'anti-centre' party: where effective communication (that is, with London) was weaker, a relatively stronger Liberal vote tended to persist in the following decades (cf. Urwin 1980).

The Labour tide after 1918 was not a flood. Until the Second World War it was very much a party of the coalfields plus some, but not all, urban areas.[27] Labour remained very weak throughout much of rural Britain, except in Wales and, to a lesser extent, Scotland and East Anglia. In the last two instances, however, its electoral popularity was due to some extent to the absence of either Liberal or Conservative intervention. Indeed, in many parts of rural England the Labour vote declined from a high point in 1918 or 1922. Since the Labour expansion generally followed the geographical pattern of industrial distribution, it too came to have a definite territorial colouration, being stronger, for example, in northern England than in the southern core. More important, perhaps, was its more rapid conquest of the economic heartlands of Scotland and Wales, which gave it a dominant position there that it subsequently lost only temporarily in Scotland.

Until the 1920s Welsh Liberalism had identified itself strongly with the issues of disestablishment and home rule. Welsh Nonconformism, however, was more hostile to industrialization and its implications (cf. Morgan 1963; Davies 1965), not least because of the large English immigration into the boom area and the erosion in the latter of much of its Welsh distinctiveness. Until the 1890s the Welsh Labour movement had few contacts with its English counterpart: three decades later the vast majority of Welsh workers were affiliated to British trade unions rather than the earlier regional associations. The organizational impulse of Welsh labour came from England (Morgan 1963: 212-55), and thereafter Welsh working-class protest against anglicization took shape within the confines of English politics and trade union organization. Labour's conquest of the Welsh coalfields therefore intensified integrative pressures. Its huge bloc vote became an important factor in its attempts to secure a British majority. In the more sparsely populated remainder of the country, it also came to pose a serious challenge to Liberalism. Paradoxically, therefore, Labour became identified, along with the Liberals, as a party of the regional interest at the same time as its weight pushed in the direction of integration: peculiarly Welsh interests, which came to focus more and more upon language, were subordinated to economic problems, being dismissed as a 'non-issue'.

By contrast, the rapid Labour expansion in central Scotland was based more upon indigenous organizational factors and nationalist sentiment: a distinctive Scottish Labour Party and trade union federation competed with the Liberals for the title of Scotland's home rule party. In the early 1920s Red Clydeside was not only anti-capitalist and anti-Conservative; it was also anti-English, and provided a soil from which sprung many of the founders of the first Scottish nationalist movements. It was not until the further Labour advance in England that the Scottish wing more or less abandoned both its separate identity and a commitment to home rule, opting for an alliance that sought a majority position at Westminster within a unified state. At the same time, this Labour expansion limited its possibilities for growth in the rest of Scotland, which looked with suspicion and even hostility upon the industrial Glasgow region with its large (Irish) Catholic minority. As in Wales, Labour growth intensified an internal division in Scotland. But whereas in Wales the Conservatives for long remained more of a third party, in rural Scotland they gradually squeezed the Liberal remnants even further: the Scottish 'periphery' had to choose between being anti-London or anti-Glasgow.

In the 1950s the nationalization of British politics reached its peak. The island had two mass parties which competed everywhere, and the swing between them, allowing for the possibility of regional variations, was remarkably uniform. The Liberals, by contrast, had reached their nadir. Contesting very few constituencies, they had strong support in a handful of scattered constituencies, most of which lay in the sparsely populated peripheries. Other parties, including the small nationalist parties, fared even more badly. Only Northern Ireland remained sui generis, with its Unionist hegemony (cf. Harbinson 1973). Even there, however, there was a glimmering of nationalization: while Irish nationalism either remained fragmented or stood aloof from competition, the Northern Ireland Labour Party began to display reasonably respectable performances in Belfast.

With the advent of Labour as a specific class party, class issues with their territorially standardizing implications were bound to increase in importance. Yet economics fitted well with earlier party patterns of relationships with the centre: geographical politics survived. Varying numbers of candidates and uncontested seats create great problems in mapping vote distributions in British politics, as does the adjustment of constituency boundaries. We have chosen to combat the problem of varying competition by taking the median vote in each constituency over a number of elections. Using the median vote counteracts to some extent the distorting effects of

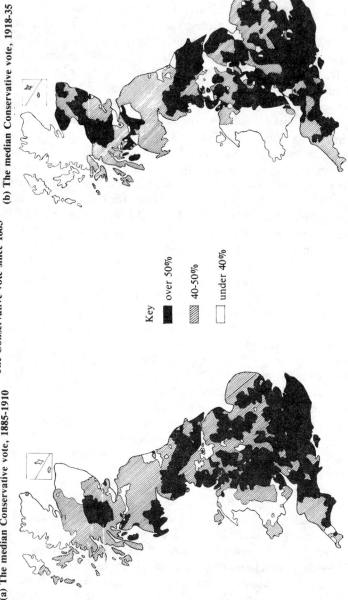

FIGURE 4

The Conservative vote since 1885

(a) The median Conservative vote, 1885–1910

(b) The median Conservative vote, 1918–35

Key

■ over 50%

▨ 40–50%

□ under 40%

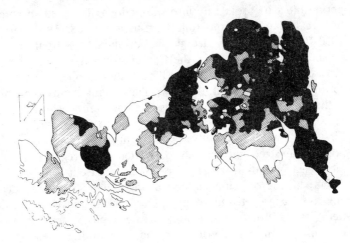

(d) The Conservative vote, 1979

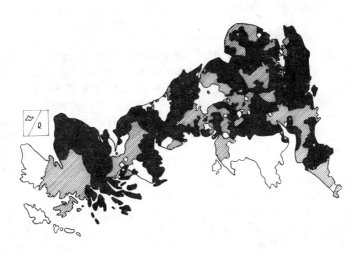

(c) The median Conservative vote, 1950-70

particular elections. Boundary reforms mean that the post-1885 period must be broken up into sub-periods. The median vote has been calculated for the following periods: 1885-1910, 1918-35, 1950-70.[28] To produce cartographic displays for all parties would be space-consuming. Since the Conservatives have been the party most closely identified with the British state and least sympathetic towards territorial aspirations, we shall examine only the Conservative vote in these three periods and compare it with the party's vote in 1979 (Figure 4). Some cautionary remarks are in order. First, the map distributions are unweighted: they pay no regard to population size, and sparsely populated rural regions stand out more clearly than urban areas. Second, because it is impossible to display all the constituencies on a single map, city constituencies have been aggregated up to an all-city level: this provides a simplified, yet potentially misleading, picture, especially in and around London. Finally, Northern Ireland has been omitted because of the distinctive nature of its party system.

However, even bearing in mind these qualifications, the maps emphasize a broad century-old pattern in voting support. The Conservative heartland has always been southern England, whereas Wales has been more or less a desert for the party. In general, the Conservative vote has tended to decline the further away from the south-eastern core. In terms of change, a steady decline in support in the northern industrial areas may be noted, though this in part may be due to extended suburbanization trends: Conservatism became a less attractive alternative within the larger cities, especially after the 1950s. Perhaps the most striking changes are those in Scotland over the whole period. There, the Conservative urban-industrial vote tended to remain static until the 1960s. Expansion came primarily in more rural areas and at the expense of the Liberals. Conservatism tended to have only shallow roots in much of Scotland, and was therefore vulnerable to the nationalist challenge that emerged in the 1960s. The Conservative vote decreased in the two 1974 elections, but not only because of a Labour growth: a major factor was the re-emergence of the Liberals who intervened effectively as a non-socialist alternative in many of the English Conservative strongholds. Even so, the historic geographical gradient persisted in the 1970s. Conservative strength remained concentrated in the South, despite large-scale migration into the industrial development within the region.

Simple electoral analysis, therefore, shows a parallel to popular images. The Conservatives have been regarded as a party of the centre; they have also been much stronger electorally in the centre regions. The reverse is true of both Liberal and Labour. What, then, of nationalism and nationalist political organization? As long as the Liberals remained a powerful political force and pledged to the idea of home rule, and/or the fledgling Labour movement expressed sympathy for the same cause, there was little political space in which nationalist organization could insert itself, except in Ireland. In the 1885-86 upheaval, home rule was obscured somewhat by the strong Welsh agitation for disestablishment and Scottish disgruntlement over a perceived lack of adequate representation of Scottish interests in London. Nationalist parties took root in the territorially redefined British state after 1922, most obviously in Northern Ireland, but also in Scotland and Wales.

In both Scotland and Wales the failure of home rule all round, disillusionment with both Liberal and Labour, and the example of Ireland convinced the committed nationalist fringe that the only means of progress was outside the traditional party structure. The success of the Irish Free State stood as a shining example, although paradoxically it was the Irish secession that permitted the British parties to remove home rule from their agenda, or at least to demote it. Both in Wales and Scotland a concern with cultural (especially linguistic) defence was a major driving force. The Welsh language has remained an important concern for Plaid Cymru, but the Gaelic protagonists were driven out of the Scottish National Party (SNP) in the 1930s. Plaid Cymru began life as a socially conservative organization, while the first Scottish party had a distinctly radical tinge, many of its founder members having come over from the socialist movements around Clydeside. In the first few decades of their existence both were truly part of the British political fringe. Their electoral activity was minimal, creating doubts over their ability to field candidates; their electoral performance was also disappointing. Like most political sects, they were introspective. Introspection led, especially in the case of the SNP, to intense debates over which strategy should be employed — electoral or not, collaboration with other parties or not — and even to convulsive purges (cf. Butt Philip 1975; Urwin 1978). While Plaid Cymru came to be more tolerant of certain forms of civic disobedience, the SNP eventually rejected all forms of direct action.

By the 1950s both were cohesive, but very small, units. Plaid Cymru had developed a more radical stance on economic issues, which it attempted to combine with its older idealistic vision that implicitly involved a rejection of the values of an industrial society.

The internal revolutions in the SNP had given it a strong conservative colouring. The party possessed a solid middle-class leadership, which sought to resuscitate and mobilize the persisting Scottish belief that London misunderstood and mismanaged Scottish affairs. The SNP's electoral activity remained almost nil until the early 1960s. Plaid Cymru had been more consistent, without achieving any marked success. The breakthrough for both came in the 1960s, as a result of single impressive by-election performances which acted, as it were, as a catalyst that generated vigour and enthusiasm. Both continued to expand before falling away again in the 1970s. The Plaid Cymru vote peaked in 1970, that of the SNP in October 1974 (cf. Table 7). Their success, especially that of the SNP, raised serious doubts about the future integrity of the United Kingdom, especially as, at the same time, the artificial structure in Northern Ireland began to crumble: 1969 marks the disintegration of unified Unionist hegemony in the province and the beginnings of a more determined minority mobilization. While Northern Ireland was, in governmental terms, brought back into the British fold, its party system, if anything, became even more distinctive. The Northern Ireland Labour Party disappeared between the polarized forces of Unionism and Irish nationalism, and Ulster Unionism itself both fragmented and ended its close relationship with the Conservatives.

Change and turbulence in Northern Ireland were generated by factors indigenous to the province. By contrast, in Scotland and Wales there was a combination of regional perspectives and British patterns. By first achieving prominence in a by-election, both parties conformed to a general picture of third-party ebb and flow in Britain. There had been two previous, albeit temporary, increases in third-party (that is, Liberal) support, in the late 1950s and early 1960s, both of which began with by-election successes. The nationalist wave that began in the late 1960s was part of the third wave, for more or less at the same time the Liberals displayed renewed life in England: Miller *et al.* (1977) point to the strong similarities between changes in the party system in both England and Scotland after the mid-1960s. Clearly, we cannot ignore the cultural distinctiveness of Scotland and Wales: it was to this that the nationalist parties appealed, and it was this that offered a foundation upon which to build electoral mobilization. Yet for some forty years patriotic appeals to national identity had fallen on exceptionally stony ground. Given the waning of the nationalist parties in the late 1970s — and at the moment it may be irrelevant to consider this as a temporary or permanent trend — it may be advisable to search for a catalyst in the British connection.

After 1922 regional politics in Britain came to be dominated by economic questions. The relative decline of industrial Scotland and Wales led to demands for special government treatment, and perhaps also to the belief that the two countries no longer had the economic resources to survive as independent units. While the interwar period marked the intrusion of class upon territorial style, the latter did not disappear completely. The Liberal role as 'defender of the periphery' was taken over by Labour. Apart from the 1945-51 period, when postwar reconstruction had been an overriding concern, Labour's joint image as a class party and as the party of the regions was, after 1931, heightened by the persistence of Conservative government hegemony until 1964. It was the straining of this dual image in the 1960s, through a combination of Labour electoral success and more adverse economic conditions, that helped loosen the two-party stranglehold. In a sense, Labour too came to be seen as having failed in the eyes of the regionalists, and, because of increasing bureaucratization and regulation, perhaps also to be as centralist in its policies as the Conservatives. There developed a malaise in British politics, expressed as an increasing loss of confidence in the managerial capabilities of the large parties. It was possible for third parties to make a much greater impact in such a climate of opinion. While a Liberal resurgence could have significance for the party system, the odds remained stacked against a decisive breakthrough without any change in the electoral system. By contrast, Plaid Cymru and the SNP had something concrete — identity — upon which to build, and were regionally concentrated. Their expansion, moreover, carried implications not just for the party system, but also for the territorial structure of the state. Scotland presented the gravest threat, for an independent Scotland would take with it the huge North Sea oil reserves upon which British calculations on future economic progress came to be based. The centre had to respond to the new challenge, and it did so in the traditional manner, seeking accommodation within prevailing structures. It was the Labour Party that faced the most serious dilemma. Not only was its weight at Westminster due in no small measure to its predominance in Scotland and Wales; to move too far in accommodation might (and did) create a backlash in those northern English regions suffering from similar economic problems, and which the party also monopolized. While the nationalist bubble burst in the 1970s, it perhaps cannot be dismissed merely as a temporary protest phenomenon. In comparative terms, the SNP and Plaid Cymru have retained a fairly respectable level of support; and, given the cultural basis of mobilization, future waves of nationalist eruptions are not inconceivable if cultural, economic and political managerial issues run in harness. In addition, the Northern Ireland sore remains. The United Kingdom is still a multicultural society, with all that that implies in terms of potential territorial mobilization, integration and accommodation.

Institutions, policies and peripheries

The early English state developed a territorial style of politics that revolved around a preference for accommodation through indirect control by the centre and direct regional representation in the centre. Actions towards and reactions by the later additions to the British state must be considered in the light of this territorial framework. With expansion and consolidation, the historical distinctions between central areas and inner periphery became less relevant. After the seventeenth century, Cornwall and the northern shires became unambiguously part of England. There was also no administrative distinction between the treatment of Wales and that of any English region. Nevertheless, because of its cultural heritage, Wales should be regarded in the same way as Scotland and Ireland.

By the mid-twentieth century regionally distinctive characteristics were regarded as little more than variations upon a basic British theme, while all conformed to standards set by the central government. The doctrine of parliamentary sovereignty came to ensure central monopoly of taxation powers, the right to revoke and retract grants of power previously conceded to other jurisdictions within the state, social and economic policy-making that was national in intention without any major or specific regard for disadvantageous regional effects, and distributive welfare policies applicable throughout the whole territory. While all these were part of the legacy of the past, the latter also included a tendency towards a positive central response to regional demands and a willingness to forgo total standardization. The centre persisted with differential treatment of Wales, Scotland and Ireland.

A consideration of this differential treatment involves a review of British institutional structures, and possible peripheral reactions to central institutions and decision-making. The institutional accommodation of territorial variations at a minimum consists of three elements: the nature of peripheral representation in the central institutions; the provision of specific institutions to safeguard territorial interests; and the extent to which there is provision for compromise between varying territorial interests. Peripheral complaints may be subsumed under the two headings of political and economic responses to central discrimination or mismanagement. One can then consider the various options available to peripheries: to work through state-wide institutions and organizations to remedy the situation; to indulge in regional mobilization; to argue that the economic, cultural and political advantages of independence outweigh the costs of secession.

A turning-point in the political reactions to, and interpretations of, the British territorial structure came in the late nineteenth century. Suffrage extension generated a need for greater organization. British party politics became more nationally oriented after the 1880s and in so doing increased the probability of regional tensions. The movement of the long-simmering Irish question, the outstanding historical failure of accommodation, to centre-stage led in the same direction. Further contributions were provided by the state's acceptance and pursuance of its claim to intervene in, legislate and administer an ever-widening circle of human activities.[29]

The peripheral response to the changing nature of British politics varied from region to region. Each response was coloured by the treatment of the periphery within the historic territorial pattern of politics. Electoral representation at the centre was gradually removed as an issue. Ireland had always been over-represented at Westminster. Scotland after 1707 might have had a grievance, for its 45 of the 568 Westminster seats was under one-half of what it deserved on demographic grounds (but also treble what would have been allotted on a taxation basis). The imbalance in Scottish representation was gradually adjusted after 1832. By the 1880s the growing demographic weight of England was producing an over-representation of Scotland and Wales, while it was deemed politically inadvisable to reduce what had become gross Irish over-representation. Since then, it has become convention that peripheral over-representation is acceptable and just. The one exception has been the Northern Ireland rump, where under-representation was justified because the province had its own legislature in Belfast.

The debate in the 1880s focused more upon influence on executive decision-making and the application of legislation. There was no particular provision for peripheral representation in government, except in Scotland, where a distinctive legal system made this necessary: but high-level Scottish governmental representation had ended in 1745. Wales, as always, had been treated the same as the English provinces. Ireland was more obviously a special case: after 1801 it retained a separate executive branch with administrative departments in Dublin, though these were controlled and directed by British ministries. Despite being treated as a separate entity, Ireland experienced perhaps a greater degree of central intervention than either Scotland or Wales.

The circumstances of the three peripheries, therefore, were very different, and so were their contributions to the territorial debate of the late nineteenth century. The overall tone and pace were set by the culturally and economically distinctive periphery of Ireland. Its increasing agitation against a perceived cultural, political and economic domination and discrimination was in the mainstream of nineteenth-century European nationalism. Compromise with London was almost impossible: the ultimate goal, backed by demonstrated popular support, was independence. The problem of Ireland and the issue of home rule made a similar policy attractive to Scotland and Wales, especially in the rural areas with their analogous economic structures. This, however, was not to be. Home rule in the Irish pattern as part of a push towards independence, though present, was not the most important component of Welsh and Scottish arguments. Both were more constrained by the integrative nature of party mobilization. Political and economic issues were less important in Wales than cultural questions, with a focus more specifically upon religion and disestablishment. Nor did Scotland raise the spectre of political nationalism: rather, it concentrated upon problems of political and administrative management of Scottish affairs. Both, therefore, were in the first instance demands for straightforward change within the existing state.

Even given Conservative/Unionist hostility towards the Irish demand the central response was fundamentally a continuation of the past — slow and positive rather than negative, seeking accommodation within a system of indirect control. There is little parliamentary or organizational evidence that suggests an intensification of pressure for Scottish home rule (cf. Hanham 1969; Urwin 1978). In any case, Scotland's specific complaints had been rectified fairly quickly, with the establishment of a Scottish Secretary and a Scottish Office in 1885, and the Scottish Grand Committee of the House of Commons in 1894. Wales was less fortunate, having to wait until 1920 for disestablishment. Even so, procrastination did not intensify or broaden Welsh pressures. And when it came, disestablishment passed, as it were, almost unnoticed: secularization, rural depopulation, linguistic decline, economic stagnation and the rise of Labour had all weakened its immediacy as an issue.

Even given the placid nature of Wales and Scotland, the Liberal leadership persisted with the grand design of home rule all round, though it was made abundantly clear that everything hinged upon a solution in Ireland. The way was not open until 1910, when the Irish Nationalists held the balance of power in the House of Commons. Concessions might have come more easily if the island had been united. Protestant Ulster, however, had mobilized just as strongly in favour of retaining the British link. The First World War forced London to shelve the whole complex issue. After that it was too late. Catholic Ireland after 1916 passed from reformism to open revolution, while the Protestant minority was also prepared to resort to military means in order to keep Ulster as part of the United Kingdom. Accommodation more or less vanished as a practical option, and it became more a question of how long the centre was willing to bear the escalating costs of maintaining territorial integrity through military control. The removal, in one way or another, of the two Irelands from the British mainstream also eliminated home rule all round as an issue. The dominant political forces in Britain were either no longer interested in or unable to press for the policy: British political passions were exhausted by the Irish dilemma, and home rule was simply shelved as a subject for discussion.

After 1918 the territorial emphasis turned to economics, not least because of the rise and geographical basis of support for the Labour Party. On a broader scale, the economic consequences of war and its drain upon manpower and resources had severely affected British industrial capacity. A host of hitherto dimly perceived problems surfaced and crystallized. Prospects of economic recovery vanished with the subsequent onset of depression and stagnation, and the definitive end of the ability of the traditional free-trade policy to service the economy. Scotland and Wales, with more specialized fields of production, were among the regions most adversely affected by economic change and decline. There was little, however, in the way of arguments that economic salvation could be achieved through independence: rather, it eventually came to be accepted as conventional wisdom that economic alleviation could not come about other than through continuance within a United Kingdom.

But regional economic policy was conceived as a nationwide design, not particularly something that catered specifically to the cultural distinctiveness and aspirations of Scotland and Wales, which held only some of the depressed industrial regions that required help.[30] The first hesitating steps were taken in the 1930s with the designation of Special Areas, though public expenditure was modest. After 1945 the stimulation of the regions continued to be pursued, more vigorously by Labour governments, less so by Conservative governments. But whatever the format, the basic assumption remained constant: that regional discontent could be satisfied by the channelling of additional financial resources. Even so, apart from a flurry of activity between 1946 and 1948, nothing much happened until the 1960s. After 1960 regional economic policies, increasing in tempo, have been in an almost continual state of flux, if not in terms of broad strategy, then at least in many details (e.g., cf. McCrone 1969; Brown 1972; OECD 1976).

The geographical area covered has steadily expanded. By 1966 the Special Areas embraced 20 percent of the British population. The revision of 1979, if we incorporate Northern Ireland, extended them to include some 45 percent of the British workforce (cf. Figure 5), a very high figure by any comparative standard. The renewed activity of the 1960s was due to a combination of factors: marked regional variations in the partisan swing in the previous elections; the accession to power of Labour; a new experimentation with central institutional structures, including decentralization; and also, of course, the rise of nationalism. While official ideology has insisted on a British planning framework, it is clear that a certain degree of flexibility has persisted, especially with regard to the non-English regions. Northern Ireland in particular has tended to be treated separately and more generously — not surprisingly, given its political and economic problems (cf. Table 6).

It would be easy to conclude from the vantage point of the present that regional policy has failed in its objectives of eliminating regional differences through the stimulation of economic growth. The problem regions remain, with the same problem industries. But failure can be assessed only in terms of what the landscape would have been like without government action. Two things are clear:

FIGURE 5
The development of regional planning:
regions designated as Development Areas

Key

▮ before 1940   ▨ 1966

▩ 1945-60   ▦ 1979

that public expenditure, both in general and with regard to the problem regions, has increased almost geometrically since the early
1960s; and that the gap between the regions, as measured by several
indicators, has tended to shrink (cf. McCrone 1969; Brown 1972;
McAllister *et al.* 1979): the economic imbalance would have been
even greater without positive government action. From the perspective of territorial structure and accommodation, two points may be
reiterated. First, there is a definite continuation of historical
gradients: a comparison of Figures 4 and 5 suggests that Disraeli's
two nations can also be interpreted in territorial terms. Second, the

centre has continued to be both positive and flexible in its response, seeking as always for gradual change and accommodation within prevailing structures. The regional economic problem may well be intractable, but in the 1960s it became linked with a more volatile party system, attitudes towards government, institutional change at the centre, and more critical English discontent at the special treatment allotted to Scotland, Wales and Northern Ireland.

Since the early 1960s the institutional framework of government has experienced turmoil in its structures and functions. Successive governments have tinkered with the apparatus of the state. Ministries have been merged, subdivided, created — even eliminated. The reports of various government committees and royal commissions have led to an attempt to restructure the bureaucratic framework and to the implementation of a new two-tier system of local government; dissenting opinions on the latter arguing for the importance of the regional dimension were not, however, followed up. The territorial consequences of all this were mixed. Most important, perhaps, was the establishment of a Welsh Office in 1964, with governmental representation, along the lines of the Scottish Office, though with a much more modest term of reference. The Scottish Office itself, relocated in Edinburgh in 1939, has continually been permitted to expand its activities (cf. Kellas 1973). There were even attempts at (administrative) decentralization within England. All these changes, however, occurred within the structure of the unitary state and did nothing to challenge the system of power and accountability.

The possibility of territorial restructuring became more prominent in 1969. As a response to the drive of Scottish and Welsh nationalism, the Labour Government established a Royal Commission on the Constitution, charged with examining 'the present functions of the central legislature and government in relation to the several countries, nations and regions of the United Kingdom' (UK 1973). Ironically, before it reported in 1973 the only major example of institutional devolution collapsed, with the breakdown of Unionist hegemony in Northern Ireland. Between 1963 and 1968 a new Unionist leadership had made the first serious attempt to reach across the abyss to the Catholic minority. The bridge-building efforts failed utterly: the major consequences were a hard-line Protestant backlash and the stimulation of an environment for rising minority expectations and potential action. After years of increasing violence, the Stormont parliament was revoked and the 'errant' province brought under direct London control, with a minister responsible for the region. Since then the Northern Ireland problem has continued to be regarded as a separate issue outside the field of partisan politics. Recognizing that an Ulster majority wishes to remain in the United Kingdom (though as an autonomous unit), that a total British withdrawal would make a grave problem even worse, that a British public has become more disillusioned with and restive over the costs of a military and political presence in the province (e.g. cf. Rose *et al.* 1978), and that there had been blatant discrimination since 1921, governments have searched for a solution that would accommodate the warring communities in a devolved framework that would again insulate Britain from the perpetual Irish problem. The Labour Government emphasized in 1975 (UK 1975: 2) that:

> The unity of the United Kingdom does not mean uniform treatment for all its parts. This White Paper is about devolution to Scotland and Wales, and its proposals are related to their circumstances. As the White Paper of September 1974 made clear, Northern Ireland is in a different category. Its history and geography distinguish it from other parts of the United Kingdom as does the presence of two separate communities. Its problems are not those of Scotland and Wales, and therefore do not necessarily require the same treatment.

The rapid collapse of the 1975 Convention and the failure of the power-sharing experiment, whereby the Catholic minority would be involved in government, reduced further the practical alternatives to direct rule. Furthermore, the years of violence have produced a more complex political mosaic after the shattering of monolithic Unionism. A more aggressive and organized Irish nationalism faces a more extreme version of Unionist loyalism, with moderate accommodative forces squeezed between. Direct rule has probably satisfied no one, yet the options are exceedingly limited, the major one perhaps being seen as even greater violence. Northern Ireland, therefore, stands in a unique position. It is a region where both majority and minority wish some form of severance of the direct links with London — a wish with which the British population, of all political persuasions, concurs — but where it has proved impossible to secure support for an option acceptable to the British government. Moreover, because of its almost anomalous position within the United Kingdom, Northern Ireland poses no serious threat to the British system and state.

The Royal Commission on the Constitution reported in 1973. It is undoubtedly the most significant document to appear on the constitutional structure of the United Kingdom. The majority opinion summarized the traditional centralist perspective, reminding everyone that the constitution 'in its essentials has served us well for some hundreds of years', as well as putting forward broad guidelines that were later followed by the Labour Government until the rejection of devolution proposals in the Scottish and Welsh referenda of 1979. It was the sensational upsurge of Scottish nationalism in the 1974 elections that seemingly made action imperative, and it was the electoral threat, especially to the Labour Party's competitiveness, rather than the Kilbrandon Report's arguments, that swung the Labour Government towards urgent action.

The Commission's recommendations, which did suggest some changes in the machinery of government, originally made little impact, in part because, while there was unanimity upon the objective, there was wide disagreement among the members as to the methods necessary to achieve it. A majority report contained in fact four different devolution outlines: a fifth was presented in a minority Memorandum of Dissent. The thrust was towards the establishment of legislative assemblies for Scotland and Wales. Since Northern Ireland had had (and, it was hoped, would have again) such an assembly, logically there should also be one for England or several English regional assemblies. This would have pushed the United Kingdom towards federalism, but this was an option rejected by the Commission's majority. Their justifications were dubious, and showed a lack of understanding of federalism and working federal systems as well as a prejudice in favour of a solution within the existing state structure. It was left to the Commission's minority to recommend also English regional assemblies which would direct the regional offices of some central government institutions as well as the numerous 'quasi-regional' public authorities already in existence. The Labour Government, however, based its tactics upon the majority recommendations.

Some basic political facts determined the devolution debate of the 1970s. The Labour leadership remained deeply suspicious of devolution: their conversion was from necessity, not conviction. They saw their Scottish and Welsh strength as essential: hence, while proposing devolution, they argued for no change in the size of Scottish and Welsh representation at Westminster. Second, they attempted to occupy a shifting middle ground between on the one hand the strident nationalists and a Liberal Party arguing the federal case, and on the other a Conservative Party that had quickly abandoned its own half-hearted approach towards devolution as well as the large majority of Labour activists in Scotland and Wales. To go too far might generate an English backlash; to hold back might simply provide the nationalists with more ammunition. In the last resort, the Labour proposals were designed to combat the SNP. They outlined a prospective Scottish assembly with a wide range of legislative powers. To accommodate a much weaker Welsh challenge, it was proposed that Wales should have an assembly with executive, but not legislative, powers. Northern Ireland was held outside the debate, while English devolution was a non-starter. Moreover, it was made clear that even a Scottish assembly would have no control over the nationalized industries and economic planning and development, and provisions for it to levy taxation were also excluded.

After numerous delays the final package was submitted to a referendum in Scotland and Wales in 1979, where it was rejected, decisively so in the latter. It was defeated not because of what the government did, but more because it did nothing. Procrastination, the consequence of uncertainty as to the most appropriate course of action, turned out to be the best policy, for by 1979 the opinion polls were indicating a sharp downturn in nationalist forces. But the proposals satisfied no one. The nationalist goal was independence, though they, like the Liberals, were willing to accept the package as a first stage. Both Labour and Conservative disliked it intensely or were extremely unhappy about it. That the electorate should also reject it should not perhaps have been surprising either. Indeed, given the demographic structure of Wales, the distribution of the Welsh language, and the nature of Plaid Cymru support, a Welsh rejection was always probable. In Scotland it was clear that voting for the SNP did not mean the endorsement of the party's goal of independence. Throughout the party's hey-day in the mid-1970s, opinion polls consistently demonstrated that this goal was supported by only one-fifth of the electorate, although a further three-fifths would have accepted some degree of devolutionary change. The question was rarely considered of how committed the electorate was to institutional reform and its saliency vis-à-vis other issues.[31] By 1979, at any rate, its saliency was clearly limited. Moreover, the Scottish electorate was not very different from those of other British regions in the distribution of opinions on devolution options. The proportion desiring 'secession' was greater in Scotland, though not remarkably so. Results from the survey undertaken for the Royal Commission are indicative of the general spread (Table 8): note the extremely low figure for Wales in the final column.

TABLE 8
Attitudes towards devolution in Britain by region: support for alternatives*

	Alternative no.				
	1	2	3	4	5
	%	%	%	%	%
Scotland	6	19	26	24	23
Wales	15	27	21	23	13
England					
North	11	27	26	20	16
Yorkshire	16	23	24	20	16
North West	8	19	30	24	15
West Midlands	15	23	26	18	17
East Midlands	12	23	26	21	18
East Anglia	16	24	17	21	20
South West	20	28	20	17	12
South	11	23	20	25	21
South East	14	29	21	20	16
Greater London	16	24	24	19	14

* Alternatives: For running (the region), which of these five alternatives would you prefer overall?
1. Leave things as they are at present.
2. Keep things much the same as they are now but make sure that the needs of (the region) are better understood by the Government.
3. Keep the present system but allow more decisions to be made in (the region).
4. Have a new system of governing (the region) so that as many decisions as possible are made in the area.
5. Let (the region) take over complete responsibility for running things in (the region).

Source: UK (1973).

The devolution debate more or less ended in 1979, but decline does not necessarily mean death. The package failed in 1979 because, per se, it had almost no committed supporters. But Welsh and Scottish nationalism will survive as long as the distinctive cultural attributes remain. Given the appropriate catalyst, identity can always be utilized as a base for regional mobilization. Perhaps the major point about the 1960s and 1970s for the future is that they demonstrated that the British party system is vulnerable to this type of challenge, and where it does arise policy-makers will be forced again to consider their own precepts of government and the territorial structure of the state.

The last old state?: flexibility and accommodation

The national identities of Wales and Scotland, which have survived the passage of time, almost oblige us to think of them as countries: the same is probably true of the warring communities of Northern Ireland. Nevertheless, they must be considered in light of the English connection and the whole structure of the United Kingdom. If the latter consists of one society, then that is a multicultural society. Moreover, the disparate cultural heritage of the various parts of the state must be measured against the persistence of a broad consensus about the nature of the overall relationship with central government. What makes the British case interesting is not the Northern Ireland dilemma as such, or the Scottish and Welsh turmoil of the past two decades, but the broader vista of three very distinctive peripheries linked to a political centre whose basic strategy was evolved almost a millenium ago. The United States was once defined as 'the first new nation' (Lipset 1963): in a sense, Britain could be said to be the last old state.

That strategy may be defined as one of territorial accommodation through the recruitment of and collaboration with regional elites who in turn were given representation in central institutions. Where conflict arose, the preferred option was compromise, but compromise at minimum cost to the centre and within the prevailing constitutional structure. This has two implications: first, in its broad strategic outlines the medieval territorial settlement developed by the emerging English state has remained very much alive; and second, the central response to regional challenges has been an ad hoc response. From the various attempts to subdue Ireland to the devolution issue of the 1970s, the British centre has treated each region and regional complaint on its own merits, a practice aided by the absence of a written constitution: the piecemeal, sectorial response to Welsh demands for linguistic parity has been typical. At the same time, the lack of any rigid institutional structure has made overall reform difficult. The introduction of devolution would perhaps have far-reaching and possibly unforeseen consequences, while a decision to accept federalism would, quite apart from the problem of what to do with England, have involved a fundamental revolution of the institutional structure of the state. In that sense, given the inertia inherent in any political system and the fact that there was no revolutionary situation in the 1970s (apart from Northern Ireland — but there 'revolution' also incorporated the majority desire to stay within the United Kingdom), a radical institutional restructuring was hardly an option.

While Anglo-Welsh and Anglo-Scottish relationships were not always historically peaceful, accommodation has been successful in that union with Wales and Scotland ultimately provided a stable political framework for elite co-operation and, after the Industrial Revolution and imperial expansion, economic integration. The paradoxical consequence was the ending of regional semi-isolation and the beginnings of a greater concentration of political power in London. Economic and political trends in and events of the twentieth century have merely intensified this centralization process. A developing party system originally reflected territorial differences, but these were still contained within the broad consensus. The crucial disrupting factor was the crystallization of these trends in the late nineteenth century and their effect upon Ireland, which helped the first 'modern' wave of regional discontent elsewhere.

The Irish imbroglio set in motion the most important political change after 1918, the everwidening cracks in and eventual rapid decline of a centrifugal Liberal alliance, and its replacement by Labour. The rise of the latter inevitably meant an increased focus upon economic issues and social class: because class differences were nationwide, class also became 'a factor making for national unity' (Pulzer 1967: 46). However, the Labour vote was also territorially skewed. Not only did it capture much of the Liberal support; it also inherited much of the traditional Liberal image, including originally a sympathy towards the regions. But by the 1930s Labour was emphasizing a programme of state planning, centralization and efficiency as a solution to regional economic problems. It was only with the decline of the Liberals and the relinquishing by Labour in the 1920s of its home rule commitment that a greater opportunity arose for nationalist parties to claim to be the sole hope for peripheral interests. But in some ways history had passed them by. Their greater concern with cultural defence and values aroused only a limited response in the bleak economic climate of the interwar years.

In contrast to the failure of the nationalist parties to evoke more than a faint response among their potential electorates, the Labour Party successfully retained an image of both underprivileged class and peripheral interests, arguing that all problems would best be solved within a centralized United Kingdom structure and redistribution of economic resources. The possible conflict between redistribution among individuals and redistribution among territories was never put to the test before the 1960s. General Conservative predominance enabled Labour to survive in this dual yet ambiguous role. The Liberals seemed to be almost extinct, while the extremely limited and spotty contesting by and performance of the nationalists in elections were not conducive to optimism. A new territorial debate and alternative solutions could come about only with an accentuation or realization of the ambiguity inherent in the Labour image. This came with the concatenation of political and economic sequences after 1964.

Recent events have reconfirmed the basic multicultural nature of the British state and the potential threat that distinct cultural identities can have as the seed-bed of political nationalism. Simultaneously, however, they also point to the benign nature of multiculturalism. In that political and economic integration with England did work, a British identity was grafted on to indigenous identity in Wales and Scotland. In many ways this dual identity survived almost untarnished until recently. In the past two decades what has emerged, because of economic problems, the end of empire and cohesive Commonwealth, and the confusion surrounding the move to Europe, is perhaps less an upsurge in 'Celtic' nationalism than a weakening of the wider sense of British identity. It is this that has thrown Scottish and Welsh nationalism into greater relief by underlining the basic tension of dual identities (and in Northern Ireland the antagonistic nature of Protestant dual identity), and that reaffirms the necessity of considering them in the light of the English connection and central strategies of integration. Membership of the European Community could either further confuse the situation or simplify it: confuse, because it creates another layer of territorial decision-making; simplify, because it may permit the accommodation of regional concerns through a bypassing of the national centre.[32] A 'Europe of Regions' is certainly not a possibility at the moment: instead, the Brussels aim is for a 'politics of concertation' (Ionescu 1975). Nevertheless, any future discussions on territorial accommodation will be influenced by the presence of a supranational level of decision-making.

Any state will pay a high price to secure the retention of its territory. Only in Ireland did the costs become too severe for the British state. Within this general constraint, the British centre, with its emphasis upon tolerance and indifference, has always been willing ultimately to respond positively to territorial demands. But that response has been determined by the concern to accommodate demands within the prevailing structure. In the last resort it has been an ad hoc attempt to resolve a specific complaint or demand: there has been no grand design. The centre, in a sense, has had no choice of standardized options, for all, including limited legislative devolution or federalism, run up against the diffuse nature of the constitutional structure and the hybrid nature of the British state. Since all would require a fundamental reformulation of the state, their attractiveness has been limited, their costs high and their outcomes uncertain. It would seem more probable that the British centre will always prefer to pursue discriminatory policies distinctive in terms of specific regional problems. Britain is not only a multicultural society. The presence of (at least) three very different peripheries, with differing concerns and issues, has produced a hybrid state in which the centre's options may be flexible, but only within very restrictive parameters.

NOTES

1. Despite the misinterpretation and lack of understanding of British history and a questionable methodological approach (cf. the critiques in Page 1978; O'Leary 1976-77), Hechter's study raises some queries that had not been answered by the conventional wisdoms. The same is true of the strongly polemical, yet highly readable, collection of papers by Nairn (1977).

2. As a piquant footnote on the variable degree of treatment across territories, it can be pointed out that several smaller islands belong to the United Kingdom. The Isle of Man, technically a Crown dependency under the Home Office, has been a self-governing unit since 1866 (cf. Kermode 1968), while the Channel Islands each received, under the 'Constitution of King John' in 1215, a form of self-government that has persisted up to the present time. While these tiny appendages to the British state raise interesting theoretical perspectives on problems of integration, they will play no part in this analysis, which concentrates upon the two islands of Britain and Ireland.

3. It might also be said that the conspicuous place that smuggling enjoys in British history does not suggest an impervious obstacle.

4. In fact, the claim was formally abandoned only in 1801 with the Treaty of Amiens.

5. We shall not discuss in any detail the geopolitical location of the British Isles and their early peripherality to European politics and economics. At the edge of Europe, they straddled the fringe of the Roman Empire. Despite the involvement of both England and Scotland with France, they were relatively minor powers and players on the European stage, as well as being at the rim of European trading systems. It was only after the long sixteenth century that the geopolitical place of Britain was radically transformed (Wallerstein 1974).

6. There are several historical indicators of the anti-central stance of the northern shires; for example, in the early fifteenth century, the powerful Percy family could be found linked to the Welsh standard of Glendower in opposition to London.

7. For example, the central Irish kingship established in the eleventh century to co-ordinate resistance to invasion soon collapsed under internal dissension. By the late twelfth century the papacy had granted the lordship of Ireland to the Angevin monarchy, and the Anglo-Irish agony had begun.

8. After all, it was through a landing in Wales that the Tudors themselves had earlier launched their bid for power.

9. If the Irish had any doubts about the force of English arms, they were dispelled a century later by Cromwell's military intervention.

10. The last monoglot speaker allegedly died around the 1780s.

11. Even so, it is worth reiterating that incorporation did not lead immediately to a purposeful and undeviating policy of standardization (e.g. cf. Elton 1972).

12. The seeds of religious pluralism, however, had been sown earlier, and played, for example, a significant role in the civil war of the seventeenth century.

13. A recent assessment paints a bleak future for the language (cf. Lockwood 1975: 33). A good picture of the politics of language can be found in Madgwick (1973).

14. The most agricultural region, Ireland, also had to struggle with an inequitous land tenure system and the changing patterns of English demand in the commodity market (Lyons 1971: 31; Solow 1971: 4-11).

15. In Ireland Protestant Belfast rose as a rival to predominantly Catholic Dublin.

16. Including Edinburgh and Glasgow in Scotland, and Dublin in Ireland. With a meagre population of 7,700, Merthyr Tydfil was the largest urban centre in Wales.

17. In addition, many rural areas also incorporated mining or textile villages.

18. Though in recent years the inner core has suffered substantial depopulation.

19. For an account of the changing influence of the northern cities in English history, cf. Read (1964).

20. It may be noted that the first, and for a considerable period of time the only, major road-building scheme was the construction of military roads in the peripheral Scottish Highlands after the Jacobite rebellion of 1745.

21. Though as a result of economic depression and structural change, the dependency of Wales and Scotland upon traditional specialization actually decreased (Brown 1972: also McCrone 1965; Nevin *et al.* 1966).

22. In 1971, of 27 United Kingdom regions, the urban population was less than 50 percent in only 4: the south-western corner of both Wales and Scotland, the Scottish Highlands, and the western half of Northern Ireland (McAllister *et al.* 1979: 33). But together these areas contained only 2 percent of the British population. The data in this and the following paragraphs are derived from the study by McAllister and his colleagues.

23. Much of the material in this section is based on the more extensive discussion

of the nationalization of British party politics after 1868 in Urwin (1980).

24. Most boroughs were small and semi-industrial: they were as likely to vote Conservative as Liberal. The corollary is that there were numerous instances of rural Liberalism. Moreover, in the larger cities the Liberal star was in persistent decline from the 1850s onwards (e.g. cf. Fraser 1976; also Pelling 1968).

25. This was especially true, and remained so, in Northern Ireland (cf. Budge and O'Leary, 1973; Stewart 1977). The importance of religion was reflected in the affiliation of candidates for the various parties (cf. Blewett 1972).

26. It is ironic to note that, but for the Great Depression and its impact upon British party politics in 1931, the electoral system would probably have been changed to that of the alternative vote before the next general election. If this had occurred, Liberal survival in some strength would almost certainly have been assured and the future course of British politics radically different.

27. Similar patterns of competition appeared in local politics. The distinctiveness of Liberalism in local politics had been especially based upon the espousal of issues dear to Nonconformism, especially education. By the 1920s these had largely disappeared owing to central policy administration. The Labour advance in competition and votes was strong in the large cities, at the expense of Liberalism, which declined particularly dramatically in London (e.g. cf. Thompson 1967; Bulpitt 1967).

28. In practice this means the omission of the 1945 election. Boundary changes make it very difficult to merge 1945 with subsequent elections, while the ten-year gap from 1935, with all that occurred within the period, make it inappropriate to combine it with interwar elections.

29. A good overview of central expansion is Parris (1969).

30. It should be noted that agriculture has rarely been involved in British regional policy. There have always been rural problems, and areas like the Scottish Highlands have received government assistance. But since the population concerned is small, the issue has not loomed large in the minds of the policy-makers or the electorate.

31. In the Kilbrandon Commission's survey of attitudes, devolution was less popular than more concrete desires for regional improvement, including the admittedly vague 'economic development'. In fact, more local democracy elicited the least number of positive responses of the 14 improvement items surveyed.

32. Some possible outcomes, now rather outdated, are reviewed by the various contributions to Kolinsky (1978).

REFERENCES

Almond, G.A. and Verba, S. (1963). *The Civic Culture.* Princeton: University Press.

Anderson, P. (1974). *Lineages of the Absolutist State.* London: New Left Books.

Bealey, F. and Pelling, H. (1958). *Labour and Politics, 1900-1906.* London: Macmillan.

Beckett, J.C. (1966). *The Making of Modern Ireland: 1603-1923.* London: Faber.

Beer, S.H. (1965). *Modern British Politics.* London: Faber.

Bindoff, S.T. (1966). *Tudor England.* Harmondsworth: Penguin.

Blewett, N. (1972). *The Peers, the Parties, and the People.* London: Macmillan.

Bolton, G.C. (1966). *The Passing of the Irish Act of Union.* London: Oxford University Press.

Bottigheimer, K.S. (1971). *English Money and Irish Land: The 'Adventurers' in the Cromwellian Settlement of Ireland.* Oxford: Clarendon Press.

Briggs, A. (1959). *The Age of Improvement*. London: Longmans.

Brown, A.J. (1972). *The Framework of Regional Economics in the United Kingdom*. Cambridge: University Press.

Budge, I. and O'Leary, C. (1973). *Belfast: Approach to Crisis*. London: Macmillan.

Budge, I. and O'Leary, C. (1977). 'Permanent Supremacy and Perpetual Opposition: The Parliament of Northern Ireland', in A.F. Eldridge (ed.), *Legislatures in Plural Societies*. Durham, North Carolina: Duke University Press.

Bulpitt, J.G. (1967). *Party Politics in English Local Government*. London: Longmans.

Bulpitt, J.G. (1975) 'The Problem of the North Parts: Territorial Integration in Tudor and Stuart England'. Coventry: University of Warwick, Politics Working Paper 6.

Butt, Philip, A. (1975). *The Welsh Question*. Cardiff: University of Wales Press.

Chambers, J.D. and Mingay, G.E. (1968). *The Agricultural Revolution 1750-1880*. London: Batsford.

Clarkson, L.A. (1971). *The Pre-Industrial Economy in England 1500-1750*. London: Batsford.

Coate, M. (1933). *Cornwall in the Great Civil War and Interregnum, 1642-1660*. Oxford: Clarendon Press.

Coupland, Sir R. (1954). *Welsh and Scottish Nationalism*. London: Collins.

Cowling, M. (1971). *The Impact of Labour, 1920-1924: The Beginning of Modern British Politics*. Cambridge: University Press.

Cullen, L.M. (1972). *The Economic History of Ireland since 1660*. London: Batsford.

Curtis, Jr., L.P. (1968). *Anglo-Saxons and Celts: A Study of Anti-Irish Prejudice in Victorian England*. Bridgeport, Conn.: Conference on British Studies.

Davies, E.T. (1965). *Religion in the Industrial Revolution in South Wales*. Cardiff: University of Wales Press.

Deane, P. and Cole, A.W. (1962). *British Economic Growth 1688-1959*. Cambridge: University Press.

Dobb, M. (1963). *Studies in the Development of Capitalism*. London: Routledge & Kegan Paul.

Donaldson, G. (1960). *The Scottish Reformation*. Cambridge: University Press.

Eckstein, H. (1962). 'The British Political System', in S.H. Beer and A.B. Ulam (eds), *Patterns of Government*. New York: Random Press.

Elton, G.R. (1972). *Policy and Police: The Enforcement of the Reformation in the Age of Thomas Cromwell*. London: Cambridge University Press.

Fraser, D. (1976). *Urban Politics in Victorian England*. London: Leicester University Press.

Gay, J.D. (1971). *The Geography of Religion in England*. London: Duckworth.

Halliday, F.E. (1975). *A History of Cornwall*. London: Duckworth.

Hammond, E. (1968). *An Analysis of Regional Economic and Social Statistics*. Durham: Rowntree Research Unit, University of Durham.

Hanham, H.J. (1959). *Elections and Party Management*. London: Longmans.

Hanham, H.J. (1969) *Scottish Nationalism*. London: Faber.

Harbinson, J.F. (1973). *The Ulster Unionist Party 1882-1973*. Belfast: Blackstaff Press.

Harris, R. (1972). *Prejudice and Tolerance in Ulster*. Manchester: University Press.

Hatcher, J. (1970). *Rural Economy and Society in the Duchy of Cornwall, 1300-1500*. Cambridge: University Press.

Hechter, M. (1975). *Internal Colonialism*. London: Routledge & Kegan Paul.

Heslinga, M.W. (1962). *The Irish Border as a Cultural Divide*. Assen: Van Gorcum.

Inglis, K.S. (1963). *Churches and the Working Classes in Victorian England*. London: Routledge & Kegan Paul.

Ionescu, G. (1975). *Centripetal Politics*. London: Hart-Davis MacGibbon.

Kee, R. (1972). *The Green Flag*. London: Weidenfeld & Nicolson.

Kellas, J.G. (1973). *The Scottish Political System*. Cambridge: University Press.

Kermode, D.G. (1968). 'Legislative-Executive Relations in the Isle of Man', *Political Studies*, 16: 18-42.

Kinnear, M. (1968). *The British Voter*. London: Batsford.

Kolinsky, M. (ed.) (1978). *Divided Loyalties*. Manchester: University Press.

Koss, S. (1975). *Nonconformity in British Politics*. London: Batsford.

Laslett, P. (1965). *The World We Have Lost*. London: Methuen.

Lipset, S.M. (1963). *The First New Nation*. New York: Basic Books.

Lockwood, W.B. (1975). *Languages of the British Isles Past and Present*. London: André Deutsch.

Lyons, F.S.L. (1971). *Ireland Since The Famine*. London: Weidenfeld & Nicolson.

McAllister, I., Parry, R. and Rose, R. (1979). *United Kingdom Rankings: The Territorial Dimension in Social Indicators*. Glasgow: Centre for the Study of Public Policy. Studies in Public Policy, no. 44.

McCrone, G. (1965). *Scotland's Economic Progress 1951-60*. London: Allen & Unwin.

McCrone, G. (1969). *Regional Policy in Britain*. London: Allen & Unwin.

Machin, G.I.T. (1964). *The Catholic Question in English Politics 1820-1830*. Oxford: Clarendon Press.

Mackinder, H. (1907). *Britain and the British Seas*. Oxford: Clarendon Press and New York: D. Appelton & Co.

Madgwick, P.J. (1973). *The Politics of Rural Wales*. London: Hutchinson.

Medlicott, W.N. (1967). *Contemporary England 1914-1964*. London: Longmans.

Miller, W.L. and Raab, G. (1977). 'The Religious Alignment at English Elections between 1918 and 1970', *Political Studies*, 25: 227-51.

Miller, W.L., Särlvik, B., Crewe, I. and Alt, J. (1977). 'The Connection Between SNP Voting and the Demand for Scottish Self-Government', *European Journal of Political Research*, 5: 83-102.

Minchinton, W.E. (ed.) (1969). *The Growth of English Overseas Trade in the Seventeenth and Eighteenth Centuries*. London: Methuen.

Morgan, K.O. (1963). *Wales in British Politics 1868-1922*. Cardiff: University of Wales Press.

Nairn, T. (1977). *The Break-Up of Britain*. London: New Left Books.

Nevin, E.T., Roe, A.R. and Round, J.I. (1966). *The Structure of the Welsh Economy*. Cardiff: University of Wales Press.

Norman, E.R. (1968). *Anti-Catholicism in Victorian England*. London: Allen & Unwin.

OECD (1976). *Regional Problems and Policies in OECD Countries*, vol. II. Paris: OECD.

O'Leary, C. (1976-77). 'Celtic Nationalism: A Study of Ethnic Movements in the British Isles', *Jerusalem Journal of International Relations*, 2: 51-73.

Page, E. (1978). 'Michael Hechter's Internal Colonial Thesis: Some Theoretical and Methodological Problems', *European Journal of Political Research*, 6: 295-317.

Parris, H. (1969). *Constitutional Bureaucracy: The Development of British Central*

Administration since the Eighteenth Century. London: Allen & Unwin.

Pelling, H. (1968). *Popular Politics and Society in Late Victorian Britain*. London: St Martin's Press.

Plumb, J.H. (1967). *The Growth of Political Stability in England, 1675-1725*. London: Macmillan.

Pollard, S. (1969). *The Development of the British Economy 1914-1967*. London: Edward Arnold.

Pulzer, P.J. (1967). *Political Representation and Elections in Britain*. London: Allen & Unwin.

Quinn, D.B. (1966). *The Elizabethans and the Irish*. Ithaca, New York: Cornell University Press.

Raftery, J. (ed.) (1964). *The Celts*. Cork: Mercier Press.

Read, D. (1964). *The English Provinces c.1760-1960*. London: Edward Arnold.

Rees, W. (1967). *The Union of England and Wales*. Cardiff: University of Wales Press.

Rose, R. (1971). *Governing Without Consensus*. London: Faber.

Rose, R., McAllister, I. and Mair, P. (1978). *Is There a Concurring Majority on Northern Ireland?* Glasgow: Centre for the Study of Public Policy, Studies in Public Policy no. 22.

Rowse, A.R. (1969). *Tudor Cornwall. Portrait of a Society*. London: Macmillan.

Rudé, G. (1964). *The Crowd in History, a Study of Popular Disturbances in France and England, 1730-1848*. New York: John Wiley.

Scarisbricke, J.J. (1968). *Henry VIII*. London: Eyre & Spottiswoode.

Schoeffler, H. (1960). *Wirkungen der Reformation*. Frankfurt: Klostermann.

Solow, B.L. (1971). *The Land Question and the Irish Economy, 1870-1903*. Cambridge, Mass.: Harvard University Press.

Stewart, A.T.Q. (1977). *The Narrow Ground: Aspects of Ulster 1609-1969*. London: Faber.

Strayer, J.R. (1970). *On the Medieval Origins of the Modern State*. Princeton: University Press.

Thompson, P. (1967). *Socialists, Liberals and Labour: The Struggle for London*. London: Routledge & Kegan Paul.

UK (1973). *Royal Commission on the Constitution I Report* (Cmnd 5460); II *Memorandum of Dissent* (Cmnd 5460-1). London: HMSO.

UK (1975). *Our Changing Democracy* (Cmnd 6348). London: HMSO.

Underdown, D. (1971). *Pride's Purge*. London: Oxford University Press.

Urwin, D.W. (1978). *The Alchemy of Delayed Nationalism: Politics, Cultural Identity and Economic Expectations in Scotland*. Bergen: INSS. Sammenliknende Politik Rapport 3.

Urwin, D.W. (1980). 'Towards the Nationalisation of British Politics?: The Party System 1885-1940', in O. Büsch (ed.), *Wählerbewegung in der Europäischen Geschichte*. Berlin: Colloquium Verlag, 225-58.

Wallerstein, I. (1974). *The Modern World-System*. London: Academic Press.

Williams, D. (1950). *A History of Modern Wales*. London: Murray.

Wilson, C. (1965). *England's Apprenticeship, 1603-1763*. London: Longmans.

Wilson, T. (1966). *The Downfall of the Liberal Party, 1914-1935*. Ithaca, New York: Cornell University Press.

3

Norway: Centre and Periphery in a Peripheral State

Frank H. Aarebrot
University of Bergen

On a map of Europe, Norway stands out mainly for its peripheral location: spanning the north-western coast of Europe, the country is indeed on the fringe of the Continent. But Norway is also marginal in terms of demography. Its small population, which barely exceeds four million, is spread relatively sparsely across the territory. Substantial urbanization has taken place over the last century, but even so the combined population of the three major cities constitutes barely one-quarter of the total population. Some of the major industries are located outside the urban centres.

Norway's economic position in the world is perhaps more significant than its political and cultural importance. Norway's share of the freight market that developed with the growth of the shipping industry after the 1850s was considerable. After 1900 cheap hydroelectricity gave rise to major industries. In the 1970s North Sea oil became increasingly important to the economy. An analysis of the country that employed only a contemporary perspective might tend therefore to underestimate its historical role as a European periphery.

Beyond this obvious peripherality, there are six further reasons why Norway ought to be an interesting case for students of territorial variations in the interplay between economic change and cultural identity.

1. The recent history of the country is a story of a territorial solution, through secession in 1905, to the problems of the Swedish-Norwegian union. An analysis of the nation-building and integration processes in the most peripheral part of that union should be of considerable interest as a 'success story' to students of peripheral minorities in other European states.

2. The juxtaposition of the politico-administrative penetration by the central authorities of a relatively unitary state, and the mobilization and counter-mobilization of markedly distinct peripheries, is one of the central themes of modern Norwegian political science, notably through the work of Stein Rokkan (cf. Rokkan and Valen 1969, 1970; Valen and Rokkan 1974; also Aarebrot and Urwin 1979). In his analyses of Norway, Rokkan offers, as a substitute for 'regionalism', a basic distinction between one central and several different peripheral areas as a means of explaining the maintenance of a territorially heterogeneous society. Centres and peripheries are interdependent and complementary antagonists, whereas regionalism can, in its extreme consequences, challenge the unity of the state.

3. Norway has experienced rapid economic growth in this century. At the same time, the cultural distinctiveness of some regions has remained relatively constant. Despite this combination of economic change and the persistence of different cultural identities in the peripheries, Norway has not been subjected to increased regional political tension — in contrast to some other areas of Western Europe, where recent regional nationalisms have been explained, at least partially, by adding the existence of traditional cultural values to the impact of economic change. Norway is interesting simply because it does not fit such explanations.

4. The Christian People's Party is one of the few Protestant parties in Europe for the defence of religious interests. Aside from the special case of Ulster, this kind of party has a tradition only in the Netherlands.[1] Moreover, the Christian People's Party is one of very few examples since 1918 of a party with clear regional origins which successfully made the transition to a national party with nationwide appeal.

5. There are at least three major examples in the last century of major victories by peripheral coalitions over central interests: the introduction of parliamentarism and the subsequent Liberal hegemony in the mid-1880s; the Prohibition referendum in 1918; and the negative popular vote on the question of Norwegian entry into the EEC in 1972. The interesting question is the extent to which a rare but major peripheral victory can contribute, within the existing political system, to a reconfirmation of the unitary state.

6. Norway has been able to afford a relatively aggressive 'district policy'. Considerable effort in terms of restrictive laws, agencies, personnel and money has gone into major projects and strong economic incentives aimed at encouraging a geographical diffusion

of industrial location and the maintenance of the scattered settlement pattern in the peripheral regions. These policies proliferated
after 1945, and the question is whether district policies have served
to anticipate and prevent regional political unrest.

The origins of the Norwegian
state and regional economic differences

The derivation of the word 'Norway' is economic rather than
political. It refers to the northern trading route, 'nordvegr', of
early Scandinavia, branching out from the central axis of the Baltic
trade from Hedeby to Gotland. The term describes the sea lane
from Olsofjord around the coast to the Lofoten islands in the
north. The route was based on a protective line of islands between
the North Sea and the mainland, which give shelter to the sea traffic. The topographical structures formed a lane, which performed
functions not dissimilar to those of rivers on the Continent.

Only slowly did the economic concept of a trading route, a 'road
to the North', change into a political entity, the state of Norway.
This took place between approximately 800 to 1350. At the Diet of
1163 the infrastructures of the state and the national church were
formally established, and the name used to describe them was 'Norway'. In the earlier part of this period (to about 1050), the Danish
North Sea empire had sought, first from Hedeby and later from
Britain, to control parts of the northern route. This was done partly
by direct territorial claims on the southern parts of the shipping
lane, and partly by seeking alliances with local nobles who controlled important western and northern segments of 'the North way'.
This imperial policy was largely successful, and the process of
Norwegian political consolidation gained momentum only after the
collapse of this empire and the subsequent internal strife in the
Danish heartlands. Thus gradually the idea of a kingdom of Norway, based upon one-half of the Scandinavian peninsula, prevailed
over the concept of a seaward empire where 'the North way'
represented an important commercial link.

The early development of towns at the extremes of the trade
route[2] did not last. But from about the eleventh century there began
a new and more lasting city development. These new towns and
trading centres were located at the intersections of the shipping lane
with supply routes from agriculturally rich inland areas. The trade
supplied business, the hinterland food. A map indicating the loca-

FIGURE 1
Norwegian agglomerations, described by terms indicating urban status in medieval sources

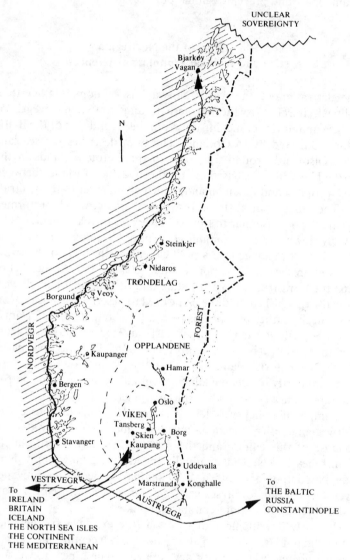

Source: Helle and Nedkvitne (1977: 191).

tion of all cities mentioned in medieval sources available to historians strengthens the impression that the early Norwegian town was a coastal phenomenon (see Figure 1).[3]

The inland areas were not part of the economic concept of 'Norway', but they were incorporated into Norway as a political state formation. Four main areas have been marked in Figure 1:

1. the coastal areas and agricultural flatlands around the inner part of Oslofjord: Viken;

2. the eastern inland flatlands around the lakes and river valleys: Opplandene (literally, 'the uplands');

3. the traditional 'germanic' peasant 'republics' in the agricultural areas around Trondheimsfjord: Trøndelag;

4. the forest area between Norway and Sweden. This was a poorer area, sparsely populated and heavily forested. Control by either realm was doubtful.

Viken and Trøndelag were both central, though often antagonistic to one another. They gave support to two viable cities: Oslo and Nidaros. The rural areas north of the Oslofjord, Opplandene, were late-comers to the Norwegian kingdom. In these areas local kings survived longer than elsewhere. The forest region was a constant source of disquiet. It was a refuge for armed bands with political aspirations. This landscape was never really controlled in the early period. By 1350, however, Norway was established as a state rather than an economic concept, although the eastern boundaries were still not firm.

The city of Bergen represents a special case; it gradually became part of the international trading network of the Hansa rather than a Norwegian city. Bergen became the centre of the trade of grain to the population of northern Norway in exchange for dried fish. The town was for long more a part of a northern European city network than a centre in the kingdom of Norway.

A second period of economic development is marked by the era of Danish domination (from around 1500 to 1800). The inland regions became more important with the development of an extractive economy based on mining and timber. Administratively, the country was divided into a predominantly inland region, 'Norway South of the Mountains', where forestry and mining activities dominated, and a coastal region, 'Norway North of the Mountains', which relied largely on the export of fishery products in the extractive economy. The export of timber in western Norway was important in the sixteenth and seventeenth centuries: it died because the extent of the activity led eventually to the deforestation of large areas in the western islands. However, the dominant economic activity in the country as a whole was still agriculture, with its products mainly consumed on the inland market. In terms of city development, the southern and eastern seaboards were particularly favoured by the Danish-Norwegian kings. The seventeenth and eighteenth centuries saw a series of new charters for towns in this area which were dependent on trade and shipping. The choice of location can be explained in terms of proximity to the Skagerakk and the shortest sea route to the Danish centre of Copenhagen. New large-scale mining activities also promoted the establishment of some chartered towns in the inland regions (Sogner 1977: 49-89).

The third phase of Norwegian economic development was a consequence of the Industrial Revolution in Britain. After the late 1840s a new transport system was built up around coastal steamers and a railroad. In the 1870s wood-based industries were founded, first pulp, later paper. Coastal cities flourished in the east, particularly those situated at the mouths of rivers large enough to float timber. The new paper and pulp industries also created a large demand for wood, thereby increasing activities in forestry. The need for forestry workers on a full-time basis, or at least as full-time seasonal labour, is an important explanatory factor accounting for the class division in some of the rural communities in the eastern inland region.

From around the turn of the century, Norwegian industrialization surged as hydroelectricity made the utilization of waterfalls economically feasible. In the early stages of this development the energy loss suffered in the transport of electrical power across long distances was high. Thus, the new industries were placed in locations close to the primary energy source, the waterfalls. This led to a geographical pattern of industrial locations quite different from that in many other European countries: large modern complexes of factories in rural and peripheral areas. Typical locations were inner parts of the western fjords and narrow mountain valleys in the eastern inland area. Urbanization was not a prerequisite for factory infrastructures. During the same period Norwegian shipping experienced a high growth rate, even though the merchant fleet had already become a considerable economic factor by the second half of the nineteenth century. The larger cities of southern and western Norway dominated this activity as ships became larger, more mechanized and, consequently, in need of more capital (cf. Hodne 1975).

In the last three or four decades the industrial tendency towards larger production units and the need for larger administrative capacity have placed increasing pressures on the economic infrastructure. The influx of population into the cities and industrial towns from the countryside, owing largely to employment opportunities and improved standards of living, has upset the traditional demographic balance. The consequent depopulation of peripheral rural areas, notably in the north, has given rise to considerable and increasing concern. Lately, governments have stepped in with rather forceful policies in an attempt to stem or even reverse the tide of internal migration, and recent accounts seem to indicate that the latter has indeed slowed down. The major incentive used to this end has been a programme tying the standard of living of farmers to the average industrial salary as a guaranteed minimum income. This increase in income in agriculture and important welfare provisions for farmers have proved to be effective demographic as well as social welfare policies.

The most recent economic development, North Sea oil, will have an impact on the territorial distribution of industries. Oil can be used as a raw material as well as an energy resource. While some older industrial installations in the Oslofjord region have been strengthened by the influx of oil, more factories and plants have sprung up in the outer coastal belt of the west, often in localities previously dominated by the primary sector. At present there is a debate about infrastructural development in conjunction with oil exploration in northern Norway. If the oil industry does expand northwards, then there too some communities with a low degree of industrialization, particularly those situated on the coast, will experience considerable economic changes in the next two decades.

We shall bear in mind two aspects of the Norwegian economy when analysing its impact on regional v. national political cleavages, and thus introduce a territorial component in the discussion:

1. the possibilities for some types of agriculture are limited by natural variations in temperature owing to northern location and altitude. This creates 'natural' differences between regions, for example in terms of the necessity to trade in order to obtain grain, and the need for seasonal labour;

2. the particularity of Norwegian industrial development: many of the largest and most important plants are situated in predominantly rural regions.

Cultural integration
and regional identities

The existence of a regionally based counter-culture, or set of counter-cultures, is a well established theme in Norwegian historical sociology (cf. Valen and Rokkan 1974). Generally speaking, the argument has been made that many localities in western Norway are marked by strong Lutheran orthodoxy, teetotalism and the usage of the 'nynorsk' written language. This constitutes a contrast to the rest of the country, particularly the centre around Oslo, where a normal secularization process has taken place and the standard Norwegian written language, 'bokmål', is used.

Throughout the Middle Ages the Roman Church mostly represented a territorially unifying force. The 'official' christianization took place in Norway throughout the eleventh century. The west probably received missionaries from Ireland and Britain, while the east was converted by the official Mission of the Church in Hamburg. But generally speaking, there were probably few regional differences of importance until the Reformation in 1536. This event caused some sharp though temporary regional dissatisfaction. The mountain and inland communities of southern Norway particularly resented the change. After a while the whole country accepted the Lutheran state church, and lasting cleavages along religious lines did not develop until around 1800. The strength of puritan Lutheranism fed on resentment created by the Danish-Norwegian state in 1751 — the year the king issued the 'Conventikkelplakaten', a decree limiting the right to preach in public to ordained ministers. This aspect of official state-building policy coincided with a low-church revival in Norway, which won a large following in the south and the west. This lay Lutheran movement has remained strong in this region ever since, despite the fact that the Conventikkelplakaten was repealed a century later. In the second half of the nineteenth century, Societies for the Inner Mission were established to defend religious orthodoxy — not as a sect, but within the church. The Inner Mission helped members to build 'prayer houses', where they could attend services given by lay 'emissaries'. While official rites such as weddings and funerals were and still are performed by the official church, in the west it is not unusual for local ministers to be highly sympathetic to the 'prayer houses'.[4] Many of the state church ministers have even obtained their degree from a separate school of divinity founded by the lay organization as a reaction to the liberal theology taught at the university.

The 'nynorsk' movement grew out of a combination of national romanticism and dissatisfaction with the Danish standard for written Norwegian. Nynorsk was created in the 1850s as an attempt to combine traits of oral Norwegian dialects into a written standard. This new language still has its strongholds in the west, but some of the eastern valleys also use the language extensively. In the primary school districts a local referendum decides which written standard will be used, and most nynorsk schools are concentrated in the non-urban and non-industrialized communes of the south, west and the east inland regions (Figure 2). The demographical pattern follows a characteristic east-west gradient: counter-cultural strength in Norway seems to be located in the west and south. This spatial distribution applies not only to the teaching of nynorsk in primary schools: the same geographical concentration can be found for acceptance by local congregations of hymn books and prayerbooks in nynorsk, and for public usage of this written standard in advertisements, signs, etc. However, because nynorsk is the main language in many of the large, sparsely populated mountain communes in the middle of southern Norway, Figure 2 tends to give a visual overestimation of the true strength of the alternative language standard.

Norway's first temperance movement was aimed at limiting the consumption of spirits and in the 1840s spread primarily along the southern seaboard. The chapters were often led by the local ministers. Some of the associations sought to facilitate and increase the consumption of wine and beer instead of distilled alcohol, as the lesser of two evils. The first teetotallers' organization was formed in 1859. All consumption of alcoholic beverages was banned for its members. The centre of this new movement was Stavanger and it spread along the west coast, both to the north and the south (Kristiansen 1974). The teetotallers' movement grew to become one of the largest popular movements in modern Norwegian history. It fought two referendum campaigns for total prohibition in 1918 and 1926, winning the first and losing the second. Ever since, the movement has acted as a pressure group to maintain the strict public monopoly on sales of wines and spirits. It has constantly fought local referenda to prevent the establishment of additional state monopoly shops, and has sought even further restrictions on the sale and consumption of alcohol. Even though the issue of teetotalism has lost some of the political saliency it held in the 1910s and 1920s, the movement is still very much alive and well.[5] That

FIGURE 2
Percentage of schoolchildren with Nynorsk as their primary language alternative

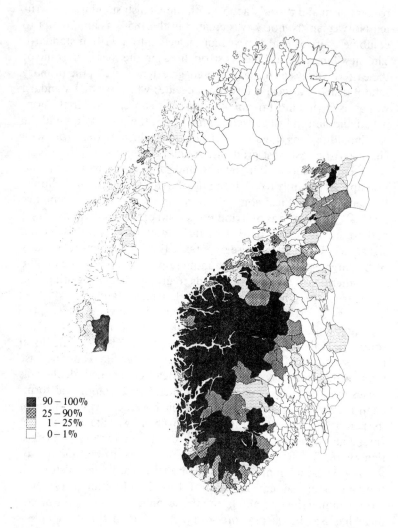

- 90 – 100%
- 25 – 90%
- 1 – 25%
- 0 – 1%

The map has been produced by the Norwegian Social Science Data Service.

this is so was indicated to some extent in the 1972 referendum on EEC membership: communes where the three counter-cultures had strength tended to vote more heavily against Norwegian membership than others.

In attempts to construct culturally based territorial partitions of Norway, three types of regions normally emerge:

1. the secularized central and urban regions, where none of the counter-cultures is strong;

2. the counter-cultural western periphery, where all of them prevail;

3. The mixed forms, in many respects the most interesting regions:

— the non-religious counter-cultural areas (e.g. the county of Sogn og Fjordane);

— the region dominated by religious and teetotalist counter-cultures, but using standard Norwegian (the county of Østfold);

— the regions where teetotalism is the only counter-cultural influence (parts of the eastern inland and the county of Nord-Trøndelag).

The relevance of these cultural regions to Norwegian mass politics will hopefully become apparent when we turn to electoral mobilization and the development of local autonomy and partisan cleavages.

The development of political institutions, local autonomy, citizenship rights and partisan cleavages

Norway emerged as a political entity with home rule in the wake of the peace negotiations following the Napoleonic wars. Under the Treaty of Kiel of 1813, Norway was taken from Denmark and annexed by Sweden as the reward for Swedish involvement on the allied side. The following year turned out to be one of intense political activity in Norway, and when the Scandinavian peninsula finally became a single political entity in 1814, the constitutional arrangement was a rather loose personal union, with extensive powers being given to a comparatively broadly based legislature, the Storting, in Norway in all fields but foreign policy. During the second half of the nineteenth century the popularly elected Storting began to assert its authority at the expense of the royal prerogative. The position of the king as an independent executive power was

further weakened by a growing sentiment among Norwegian voters
that their sovereign was not a national ruler, but rather a political
actor whose decisions were determined mainly by Swedish interests.
The introduction of parliamentarism in 1884 and the declaration of
full independence in 1905 were the two major steps in this process.

Throughout this period the Storting was a body with two
interlocking and overlapping, yet principally different functions: a
constitutional role as a legislative body; and a legitimizing role as
the ultimate expression of popular sovereignty — as the voice of the
nation — locked in constant political combat or even constitutional
struggle with a 'foreign' king. The importance of the latter function
made the popular base of Parliament extremely important. One
explanation of the absence of serious regional challenges to the
constitutional unitary state is the early over-representation of the
rural population: the 1814 constitution gave the franchise to all
farmers, leaseholders as well as freeholders. No similarly broad
social stratum in the cities was given the right to vote until the suffrage extensions of 1884 and 1897.

Originally the constitution also included a provision to insure a
rural majority. This so-called 'farmers' paragraph' specified that
two rural representatives should be elected for every urban
representative. In the first decades after 1814 this paragraph further underlined the dominance of the peripheries over the centre.
The early political system can be described as an elitist unitary state
seeking legitimacy primarily from the peripheries. This system was
later taken over by a coalition of peripheral and marginal groups.
Because the early franchise penetrated rural Norway more
thoroughly than the cities, regional devolution could never be an
issue since the system permitted a peripheral penetration and subjugation of the centre.

But in order to understand the impact of this, it is necessary to
examine the kind of Norwegian society that emerged from the
Napoleonic wars. The social structure was remarkably egalitarian,
certainly in the rural areas and especially around the coast of
southern Norway. No nobility existed after 1820, and the urban
liberal middle class was still very weak. In this situation the
bureaucracy, with its legacy from an absolutist regime, functioned
very much as the upper echelons of society, on both the national
and local level. This elite, however, was oriented towards using
achievement rather than ascriptive norms as a basis for elite recruitment. In practice, the 'class of officials' tended to be self-recruiting

to a large extent. The importance of their achievement orientation lies more in the emphasis on improving the school system, on enlightenment in general as an ideology and on the willingness to grant political rights to a large segment of the population. During the first half of the nineteenth century the Norwegian economy was still dominated by the primary sector. And when the suffrage in the rural areas was tied to the condition of *any* kind of land ownership, it is rather evident that voting rights were much more widespread at this early stage than they were in most European countries. An estimated 45 percent of the male population of more than 25 years of age was enfranchised in 1814 (Kuhnle 1972).

The territorial distribution of the franchise can be further illuminated by considering the regional consequences of the next suffrage extension in 1884. Figure 3 shows the composition of the electorate by commune in the first election after the major suffrage extension of 1884. It indicates the percentage of the electorate who held their franchise only through the new census criterion, that is, who formerly would have been excluded from voting. The larger cities (population size is proportional to the area of the 'city-boxes' around the map outline), the eastern towns and the agricultural heartlands in the east and mid-Norway (Trøndelag) gained most in terms of new voters. As the census was related to income, it is clear that communities with the most monetarized economies and the best possibilities for capital accumulation would receive most new voters. The effect of the extension was a more homogeneous electorate in terms of geographical location. The new electorate of the 1880s lived in communities previously under-represented. In other words, the cities and central regions were strengthened. However, the preceding period had established a political style which meant that peripheral demands had a tendency to be taken seriously in the centre. Increased local autonomy was one way of catering for this peripheral interest (cf. Kaartvedt et al. 1964).

Early electoral over-representation of the periphery is not the only institutional characteristic that may account for the absence of a tradition of regionalism or federalism in Norway. Almost equally important was the administrative and political decentralization caused by the implantation of a relatively high degree of local government autonomy in the 1830s. By retaining a centralized, unitary state, and by combining this with a strong local government on the parish level, the need for a more or less autonomous regional authority was by-passed. The directly elected regional council is a child only of the 1970s.

FIGURE 3
Newly enfranchised voters under the Income Census criterion
as a percentage of the total electorate in the 1888 election

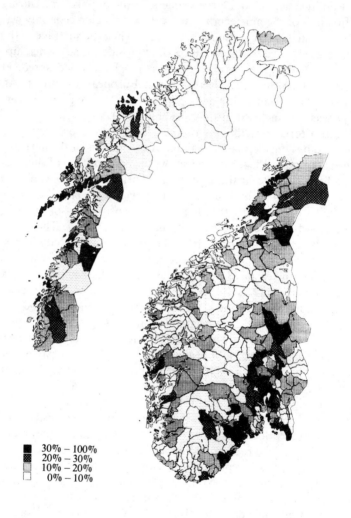

■ 30% − 100%
▨ 20% − 30%
▨ 10% − 20%
□ 0% − 10%

This map has been produced by the Norwegian Social Science Data Service.

Table 1 indicates that local government reform in Norway led to a long period of expansion after the passing of the 1837 Community Council Act (NKS 1972). This development was broken in 1920, following a number of municipal bankruptcies caused by financial overextension during the construction of hydroelectric plants. Since 1920 the communes have tended to become increasingly burdened by mandatory administrative tasks defined by laws made in the national centre, Oslo. Communal autonomy has also been seriously restricted by the removal of instruments of choice, such as the option for communal councils to determine the level of local taxes independent of central decisions.

TABLE 1
The development of Norwegian local government

Time span	Institutional characteristics of the period
Phase I Pre-1837	Special-purpose local government for certain sectors based on the parish as a territorial entity. Port authorities by royal decree in 1735. Poverty commissions and church elders were made mandatory by law later in the century.
Phase II 1837-1900	Local autonomy. Limited suffrage and relatively low electoral participation. Local taxing authority. Local roads, primary school system were the major expenses. The representatives of most rural councils considered saving public means and low taxes a very important political aim in itself.
Phase III 1900-1920	Communal expansion. Heavy involvement by some local councils in industrialization, notably the development of hydroelectricity, often in co-operation with local banks.
Phase IV 1920-1950	Central government intervention, often on an ad hoc basis. Communal bankruptcies lead to central government administration of the bankrupt communes.
Phase V 1950-	Planned central government regulation and control. Incorporation of the local governmental level in the implementation and administration of new government enterprises and programmes initiated at the central level of government. The communes are often obliged to take action in a new field as directed by the ministry, and are hence deprived of the possibility to change priorities between fields. More means have been channelled to the local level, but at the same time a growing proportion of the means are bound to specific central government programmes.

Generally speaking, the present trend is towards increasing standardization across communes. In the 1960s and 1970s large government programmes led towards a national comprehensive school system, the establishment of general plans for all communes, and the obligation to open social welfare agencies in each community. All these new agencies have been established by mandatory requirement upon the communal administration, in most cases supplemented by an increased transfer of financial means from the central level of government. These transfers have, however, been differentiated to secure optimal equalization across communes.

Another important present-day trend is the penetration of the idea that localization of decision-making to a given level of government should imply that economic responsibility be placed on the same level. This philosophy has particularly contributed to a strengthening of the provincial level. In 1973 this fact was also acknowledged as part of political reality: direct elections were held to all 19 provincial assemblies for the first time. Previously, the mayors and some communal committeemen from all local governments units in a province had constituted the provincial assembly.

The first mass-based political parties came into being in a situation of ongoing constitutional crisis fought with increasing vigour and bitterness by the groups involved. Parliamentarism was introduced in 1884 and the role of the Swedish-Norwegian king came under more or less continuous attack. In this situation new political organizations, the parties, emerged as excellent instruments for the aggregation and prolongation of political conflict. Their role as mobilizing agents should not be underestimated. Figure 4 provides a summary of these developments in the form of a genealogical 'tree' of Norwegian political parties.

From 1883 to 1905 political conflict in Norway was centred around the polarity of Liberals (V) and Conservatives (H). The former were the force behind the introduction of parliamentarism and early demands for full independence from Sweden. The latter clung to the principle of the division of powers, but never attempted to reintroduce it once parliamentarism had been adopted. The Conservatives favoured negotiations to demonstrations vis-à-vis the Swedes. They were, however, driven back on the Union issue, and in 1905 fully supported Norwegian independence. In fact, ideologically the Norwegian Conservatives had, and have, much more in common with Continental urban-based Liberal parties than with their British or German counterparts.

FIGURE 4
Changes in the formal structure of the Norwegian party system, 1883-1978

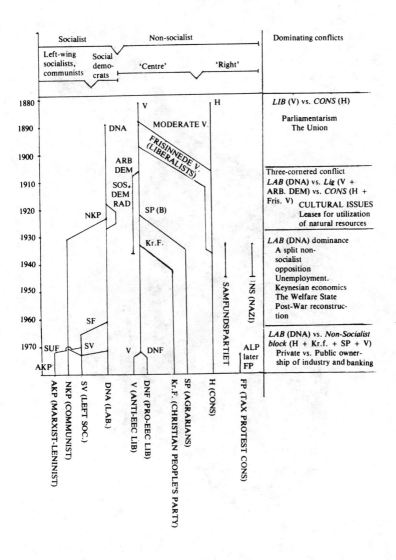

This information is based upon Naerbøvik (1973) and Furre (1971).

FIGURE 5
Conservative votes as a percentage of all
votes cast in 1882, southern Norway only

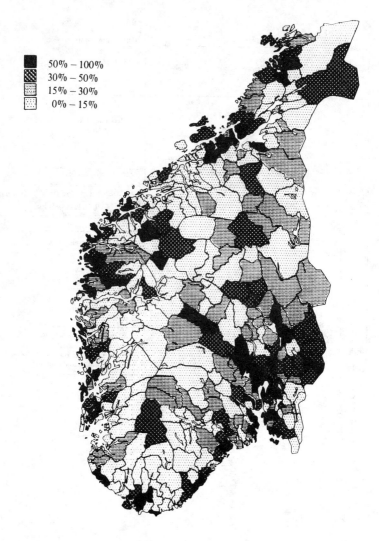

This map has been produced by the Norwegian Social Science Data Service.

FIGURE 6
Conservative votes as a percentage of all votes cast in 1973

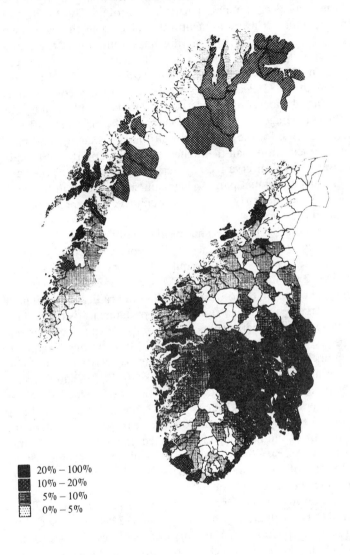

■ 20% − 100%
▨ 10% − 20%
▤ 5% − 10%
▦ 0% − 5%

This map has been produced by the Norwegian Social Science Data Service.

Figures 5 and 6 show the geographical pattern of Conservative support in the 1882 and the 1973 elections. In both elections the party gained most of its support from the cities, particularly in the east, the coastal communes, particularly around Oslofjord, and central communes in the agricultural heartland in the east. Two aspects of Conservative support are important. First, there is a relative stability in regional support for the party across the 89-year interval. In 1973 the support may seem weaker in terms of percentages, but the 1882 election was a strictly two-party competition. Thus a map of Liberal support would be simply complementary. In 1973 the percentages are generally lower, but the map may reflect a stronger position for the party in a multi-party system with proportional representation than the higher percentages of 1882. Second, even though the Conservatives are very strong in the east, they are by no means a regional party. They gain support also in the west and north, and the party is strong enough to elect representatives from all 19 provinces.

The Conservative Party (H) has been one of the most stable factors in Norwegian politics since 1882, representing almost continuously the second largest group of voters in the Storting. Around 1912 the Labour Party (DNA) became a serious adversary. The Liberal coalition (V) of farmers and urban radicals, which had been the force behind the introduction of parliamentarism, dwindled away in frequent party splits, and from 1933 the stage was set for a period of Labour Party dominance. After a strong beginning the Labour Party experienced no less than two party splits, in 1918 and 1923, and a merger in 1927. In the wake of the Russian Revolution the party convention decided to join the Comintern. This caused the moderate wing to secede to found a Social Democratic Party. When the party convention voted to leave the Comintern in 1923, the pro-Comintern faction formed the Norwegian Communist Party (NKP). When Labour and Social Democrats merged in 1927, the crisis of the left was over and the party has been the largest in the Storting ever since.

Figure 7 shows the percentage of the votes cast for the Labour Party (DNA) and all groups to its left in the 1973 election. Socialist voting is a real phenomenon in Norwegian politics. Parties to the left of Labour would never seek positive co-operation with the non-socialist parties which would be necessary to support a government. Occasional negative coalitions do, of course, occur, and have caused governmental crises. But generally speaking, Labour can form minority governments if representatives from all elected socialist parties constitute a majority of the seats in the Storting. On the other hand, Labour has never attempted to pull left-wing socialist parties directly into government by offering them ministerial posts. Socialist voting is particularly strong in the east, with the exception of the coastal communes around the Oslofjord, and in the north.

Three aspects of the regional support of the socialist parties should be noted. First, even though the left-wing socialists may sometimes be pivotal in the election of a Labour government, the latter party secures a vast majority of the votes (see Figure 8). Second, contrary to the Conservative case, Labour's urban support is dependent on its regional strength in the area where a given city is situated. Moreover, like the Conservatives, Labour has a regionally uneven pattern of support, without being a regionalist party. It too has elected representatives from all 19 provinces.

Figure 7 tends to give a false impression of a more homogeneous socialist political support than is the case. An attempt to correct this is made in Figure 8, which shows the anatomy of the socialist vote in 1973. The left here is composed of three parties: (1) the Socialist Left Party (SV), composed of splinter Communist and anti-EEC Labour Party members surrounding a core constituted by the former Socialist People's Party (SF), a group of anti-NATO socialists excluded from the Labour Party in 1961; (2) the Moscow-oriented Norwegian Communist Party (NKP); (3) the Workers' Communist Party (AKP), a splinter from the old Socialist People's Party of pro-Chinese Communists. Table 2 shows the content of the typology used in Figure 8, which indicates that, in most communes with a large socialist percentage of the vote, the Labour Party is dominant. The left-wing socialists form a strong faction in several northern communes, and in a belt of communes running alongside the Swedish border in southern Norway. The greater success in the former may be due to the relative uncertainty surrounding fishing as a way of life. The latter, by contrast, are traditional forestry areas, and radical Scandinavian syndicalism was strong in forestry regions in the 1920s. Some agricultural communes in the east, in a semi-circle around the Oslo region, also display strong left-wing electoral support. This has been attributed to a relatively high degree of class division in these richer farmlands — an uneven distribution of land, and traditional, normative distinctions between farm owners and crofters. The left-wing socialists are strong in the south-west and north-west, but in these areas Labour is

FIGURE 7
Socialist (DNA + SV + NKP + AKP) votes as a percentage of
all votes cast in 1973

50% − 100%
40% − 50%
20% − 40%
0% − 20%

This map has been produced by the Norwegian Social Science Data Service.

FIGURE 8
The anatomy of the socialist vote, 1973

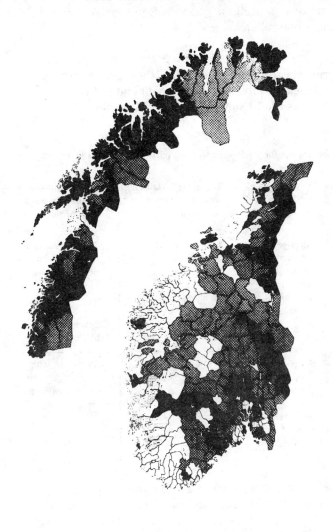

■ All soc. strong and left-wing soc. strong
▦ All soc. strong and left-wing soc. weak
▤ All soc. weak and left-wing soc. strong
☐ All soc. weak and left-wing soc. weak

This map has been produced by the Norwegian Social Science Data Service.

weak, a trend related to the rapid growth of industries in some rural western communes. Most of these communities are recent agglomerations housing a ghetto of industrial workers in a predominantly rural area.

TABLE 2
A cross-classification scheme for Norwegian communes, based on total socialist strength and left-wing socialist support in the 1973 election

		Votes for the three left-wing socialist parties as % of all votes cast	
		Left-wing socialist weak (less than 10% of the vote) ↑ 10%	**Left-wing socialist strong (more than 10% of the vote)**
Votes for all socialist partners as %	**All socialist parties strong (more than 40% of the vote)**	Labour strongholds, weak left-wing socialists	Socialist strongholds with large proportion of left-wing socialists
		Shade on map: dark grey	Shade on map: black
of all votes cast	40% ← **All socialist parties weak**	Weak socialist *and* left-wing socialist support	Strong left-wing socialist, in weak socialist environment
	(less than 40% of the vote)		

The lack of a Labour Party tradition and the propensity for a polarized situation to develop have proven to be fertile grounds for the small socialist parties to the left of Labour. All kinds of socialists are weak in the inland areas of the west, particularly around the fjords. In some cases, however, socialist strongholds stand out in the western fjordland. Without exception, these are industrial enclaves depending on hydroelectricity for their power. Some of the most radical left-wing trade unionists in Norway represent these industrial communities, situated in primarily rural surroundings.

But the west is not only the weakest region for Norwegian socialism: it is also the stronghold of the final party to be discussed in some detail here, the Christian People's Party. This is the party for the defence of Lutheran fundamentalist values and also, to some extent, a party rendering support to the three counter-cultures. Figure 9 displays the regional support for two election results, 62 years apart, for two parties both of which were established by lay Lutheran groups with fundamentalist theological views. In 1886 the Moderate Liberals split from the Liberal Party on the issue of direct elections to the councils of parish elders. Conservatives and Liberals alike wanted limited powers for these bodies, whereas the Moderates saw them as important controlling institutions and channels of influence for laymen. Towards the end of the century the Moderates in most places merged with the Conservatives. The Christian People's Party (Kr. F.) grew out of a local party split of fundamentalist Lutherans from the Liberals in the western province of Hordaland just before the 1933 election. Before 1940 it remained a western party, but since the war it has presented candidates over the whole country. A comparison of the two maps shows striking similarities. The strength of counter-cultural movements in the west is manifested in support for the Moderates as well as the Christian People's Party. Comparing these parties with the Conservatives and Labour, there is also in this case a clear regional support base. The Christian People's Party is, however, no more a regionalist pressure group than the other two, even though it grew from being the splinter of a provincial branch to a national party in about 20 years.

All the parties discussed here are national parties, despite the documented regional variations in electoral strength. All obtain votes in all parts of the country, and none of the party programmes contains specific regionalist demands. The left-wing socialists, Agrarians and Liberals include redistribution of wealth to the peripheries as a major plank in their platforms, but no party advocates the interests of one particular periphery as being more important than any other. All parties advocate district policies, and no party represents a single district. The structure of the contest is nationwide; that is, most parties campaign in most provinces. But part of the answer is also that the size of the partisan support varies across regions. The yoke of being in a more or less permanent minority lies upon the socialists in the west and upon the non-socialists in the north. Figure 10 shows the relative weighted

FIGURE 9
The strength of the moderates in 1891
and the Christian People's Party in 1953: southern Norway only

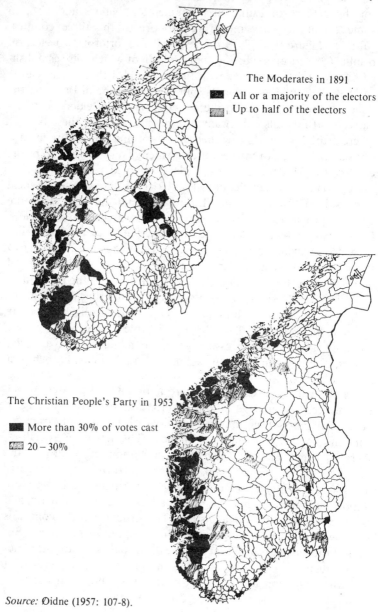

Source: Øidne (1957: 107-8).

FIGURE 10
The weighted percentage of all votes cast, given to the Conservatives, Labour and the Christian People's Party, in all elections from 1906, in the north, the west and the east

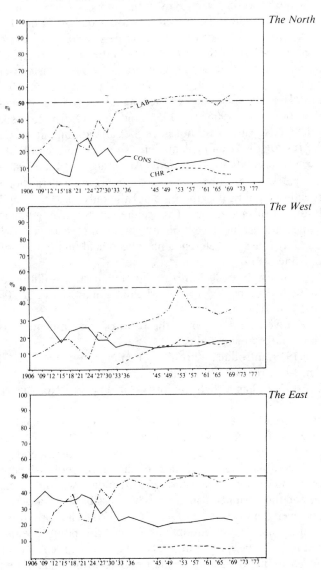

percentage of the votes for the Conservatives, Labour and the Christian People's Party in elections since 1906. The graphs show the general development of electoral support for the three parties in the east, the west and the north. The time-series plots for the east and north show Labour in the lead with around 50 percent of the vote in the postwar period. Given the support from the left-wing socialists, this indicates a socialist majority of votes since 1945. In both regions Labour has been the largest party since 1927. The Conservative Party is the second party in the north as well as in the west, and in both areas the Christian People's Party oscillates around 6 – 7 percent of the votes. The west presents a very different picture. Here the socialists have held a majority of the votes in only one election since 1906, the 1953 landslide. On the non-socialist side the Conservatives do not have the same dominant position as in the rest of the country. The vote is much more evenly divided between Conservatives, Christians, Agrarians and Liberals.

But party vote, albeit the most important institutionalized expression of political opinion, is not the only one. Norway has a tradition of using consultative referenda to give guidance to parliamentarians on exceptionally important issues (cf. Nilson 1972). Only five such referenda have been held in modern political history since 1814:

1. in August 1905, legitimizing the unilateral dissolution of the union with Sweden by an overwhelming majority (362,208 yes votes; 184 no votes);

2. in November 1905, over the form of government: constitutional monarchy or republic;

3. in 1919, introducing prohibition in Norway;

4. in 1926, rejecting prohibition;

5. in 1972, when a slight majority voted against Norway's entry into the EEC.

These referenda may provide one of the keys to explaining the absence of regional parties in the country. Two of the five referenda, in 1919 and in 1972, were victories for peripheral groups over the combined forces of most of the central political elite and major corporate interests. The 1926 referendum was only a narrow defeat for the peripheral interest. This possibility of a peripheral victory over the centre is in fact a sustained belief in Norwegian political culture. It is essential to appreciate this point if we wish to understand why peripheral interests have allowed themselves to be organized by national movements rather than mobilizing under a regional banner. The formation and success of these coalitions of

peripheral interests have been more dependent on a national network than on regional strongholds.

The anatomy of the anti-EEC vote will serve as an example.
Figure 11 shows the distribution of communes with a 'no' majority.
The resistance was strongest in the western, central and northern
peripheries, but the agrarian communities in the east also voted
against Norwegian entry. Only the most central areas around the
Oslofjord and some major cities had clear majorities in favour of
Norway becoming a member of the EEC. The result is surprising,
given the relative importance of the forces facing one another in the
campaign. A clear majority of leading groups and organizations in
the political centre were in favour of Norwegian membership: the
Labour Party, a vast majority of parliamentarians, the leadership
of the two largest parties, the Trade Union Congress, the
Norwegian Federation of Industry and the entire establishment of
the Foreign Office. The 'No to the EEC' campaign was organized
by a series of more peripheral groups: one strong central organization — the Norwegian Farmer's Organization — and a number of
less influential organizations — left-wing socialists, northern
Norwegian fishermen, fundamentalist Lutherans, the Agrarian
Liberals. Yet again, David bested Goliath — and if loose interest
coalitions work, why try an alternative regionalist strategy when
the former demonstrates a proven ability to change the policy of
the entire national system?

The impact of district policies

The importance of the peripheries as a base for political mobilization may well be the structural background for the relative consensus within the Norwegian political system about the necessity of a
district policy. The size of this transfer of resources and the form of
redistribution lies, however, within the realm of valid political
dispute. It is almost self-evident that an aggressive district policy by
the central authorities ought to be to the detriment of any attempt
to establish a regionalist political base.

The area of the politics of redistribution demonstrates perhaps
most clearly the difference between, on the one hand, a territorial
identity tied to the idea of a peripheral location and, on the other,
the assertion of regional autonomy. In the former case the
spokesmen for the peripheries cherish their links with the centre
and try to extract maximum resources for their own area; by con-

FIGURE 11
'No' votes cast as a percentage of all votes cast in
the 1972 referendum on Norwegian entry into the Common Market

■ 70 – 100%
▨ 50 – 70%
☐ 0 – 50%

This map has been produced by the Norwegian Social Science Data Service.

trast, true regionalists would seek to sever the bonds between the region and the heartland through demands for autonomy, home rule or even independence. Norway's case fits the first description.

Government attempts at territorial equalization have a long tradition in Norway. As early as the nineteenth century, special funds were set up to accelerate the establishment of a school system in northern Norway. The devastation of the northernmost provinces during the German retreat towards the end of the Second World War had a major impact on central government involvement in district policies. In the late 1940s and early 1950s, area planning was started in northern Norway as an integral part of the reconstruction programme. After the official reconstruction had been completed in the early 1950s, the programme was continued through an industrial development scheme and a development fund for northern Norway. In 1956 a similar development fund was set up for areas of southern Norway with high unemployment, and in 1961 these institutions were merged into a general District Develop-

TABLE 3
The relative distributions of resident population in 1970 and loans given through the District Development Fund in 1973, by six Norwegian regions

Region	Percentage of all loans and loan guarantees given through the District Development Fund in 1973 (1)	Percentage of total resident population in 1970 (2)
	%	%
Oslofjord	0.8	30.8
Eastern inland	19.3	18.2
The south	7.0	5.3
The west	26.3	24.9
Trøndelag (mid-Norway)	11.6	9.1
The north	35.0	11.8
The whole country	100.0	100.1

Sources: Col. (1), Stortingsmelding (1973-74: 90); col. (2), NOS (1972: 6).

ment Fund. The relative distribution of loans from this fund across Rokkan's six-way regional partition of Norway is shown in Table 3, together with the relative distribution of the resident population. The Oslofjord region is the only one that receives a smaller share of the loans relative to its population size. The table gives a general impression of a channelling of resources away from the centre to the peripheries, with northern Norway as the main recipient of redistributed resources. There is little doubt about the redistributive effect of the fund at a regional level.

There is, however, a common criticism of the District Development Fund, which points to the lack of distributive effects within these broad regions. The most deprived communes do not benefit as strongly from the fund as those communities with some already existing industrial infrastructure. In short, it is local centres in the peripheries that benefit most. This might not be desirable from the point of view of economic development, but from that of an analysis of the probability of regional political unrest it is most interesting. If a not necessarily intended side-effect of industrial development serves to 'buy off' local centres in the peripheries, we have another partial explanation of the failure of Norway to produce a regionalist party.

But a vigorous district policy can in itself limit the political appeal of regional dissatisfaction. The range of political means available to politicians and planners to improve peripheral conditions has increased since the early 1960s. Loans, guarantees and other types of financial aids in establishing industries have been supplemented with other positive incentives as well as with restrictive policies vis à vis the central areas. Labour market incentives, retraining programmes and encouragement of new enterprises are means used in addition to funding. Direct investment aid, selective transport subsidies, exemption from community infrastructure costs and differential employers' tax rates have all been added to the arsenal of positive incentives in the peripheries throughout the 1970s. On the other side of the coin, new laws have been introduced to prevent uncontrolled industrial establishment in the centres and 'pressure areas'. This Industrial Establishment Control Act allows the authorities to refuse permission for medium- and large-scale enterprises to establish themselves in the centre. Furthermore, governments have attempted to contribute towards increased employment opportunities in the peripheries by moving government institutions away from Oslo. The impact of the latter, however, is probably more symbolic than real.

Overall, the political centre today has a wide range of instruments with which to conduct an aggressive district policy and thus avoid too much peripheral unrest. By contrast, it is a disadvantage for government increasingly to be held responsible for the failure of development in some regions. Since peripheral leaderships are aware of the powerful tools available, governments are criticized for not using them. Today, some geographical redistribution is expected and accepted in the Norwegian political system. The net result has been a growing interdependence between centre and periphery.

Conclusion: the missing regional link

One inescapable question remains: why has Norway not experienced serious regional unrest? Admittedly, many conflicts are strengthened by their relationship to a centre-periphery cleavage; yet none has found regionalist advocates within the political system, even though the combination of considerable economic growth in the peripheries and territorial concentrations of counter-culture identities in a European context often seems to predict regional unrest.

An attempt to answer this question is made in Figure 12. Norway's economic growth and territorial identities did not emerge as fully-fledged regional political problems for three reasons: (1) the regional level has very little institutional infrastructure in the Norwegian political system; (2) cultural identities and aims are not tied to specific regions; and (3) the basis for Norwegian economic development has emphasized the local establishment of industries rather than regional plans, at least until recently.

Two forms of economic infrastructures have been built up through the process of urbanization and industrialization. In the former administration and consumer-oriented industries have been located near the cities, notably in the Oslofjord area. In the latter, dependence on cheap hydroelectricity has caused many large industries to be placed in rather remote locations. But the point is that this development, and the current development of North Sea oil resources, has caused local industrialization rather than regional growth in manufacturing. Norway has no regional resources similar to the brown coal belt in Germany or the iron ore of northern France or Wallonia. Particularly in the early half of this cen-

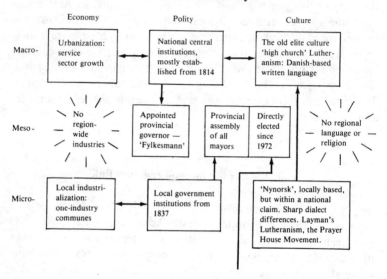

FIGURE 12
**Regional institutions between national centre
and local authority**

tury, those communes that had the waterfalls also obtained the industrial infrastructure.

Similarly, in the field of symbolic systems, no regional rallying points seems to exist. The central culture of high-church Lutheranism, a Danish-related written norm and an acceptance of alcohol has been opposed by the three counter-cultures of lay Lutheranism, nynorsk and teetotalism. Albeit often regionally skewed in terms of their support base, none of the latter has even tried to create a base of regional identity. Since all three have carried the ambition of taking over the national centre, none has been willing to settle for a regional infrastructure.

But the lack of regional infrastructure is most evident in the political institutions. Parts of Norway's central state bureaucracy date back to the seventeenth century. The 1814 constitution established a Parliament, a Cabinet and a High Court of Justice, all located in the capital. As early as 1837 Parliament passed a Communal Government Act giving the local level considerable powers, the most important being that of taxation. The suffrage for the local assemblies was gradually extended parallel to the franchise for general elections. For the regional level of government — the province — directly-elected regional assemblies were established only in 1972. Their income from taxation is negligible compared with that of the communes and the state. The senior official in the province, the 'fylkesmann', is not elected or appointed by the regional assembly, but is a civil servant appointed by the government. The regional assembly, without powers of taxation before 1972, consisted of all the mayors of communes within the region. Introducing direct elections has not changed the strange dualism of Norwegian regional government: a civil service that owes its allegiance to the centrally appointed fylkesmann, and with considerable administrative decision-making powers, juxtaposed against a provincial assembly with some budgetary powers, but whose control of the provincial bureaucracy is very much in doubt. In modern Norwegian history it is not unreasonable to refer to an institutional lacunae of authority on the provincial level. However, the provincial election is a recent source of popular sovereignty in the Norwegian system of government, and only the future will show if regional solutions to national problems will be advocated by a new class of regional politicians whose careers lie in this level of government. Today, the local level prevails as the important subnational political forum in Norway.

Other factors may also account for the lack of regional difficulties in Norway, bearing in mind that peripheral problems of distribution do exist: the relative youth of the country since its successful secession from the Swedish centre and the prominence of district policies in Norwegian politics. In a comparative perspective, each of these factors appears to be rather idiosyncratic. The secessionist states of east-central Europe did not avoid serious regional problems. Economic redistribution to the peripheries, through industrial location, capital influx and other 'district policies', did not prevent the rise of the SNP in Scotland.

The relatively weak regional government in Norway has been made even more vulnerable by the lack of region-wide industries and by the fact that specific identities tied primarily to the regional territory have been absent. Peripheral problems and counter-cultural identities have channelled potential territorial problems away from the region as a political base. At the same time, no class of politicians has a vested interest in the region as a career base for their political life. Whether this situation will continue in the future is another matter altogether.

NOTES

1. During the last decade Christian parties of this kind have been formed in Denmark, Sweden and Finland, but these parties are considerably weaker than their Norwegian counterpart.

2. Bjarkøy in the North and Kaupang in Oslofjord.

3. Recent research on the formation of Scandinavian cities is reported in Blom (1977).

4. An interesting discussion of these regional differences can be fund in Øidne (1957).

5. For an interesting hypothesis of the role of teetotalist organizations in elite circulation, see Gusfield (1963).

REFERENCES

Aarebrot, F.H. and Urwin, D.W. (1979). 'The Politics of Cultural Dissent: Religion, Language, and Demonstrative Effects in Norway', *Scandinavian Political Studies*, new series 2: 75-98.

Blom, G.A. (ed.) (1977). *Urbaniseringsprosessen i Norden*, 3 vols. Oslo: Universitetsforlaget.

Furre, B. (1971). *Norsk Historie, 1905-1940*. Oslo: Det Norske Samlaget.

Gusfield, S.R. (1963). *The Symbolic Crusade*. Urbana, Illinois: University of Illinois Press.

Helle, K. and Nedkvitne, A. (1977). 'Norge. Sentrumsdannelser og byutvikling i norsk middelalder', pp. 191ff. in C.A. Blom (ed.), *Urbaniseringsprosessen i Norden*, Vol. 1. Oslo: Universitetsforlaget.

Hodne, F. (1975). *An Economic History of Norway, 1815-1970*. Trondheim: Tapir.

Kaartvedt, S., Danielsen, R. and Greve, T. (1964). *Det Norske Storting gjennom 150 år*, 4 vols. Oslo: Norges Storting/Gyldendal.

Kristiansen, B. (1974). 'Organisasjonsdannelse og organisasjonseffekt: En studie av totalavholdsbevegelsen i Norge 1859-1894'. Paper presented at the Nordic Political Science Association Convention, Århus.

Kuhnle, S. (1972). 'Stemmeretten i 1814: beregninger over antall stemmerettskavalifiserte etter Grunnloven', *Historisk Tidsskrift*, 51: 373-90.

Nærbóvik, J. (1973). *Norsk Historie, 1870-1905*. Oslo: Det Norske Samlaget.

Nilsen, S. Sparre (1972). *Politisk avstand ved norske folkeavstemninger*. Oslo: Gyldendal.

NKS (1972). *Selvstyre og Samarbeid. Et historisk tilbakeblikk ved konstitueringen av Norske Kommuners Sentralforbund*. Oslo: Norske Kommuners Sentralforbund.

NOS (Norges Offisielle Statistikk) (1972) *Statistisk Årbok 1971*. Oslo: Central Bureau of Statistics.

Øidne, G. (1957). 'Skilnaden på Austlandet og Vestlandet', *Syn og Segn*, 63: 107-08.

Rokkan, S. and Valen, H. (1969). 'Regional Contrast in Norwegian Politics', pp. 190-247, in E. Allardt and S. Rokkan (eds), *Mass Politics*. New York: Free Press.

Rokkan, S. and Valen, H. (1970). 'The Mobilization of the Periphery', pp. 181-225, in S. Rokkan, *Citizens, Elections, Parties*. Oslo: Universitetsforlaget.

Sogner, B. (1977). 'De "anlagte" byer i Norge', pp. 49-89 in G.A. Blom (ed.), *Urbaniseringsprosessen i Norden*, vol. II. Oslo: Universitetsforlaget.

Stortingsmelding (1973-74). *Om virksomheten til Distriktenes. Utbyggingsfond i 1973*, no. 92.

Valen, H. and Rokkan, S. (1974). 'Norway: Conflict Structure and Mass Politics in a European Periphery', pp. 315-70 in R. Rose (ed.), *Electoral Behavior*. New York: Free Press.

4

Finland: An Interface Periphery

Risto Alapuro
University of Helsinki

Two central characteristics concerning the formation of the Finnish nation seem to be due to the fact that Finland was consolidated in a territory between two established members of the European state system, Sweden and Russia. This position resulted, first, in an exceptionally calm and steady consolidation of a bourgeois national culture during the nineteenth century, in comparison to other minor nationalities within the great multinational empires of the time. Second, it affected the basic configuration of political integration in the early twentieth century. Finland was not only a minor nationality within Russia: in a larger perspective, the Finnish state developed in the interface between two centres, one of which was economically and culturally dominant, the other politically dominant but economically backward. This interface position has retained much of its importance in the late twentieth century. The character and manifestations have changed in several respects, but it is nevertheless useful to view Finland's present position from the state-making perspective, in relation to both Sweden and the Soviet Union.

One can focus upon interfaces at several levels. In his analyses on interfaces Stein Rokkan has dealt with regions where conditions for independent core formation have been small and where the interfaces have not eventually survived, for example, Alsace or Silesia. The periphery of Finland progressed from an interface position between Sweden and Russia towards development of its own strong centre.

European perspectives[1]

The dual nature of nationalism in nineteenth-century Europe

In an influential essay, Ernest Gellner (1964: 166) has portrayed nationalism as 'a phenomenon connected not so much with industrialisation or modernisation as such, but with its uneven diffusion'. The tidal wave of modernization has struck various parts of the world in succession, mobilizing the late-comers to a nationalistic defence against those territories already modernized, and bringing about struggles for independence in the territories defined by nationalist criteria. In the nineteenth century, this wave moved from the west to the marchlands of Europe (Gellner 1964: 164-72).

Gellner's definition of nationalism covers especially movements of national self-assertion and liberation, particularly those not linked to an existing or even historically remembered polity (the 'unhistoric nations'), where nationalism is not the awakening of nations to self-consciousness but, rather, the inventing of nations that did not exist. Nationalism arose with the introduction of mass education in those late-comer regions where language or other ethnic differentiae provided a strong incentive and means for the backward population to think of itself as a separate 'nation' and to seek independence — a liberation from second-class citizenship. Several nationalist movements in the nineteenth-century multi-national empires are appropriate examples (Gellner 1964: 171-2; Hobsbawm 1972: 395-401).

However, besides being a mode of confronting the consequences of later modernization, nationalism has another aspect. It is, to cite Eric Hobsbawm (1972: 392, 404), a 'civic religion' for the modern territorially centralized state. A territorial state that functions through a direct linkage between the individual citizens and a strong centre must develop a set of motivations in the citizens that gives them a primary and overriding sense of obligation towards it and eliminates the various other obligations that they feel towards other groups and centres within or without the territory. In an era of capitalist economic development and mass participation in politics, nationalism has functioned as the ideology by which the population has established a sense of identity with the modern state.

The two sides of the phenomenon — or, rather, the two phenomena — are historically linked, at least in the sense that, after a successful national liberation struggle, nationalism has functioned as the 'civic religion' of the new state (Hobsbawm 1972: 404; Kiernan 1976: 115-6). However, in nineteenth-century Europe the importance of these two aspects differed in different parts of the Continent. This has been explicitly suggested by Nairn (1977: 177-8, 184). In the older states of western Europe the nineteenth century maximized the ascendancy of the dominant nationality, whereas in east-central Europe nationalism arose as the protest of underdeveloped peoples. For the latter, nationalism became the way of mobilizing and trying to catch up with the already industrialized areas in the west.

Nairn (1977: 153-4, 339-40) links nationalism among the late developers of east-central Europe to the uneven development of capitalism. It was in essence a forced reaction to the spread of capitalism. The majority of the better-off groups saw themselves excluded from the material progress of the advanced lands, and mobilized against this 'progress'. In the mobilization of the people a militant, inter-class community was consciously formed and made strongly, though mythically, aware of its separate identity vis-à-vis the outside forces of domination. A nationalist mobilization against 'progress' was the only way for the backward, dominated lands — or, more precisely, for certain social strata in these regions — to seek access to this progress.[2] Nairn's analysis demonstrates the close relationship between the two nationalist phenomena. What is also essential in nationalism as the protest of underdeveloped peoples — though in an embryonic form — is mobilization across class boundaries, the creation of an inter-class community.

Citing Miroslav Hroch's study of nineteenth-century nationalist groups (1968), Nairn (1977: 117) states that nationalism as a reaction to underdevelopment normally involved, first, the intelligentsia, then wider strata of the middle classes, and finally the masses. Hroch's study is a rather unique piece of comparative research on the structure of 'patriotic' groups in the phase immediately preceding nationalist mass mobilization among seven small European nationalities which are also, for Hroch, 'repressed peoples', repression meaning the unequal cultural and political position of a group in a larger political unit (Hroch 1968: 16).

The overwhelming importance of intellectual groups was, understandably enough, typical of nationalism in these cases. Moreover, petty officials, in contrast to higher bureaucratic strata, and small merchants and artisans, in contrast to entrepreneurs and

large merchants, tended to provide activists and supporters to the
nationalist movements in this phase. The 'patriots' studied were
predominantly upwardly mobile, the sons of parents from the
lower ranks, who had risen just as far as was possible for persons of
such parentage (Hroch 1968: 125-37; 1971, 129-30; see also Plakans
1974; Koralka 1971: 57-8, 62-7; Portal 1971; 97, 100).

It was particularly the eastern European cases in Hroch's study
that displayed these traits — which fits in well with the suggestions
by Gellner and Nairn that nationalism is a reaction to underdevelopment. At the same time, the social structure in these cases
bore strong marks of the feudal past. The activists were recruited
from outside the nobility and high bureaucracy, or from outside
the ruling class of the feudal society (which largely identified itself
with the repressing culture). They also came from outside the new
rising bourgeoisie, which was likewise culturally alien to the nationalist groups. It was these who were able to spread the nationalist ideas to the masses and to mobilize them in the next phase
(Hroch 1968: 16-17, 33). On the other hand, in the core areas of
capitalism in western Europe, the main bearers of nationalism were
intellectual groups linked to the rising bourgeois ruling classes.
What was important there was nationalism as a civic religion, the
need for a unifying ideology which would overcome the unstabiliz-
ing effects of class conflict (Kiernan 1976: 111-2).

Finland in a European perspective

At first glance, the position of Finland — or, to be precise, of its
great Finnish-speaking majority — in this two-fold division of
nationalist phenomena seems fairly clear. Finland was one of the
'unhistoric nations'. The Finns were also one of the ethnically
distinct minority groups of the multinational empires of the time,
all of which gave rise to national movements. In Hroch's study the
Finnish case appears as one example of national self-assertion by a
repressed people, and being more reminiscent of the eastern than
the western cases. For Hugh Seton-Watson (1977: 72, 430), the Finnish national movement resembles closely the national movements
that arose in central-eastern Europe and the Balkans.

There is no doubt that in all these respects Finland was one of the
late developers of the east. However, it is not quite correct to picture Finland as a colonial territory or as an eastern European
periphery struggling through nationalism to free itself from the

dilemma of uneven development. It may be hypothesized that the Finnish 'deviations' from this pattern explain much of the steady advance of national consolidation and nationalism in Finland.

Up to 1809 most of the mainly Finnish-speaking regions, which later came to make up Finland, remained more or less on the periphery of the Swedish state. In the Napoleonic wars they were transferred to the control of Russia, which made the Finnish-speaking regions an autonomous grand duchy in the Russian Empire. Russia was economically more backward than Finland. This relative advantage fostered autonomous Finnish development in the nineteenth century, so that Finland grew up as a state with an autonomous economic core (see Pihkala 1970: 241-4; Schybergson 1973: 167-8; Suni 1979: 34-47). Its independence in 1917 was a consequence of the Russian Revolution.

There are at least three important characteristics of this development. First, the autonomous Finnish state was founded some decades before the politicization of ethnic differentiae, that is before the rise of nationalism. In a more fundamental sense than was true for any of the other small nationalities in Hroch's study (with the exception of Norway), Finland was an autonomous unit, with its own state apparatus. Autonomous institutions developed gradually during the nineteenth century, which was much more decisive for the consolidation of Finland than imperial declarations or juridico-formal statements (cf. Schweitzer 1978: 4-5, 18-30; Jussila 1979a). The necessity of fighting for the formation of a separate state, which faced nationalist movements elsewhere, did not occur in Finland.

During the late nineteenth century, simultaneous with the increasing penetration of nationalist ideas beyond small intellectual circles, it was also important for the development of a Finnish national economy that Finland was able to benefit from the large Russian market. This importance was accentuated because, in its trade with the far more industrialized western Europe, Finland served as a source of primary products, sawn goods and raw timber, which were highly vulnerable to international price fluctuations. Like interface peripheries in general, Finland has some level of choice between alternative centres in its strategy of growth. In the Russian Empire Finland was a region of 'relative over-development' (see Nairn 1977: 185-7). In short, the formation of a national economy as a necessary condition for the consolidation of the state had advanced exceptionally far by the end of the century. As regions of 'relative over-development' in eastern Europe, only Bohemia and

Croatia seem comparable to Finland (see Nairn 1977: 185-7).[3]

Third, the Finnish state was consolidated in the interface between the Swedish and the Russian centres. Because of the earlier history under Swedish domination and the subsequent transfer to Russia, political domination on the one hand, and cultural and economic domination within the country on the other, were not superimposed one on another. Political domination was ultimately in St Petersburg, whereas domination in the economic and cultural sphere belonged to the Swedish-speaking upper class. Furthermore, this consolidation between Sweden and Russia involved a combination of a non-feudal class structure (with a large freeholding peasantry that had emerged during the Swedish period) and a subordination to a great multinational empire. In these respects Finland displayed only superficial similarities to other eastern European regions.

In both types of nationalist phenomena, mobilization across class boundaries, or the creation of an inter-class community, played a central role. The function of this mobilization — whether as a civic religion for the state or as the protest of underdeveloped peoples — was closely connected with the position of the country or region in the international (capitalist) system. In the western core areas of capitalism, nationalism was closely linked to the ruling class; in the periphery, it was linked mainly to middle-class groups seeking popular support against alien economic and political domination.

The early development of the Finnish case points straightforwardly to the latter alternative. During the Swedish period there arose a non-feudal class structure and a cultural division of labour (Hechter 1975: 35-43) with a Swedish-speaking upper class. At the same time, the Finnish-speaking regions were backward in comparison both to the core of Sweden and to western Europe in general. But immediately before the rise of nationalism, this picture was radically altered. Finland was established as a state, and it became politically dependent on an empire, in contrast to which it was economically 'over-developed'.

The consolidation of a national culture

All this resulted in a nationalism in which both aspects seem to have intertwined exceptionally closely. In other words, nationalism in Finland did not play the role of a liberating force in the typical eastern European way; it also displayed, practically from the very beginning, strong elements of nationalism as a 'civic religion' for the territorially centralized state.

The political dependence on a great feudal state points to nationalism as a mode of self-assertion and liberation. On the other hand, the fact that Finland was a state that emerged and consolidated itself during the nineteenth century points to the possible importance of nationalism as a 'civic religion'. Economic overdevelopment' contains, at least potentially, varying elements. Nairn (1977: 185-7) cites it as an example of a situation causing nationalist movements for liberation and self-assertion. But if looked at as a factor in the consolidation of the emerging state, as in late nineteenth-century Finland, it may also strengthen tendencies for inculcating the sense of obligation to the state. Which aspect predominates depends on how tight or loose the political dependence is, and the extent to which it limits economic freedom of action. In Finland the economic limitations were few (see Pihkala 1970).

This dual background may serve as a starting point for an analysis of the consolidation of a rather unified bourgeois national culture in Finland. There were exceptionally strong incentives for the upper classes, with their Swedish culture, to adopt or accept rather easily the language and culture of the great majority of the people. It was important, both because of the country's political dependence on Russia and because of the need to establish a sense of obligation to the Finnish state. In this respect Finland differed from the other ethnically distinct regions of the multinational empires of the time, where the upper classes identified themselves with the power on which the region as a whole was dependent politically and/or economically, and therefore had little incentive for nationalist mobilization (cf. Molnar 1971: 221-7): what was central for them in the maintenance of their position was not so much the efforts to strengthen solidarity towards the state as the maintenance of feudal class domination. Instead, in Finland the class structure was basically similar to the Scandinavian pattern, and therefore it may be hypothesized that the need to establish a sense of identity towards the state was also a primary concern for the upper classes.

This suggestion may be formulated in another way by saying that in Finland, unlike anywhere else in eastern·Europe, there existed strong incentives for nationalist mobilization not only among the middle, but also among the upper, classes. This view may be supported by referring to some traits in nineteenth-century nationalism in Finland. Information on the structure of early nationalist groups may be found in Hroch's study. He focused on the development of nationalist movements in the phase when a group of 'patriots' had already attempted systematically to spread 'the national idea', but without as yet penetrating the masses to any extent. In Finland this phase was reached in the 1840s and 1850s. In those decades the nationalist movement, which originated within the Swedish-speaking upper class, began to take on a clearly socio-political character. At the same time, the estate-based political system (with representation of the freeholding peasantry) remained dormant. The estates convened for the first time after a long delay in 1863. The presence of the upper classes among the nationalist activists of the period was larger than it was among the corresponding activist groups elsewhere in eastern Europe. The proportion of nobles and high bureaucrats was particularly high. This picture is completed with information on the recruitment of the activists. In Finland they were less often sons of parents from the lower ranks and less often geographically mobile than elsewhere (Hroch 1968: 83-167 passim; also Klinge 1969: 189-98).

The aim was to establish national unity on a Finnish cultural base. Finns with university training were urged to adopt Finnish as a working language, and to create cultivated literature that was both popular and patriotic. Over a longer period, national unity and patriotic indoctrination were to be achieved through revision of the school system. A Finland united in language, culture and loyalty might resist the dangers resulting from the dependent position of the country, and might also develop further.[4]

The cultural demands were not accompanied by demands for any thorough reforms in the structure of society. On the contrary, by the end of the century the so-called Fennoman movement had developed increasingly into a movement in which conservative tendencies gained an upper hand:

> The fact that the Finnicization movement was directed against the exclusively privileged Swedish-speaking upper class of that time, did not imply that the upper class should have been eliminated in order to found a democratically organized society, but that the upper class speaking Swedish and oriented to the Swedish culture should have been replaced by an upper class speaking Finnish and oriented to the Finnish culture. [Wuorinen 1935: 273; also Klinge 1968: 74, 114]

On the other hand, by urging linguistic reform and the broadening of the social basis for school attendance, the Fennoman movement did contribute to the recruitment of new groups to the upper classes in the latter half of the century.

It may be argued that the Fennoman movement acted for

national self-assertion against Swedish cultural dominance and against the dangers arising from the politically dependent position of the country. On the other hand, it clearly strove for the creation of an integrating ideology for the emerging Finnish state. This latter characteristic became increasingly dominant in the last decades of the 1800s, simultaneously with the reactivation of the estate-based political system. At that time the nationalist movement first gained support from outside intellectual circles. Besides the Evangelical Lutheran clergy, it was backed by the wealthy freeholding peasantry, which had representation in the political system. This group greatly benefited from the simultaneous capitalist transformation in agriculture, which created a peasant upper class and enormously widened the gap between the large peasants (becoming capitalist farmers) and the landless proletariat.

In developing into a political force based mainly on the clergy and wealthy peasantry, the Fennoman movement began, in a pronounced manner, to function as the introducer of an agrarian and religious ideological alternative for the integration of the emerging nation. This was to be conclusively proven after political mass mobilization: the Finnish party — the heir of the Fennomen — became the main opponent of the Social Democrats in the countryside, where the latter gained a larger turnout than in the cities, not only absolutely but also relatively.

Also characteristic of Finnish national consolidation is the attitude of the Liberals. From the 1860s onwards they were the main opponents of the Fennomen. True, in the 1860s there also rose a Swedish nationalist movement which leaned in its ideology on the Swedish-speaking agrarian and fishing population in the coastal areas. Its significance remained limited, however. The Liberals derived their support from the rising bourgeois groups and from the bureaucracy. While being mainly Swedish speakers, they did not direct their main opposition against the cultural programme of the Fennoman movement. They opposed the creation of a unilingual Finnish national culture, but more important for them than language was the preservation of political institutions vis-à-vis Russia and the continuity of the cultural heritage of the Swedish period (Klinge 1975: 20). Consequently, they were prone to stress the position of Finland as a separate state unit more sharply than the Fennomen did. It was particularly Liberal circles in which some central national symbols, for example the personification of Finland in the figure of the so-called Maiden of Finland, were first put forth (Reitala 1980).

In these circumstances there were central common elements in both the Finnish and Swedish nineteenth century upper-class culture. The Fennomen were largely linguistic converts, and the whole educated class attended the same university (see Klinge 1975: 92-3; 1977: 51, 62). By the end of the century Finnish had arisen to a strong or even predominant position in the central institutional spheres of the society. The aim, crystallized in the Fennoman movement, to create an upper class culturally united with the majority of the people, largely by linguistic conversion, was materializing rapidly (Klinge 1968: 331-2). While there was some upward mobility into the elites, the old upper stratum — consisting of the old noble, burgher and particularly clerical families — was mainly to remain in charge of Finland up to independence in 1917, and far beyond (Klinge 1975: 17).

All this resulted, in the long run, in the rise of a comparatively united nationalistic culture among the upper classes and such middle-class groups as teachers and lower civil servants, despite the linguistic division. The development was strengthened by the heightening of class conflict in the first decades of this century (see, e.g., Murtorinne 1967: 90-104, 222, 229, 231; Sarajas 1962: 135-80). In the interwar period most support for nationalist campaigns against the former linguistically distinct upper class came from among the students and the young educated class, but they had no real base in other groups. By the late 1930s at the latest, the linguistic fervour had exhausted its strength even among the students. The upper classes were, after all, overwhelmingly Finnish-speaking, or at least bilingual in this period.

On the other hand, because Swedish did not lose its significance among the upper classes, the Swedish-speaking agrarian and fishing population never came into a situation in which they would have been forced to fight for their language. The position of the Swedish language was guaranteed in the constitution and the language Acts immediately after the gaining of independence. Owing to the specifically Finnish situation where the linguistic minority had roots both in the 'centre' and in the 'periphery' (Allardt and Miemois 1979: 50), the Swedish-speaking small farmers, fishers and workers were not to face a situation of several other linguistic minorities which has given rise to strong ethnic movements.

Territorial integration in
the nineteenth century

The Finnish-speaking regions up to 1809

To assess the nature of political mass mobilization, and especially its main regional variations, state-making has to be viewed from another perspective. Up to 1809 most of the Finnish-speaking regions that later came to make up Finland remained more or less on the periphery of the Swedish state. There were only weak interactions among various Finnish-speaking regions; they were in no sense an autonomous economic unit (cf. Åström 1977). The Finland that the Russians took over in 1809 was a country without a centre. It had never before been an administrative unit, although its main areas had temporarily been under a united administration (Klinge 1975: 10, 28). It is indicative of the pre-1809 Swedish stance that in certain symbolically central contexts the provinces are presented, not separately as 'Swedish' and 'Finnish' provinces, but in an order starting from the regions around the core and arriving at the more peripheral regions in various corners of the state (Klinge 1975: 26).

Figure 1 presents the basic divisions that are commonly used in accounts dealing with socioeconomic conditions in the nineteenth century.[5] These regions developed gradually through the centuries, and were to be influenced in different ways by nineteenth-century capitalist integration. They have a rough correspondence with the historical division of provinces.[6] Also, the regions correspond rather well to a division by county or groups of counties in the beginning of the twentieth century, the only major exception being Ostrobothnia. To south-western Finland belong the counties of Uusimaa, Turku and Pori, and Häme, and to eastern Finland the main areas of the counties of Kuopio and Mikkeli. Ostrobothnia is composed of the western parts of the county of Vaasa and of the adjacent coastal areas of the county of Oulu. Northern Finland is coterminous with the bulk of the county of Oulu. The regional statistics used in the subsequent pages exist in most cases only for the counties. Since the Second World War there has been some reorganization, and the number of counties has increased (see Figure 10 below).

Both geographically and socially, the south-western region was closest to the core of the Swedish state. From there were also established trade connections with Stockholm. The southern coast

FIGURE 1
The regions and the line between the
Reds and the Whites in the civil war of 1918

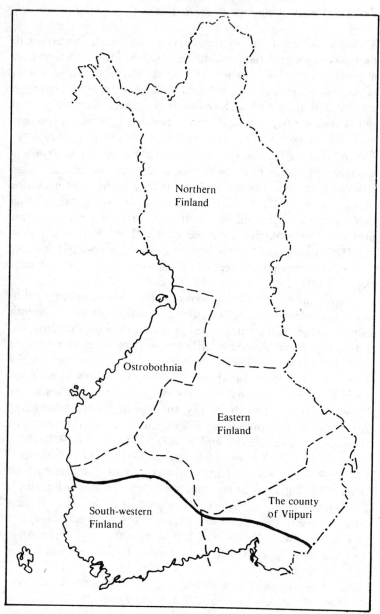

is the area with the longest history of Swedish domination and influence. In this region were the largest trading centre, Turku (Åbo), and the majority of the manors in Finland. Besides the gentry, the wealthier peasants also began to penetrate the agrarian upper class after enclosures and other reforms in the late 1700s. In the early nineteenth century the number of crofters, and especially of landless labourers, rapidly increased.

The other region with close links to the state core was Ostrobothnia. Up to the mid-eighteenth century it was dominated by Stockholm merchants and specialized in tar export. It was a part of the regional whole surrounding the upper end of the Gulf of Bothnia. Geographically, it displayed a 'strong ecological independence' (Sarmela 1969: 253), with a ridge separating it from the other regions, particularly in the south and south-east. Coastal cities served as way-stations in the trade between the Ostrobothnian hinterland and Stockholm. The countryside was dominated by independent peasants, and it seems that the tar-selling peasants had considerable bargaining capacity in their relations with the merchants in the cities. The villages were compact peasant communities with no manors. In Ostrobothnia the wealth of the peasant upper class was not primarily dependent on the exploitation of the crofters' labour power, as it was in the south-west. The crofters were in a kind of partnership with the landowners, who needed labour for tar production.

Early eighteenth-century developments definitely separated and differentiated the area surrounding the city of Viipuri from the south-west. Up to that time Viipuri had been, besides Turku, the most important centre for overseas trade in the Finnish-speaking regions. In losing its position as a great power, Sweden was forced to surrender this area to Russia. Russian nobles dominated large areas of land, and feudal relations in the countryside were strengthened. At the same time, the proximity of St Petersburg linked small peasants directly with its market. In 1800 the Russian capital, with 220,000 inhabitants, was among the largest cities of Europe. The small landholding peasants transported their own products to St Petersburg and worked at other transport jobs both during and outside the farming year.

In the east, or the so-called Savo-Karelian slash-and-burn region, the linkages with the dominant external centres were tenuous. Slash-and-burn cultivation meant that people were mobile and lived in dispersed settlements: there were no strong exploitative relations within the peasant population and, therefore, no strong peasant upper class. The small local gentry had only a minor role in agricultural production. Up to the eighteenth century the region had remained an economic hinterland of Viipuri, exporting mainly tar. Here, merchant-peasant relations seem to have been exploitative by nature. After the annexation of Viipuri and its sur-roundings by Russia in 1721 and 1743, peasant trade with the southern centres declined. The scant grain and tar trade that existed was forced to turn to the Gulf of Bothnia, benefiting the Ostrobothnian economy.

The extremely sparsely populated north of Finland, including Lapland, corresponds roughly to the 'backwoods Finland' (Luonnon-Suomi) in the geographers' terminology. During the Swedish period and later it was an area of colonization. On the other hand, there existed old market connections based on salmon, furs and meat, through the rivers running to the upper end of the Gulf of Bothnia. The eastern areas were involved in the market through Russian centres.

To sum up, in 1809 the regions of the newly founded Grand Duchy of Finland had a tradition of linkages to several outside cen-tres. The south-west and Ostrobothnia had closest connections with the core of Sweden; the future county of Viipuri had establish-ed close links to St Petersburg; and in the east the ties with Viipuri had been partly replaced by connections with Ostrobothnia. Therefore, in an analysis of trade routes, the eighteenth century (the end of the Swedish period) has been characterized as a period of north-west orientation (Ajo 1946: 22-3). The weakest links were between the east and the south-west: even by the mid-nineteenth century, their commercial ties were 'quite negligible' (Wirilander 1960: 753, 754).

Re-orientation from Stockholm
to St Petersburg

In the first decades of the autonomous Grand Duchy, Stockholm was in some respects replaced by St Petersburg, even though many economic and other connections with Sweden continued to exist. In eastern Finland the peasant trade with the southern coast and with St Petersburg was soon revitalized. Butter became the main export to the Russian capital. Timber and some iron was also exported from the region. Ironworks were owned mainly by merchants and industrialists in St Petersburg. From both Viipuri and eastern

Finland, there was a considerable flow of seasonal and permanent migration to St Petersburg, which stagnated only in the 1880s (Engman 1978) (see Figure 2). A very significant connection between eastern Finland and the south, and at the same time between eastern Finland and St Petersburg, was created when the Saimaa Canal from the eastern watercourse to Viipuri was completed in 1856 (see Figure 3).

In other words, both regions in the east had close economic connections with the St Petersburg area or belonged to it. To some extent the same development was true of other regions. Over the whole country trade turned increasingly to Russia. An important partial exception was Ostrobothnia, which continuously maintained active trade connections with Sweden (Joustela 1963: 304, 342-3).

An indication of the conscious efforts to weaken the orientation towards Sweden is the promotion of Helsinki as the capital city of Finland. During the Swedish period the most important Finnish centre had been Turku, where the university, the archbishopric, and the court of appeal were situated. But the role of capital was granted not to Turku, which both economically and culturally was oriented towards Sweden, but to Helsinki, which was situated nearer St Petersburg on the southern coast. Helsinki was a small city which was now deliberately built up as a prestigious administrative centre of the country. For several decades it was a 'parade capital' (Klinge 1975: 29) with huge and continuous construction works for the central administration, the church and the university (which was transferred to Helsinki after a fire had devastated Turku in 1827).

Territorial integration in
the late nineteenth century

Another tendency was, however, gaining ground. After the middle of the century a development gained momentum which united Finland economically around a domestic core. Capitalist transformation created a regional division of labour that tied the various regions to each other in a more fundamental sense than ever before and accentuated many regional inequalities. Territorial integration was affected by Finland's dependent position; but above all it was affected by the varying internal characteristics of the regions consolidated earlier. This pattern holds for western Europe in general.

FIGURE 2
Finns resident in Russia on a passport in 1881,
by commune. The Countryside. (As a proportion
per thousand of the population entered in
the census registration)

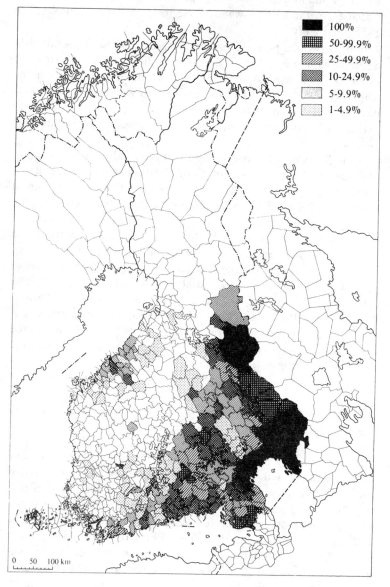

FIGURE 3
The Finnish Railway network in 1918, and the Saimaa Canal

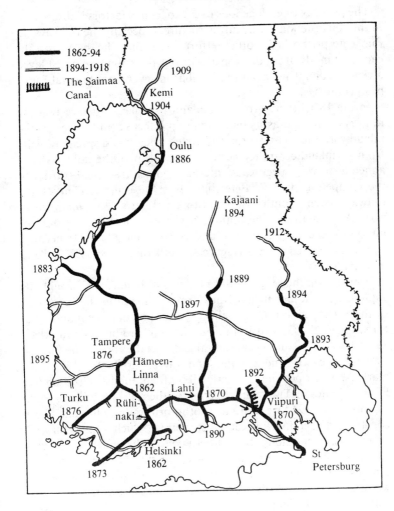

Source: Based on the map in Jutikkala (1968: 172) and on *Suomen valtionrautatiet 1862-1912*, II (1916).

Industrial growth in the rise of the capitalist mode of production concentrated on those regions that initially were the more advanced.

The primacy of endogenous over exogenous factors is discernible in the construction of the infrastructure, especially the railways. In this enterprise, as in other efforts at laying the basis for the capitalist mode of production and its reproduction, the state played a central role. This, again, is familiar in several continental European countries.

In the 1850s there arose in Finland two plans for directing the railways. One had as its main point an effective unification of the inland with the coasts: a railway system that would spread radially from an inland centre to the coasts. In this plan the railways were aimed at serving specifically internal Finnish trade and traffic. In the competing plan St Petersburg was the central point of the railway system: south-east – north-west was to be its main direction. At first the former gained a victory (the Russians did not interfere), and as a first step a railway from Helsinki to Hämeenlinna was completed in 1862 (Figure 3) (Jutikkala 1968: 169-73; Tommila 1978: 17).

However, in the long run this plan was also defeated through the capitalist development that transformed Finland in the latter half of the nineteenth century. In this process 'Finland's economic face turned from the Gulf of Bothnia to the Gulf of Finland' (Jutikkala 1950: 164). The south became established as the core of the Finnish state. There was the capital, and there, closer to the export markets, developed the gravitational centre of industrialization (see Figures 4 and 5). The main direction of the railways became south to north. By 1894 three parallel south-north railway lines had been constructed; a connecting east-west railway (which went as far as St Petersburg) went along the southern coast. This indicates the consolidation of the south as the core, but at the same time it is reminiscent of the division between the two eastern and the two western regions: link lines connecting the east and the west were constructed only later.

Central in the capitalist transformation of Finland was the sudden rise of the sawmill industry in the 1870s. The Finnish linkage to the capitalist world market in the late nineteenth century became overwhelmingly based on forestry; between 1900 and 1909 69 percent of Finnish exports were based on lumbering. At the same time, the Finnish economy was oriented to western Europe as an exporter of this primary product. The leading industrial sector was a

Risto Alapuro

131

FIGURE 4
Urban industry in Finland, 1884-85 and 1938

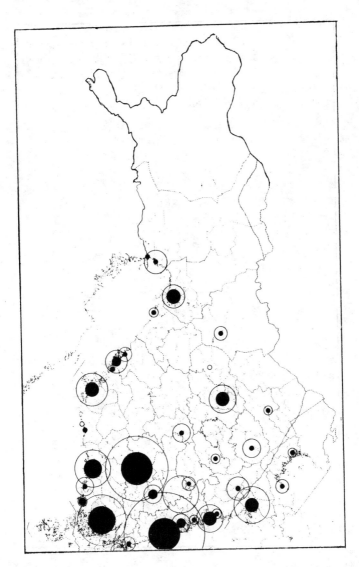

Solid circles indicate the size of the industrial workforce in 1884-85, the open ones, its size in 1938. Indicative of the scale is that in Helsinki in 1938 (the largest open circle), the number of workers was 31,000.

Source: Jutikkala (1959: 61).

FIGURE 5
Rural industry in Finland, 1884-85 and 1938

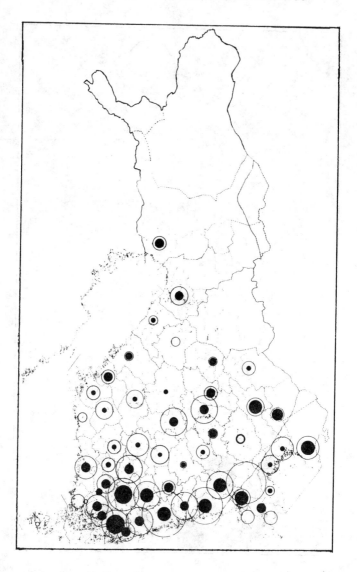

Solid circles indicate the size of the rural workforce in 1884-85, the open ones, its size in 1938. On the scale see n. 1 in Figure 4.

Source: Jutikkala (1959: 61).

primary sector, based on the demand of the more industrialized western states. In the Russian trade, instead, Finland exported more processed products: by 1914 these dominated exports to Russia.

Because of the predominant position of forestry, Finnish industrialization had few linkages with other sectors of the economy. The employment structure remained agrarian: in 1910, for example, 74 percent of the labour force was engaged in agriculture, and only 12 percent in industry.

Significantly, however, export trade was based on a product having a strong and very direct linkage with the countryside because the peasants owned the bulk of the forests in Finland. Therefore, the sudden linking of a thoroughly agrarian Finland with the developed capitalist market system was felt immediately and profoundly, but with great regional variations, among the agrarian population.

Core – periphery interaction between
the county of Viipuri
and eastern Finland[7]

Perhaps the most striking indication of the transformation in the nature of regional interconnections was the new linkage between Viipuri and eastern Finland. The latter became a periphery in the emerging national economy of Finland. At the same time, the coastal area of Viipuri developed into a part of the core. In the mid-nineteenth century the sawmill industry was, besides a modest iron production, the only industry in the east. The sawn goods were transported to the coast where the commercial houses in Viipuri dominated the trade. In the 1850s about two-thirds of Finland's sawn goods were produced in eastern Finland while the province of Viipuri was the 'least developed' sawmill region in the country. Simultaneously with the rapid expansion of forestry, the location of the sawmills changed completely. By the end of the century, those in the eastern region produced only 12 percent of the total production, whereas Viipuri became the main region for the sawmill industry, producing about 25 percent of the country's sawn goods (Table 1).

In the redefinition of the regional division of labour, the introduction of the large steam-operated sawmills played a central role. They were built at the mouths of rivers on the southern coast,

TABLE 1
Production of sawn goods in 1860 and in 1900,
by county

County	1860		1900	
	1000 std	%	1000 std	%
Uusimaa	1.6	4	55.2	9
Turku and Pori	2.1	6	131.0	22
Häme	1.9	5	69.8	12
Viipuri	1.3	3	151.4	25
Kuopio	12.0	31	44.9	8
Mikkeli	13.7	35	24.2	4
Vaasa	1.6	4	50.8	9
Oulu	4.8	12	65.3	11
All counties	38.9	100	592.6	100

Source: Hoffman (1978: 171)

mainly by the commercial houses in Viipuri and other centres. The timber they required was mainly transported from the eastern region. There the new steam-operated sawmills remained small, and the old sawmills, which were run by water power, served local needs to a greater extent than before: 'The small firms fell into the possession of the big firms, and the concentration of production in the coastal areas made the former inland centers of industry into mere sites for delivery of the raw material' (Lakio 1975: 105).

In other words, industry in the eastern region stagnated in relation to the south. Not only did the proportion of the industrial labour force living in this region decline at the turn of the century, but its numerical strength also declined temporarily in large areas. The prevalence of agrarian occupations was actually accentuated at this time, as was to some extent the lack of an urban population (Table 2).

Perhaps the main indication of a peripheral position is, however, the fact that the timber companies purchased large areas of peasant land. This development was limited almost exclusively to eastern Finland and to areas adjacent to it in the north. In other regions, particularly in the south-western counties, the peasants sold timber from their forests, but little land. Most of the land owned by the companies lay in the eastern region and the bordering areas further

TABLE 2

Population and migration in Finland during the late nineteenth and early twentieth centuries, by county

County	Population		Percentage growth in population, 1865-1910	Percentage of population engaged in agriculture		Percentage of urban population		Migrants from the rural communes during 1891-1900 as a percentage of the rural population in 1891
	1865	1910		1865	1910	1865	1910	
	('000)	('000)	%	%	%	%	%	%
Uusimaa	173.6	376.2	117	64	43	19	43	3.5
Turku and Pori	305.7	499.3	63	73	63	11	16	4.6
Häme	196.0	342.3	75	80	61	4	17	5.5
Viipuri	275.0	521.5	90	85	65	5	9	-5.1*
Kuopio	227.6	333.8	47	85	79	3	7	6.4
Mikkeli	159.6	198.8	25	85	82	2	5	5.7
Vaasa	312.9	514.9	65	84	74	5	8	3.0
Oulu	188.7	328.3	74	78	71	6	8	0.8
All counties	1,839.0	3,115.2	69	79	66	7	15	

* i.e., more in-migration than out-migration.

Sources: *Official Statistics of Finland*, VI: 1 and 45; Jutikkala (1963: 386); Kilpi (1913: Appendix, Tables 1 and 2).

to the north. By 1915 the companies had, in large areas of the region, purchased 20 percent or more of the land not owned by the state. The boom in land purchases began in the 1890s.

Finally, an important indicator of the intensity and quality of the new interconnection was the mass migration to the south (see Figure 7 below). St Petersburg had previously been the main target for migration, but towards the end of the 1800s Viipuri took its place. At the same time, the volume of migration greatly increased, apparently accelerated by the construction of railways from the south to the region in 1889 and 1894 (Lento 1951: 168-9). The migrants largely moved to the new centres in Viipuri, outside the established towns: hence the prevalence of rural in-migration over rural out-migration in this province.

The intensity of migration was greater in eastern Finland than in any other region, and therefore the increase in population was slower there than elsewhere. It demonstrates the increasing permeability of regional boundaries with economic integration, and also shows an aspect in the redefinition of the division of labour between the regions. About half of the sawmill workers in the county of Viipuri were born elsewhere, mainly in eastern Finland (Snellman 1914: 35).

In the development of the eastern region there are significant parallels with the so-called dependent industrialization depicted, for example, by Michael Hechter (1975: 33). On the other hand, this change may be portrayed as a move from the dominance of merchant capital to that of industrial capital. During the Swedish period the merchants had dominated the tar-based linkage. They controlled the export of tar, but production was in the hands of the peasants. The same held true for the butter trade. From the late nineteenth century onwards, however, the owners of capital (still largely merchants from Viipuri) dominated the linkage. Capital was required to construct modern sawmills and buy land. The peasant population was required mainly as wage labour in felling and floating timber, and in the sawmills on the coast. In this way the territorial division of labour was redefined, accentuating regional inequalities.

Despite the fact that capitalist transformation seized the region, it did not make it possible for the increasing rural proletariat to move to the towns or industrial centres in the region, as was the case with industrialization in the core. It seems that even the intense migration of the late nineteenth century did not decrease the proportion of the landless labourers compared with the other regions.

And more importantly, the structure of the landless in eastern Finland differed greatly from that elsewhere. In the east rural poverty was much greater, and the relative poverty of the eastern landless proletariat was actually increasing in the last decades of the 1800s (Haatanen 1968: 142-3).

It is also significant that here the timber boom did not create a strong peasant upper class. True, the landowning peasants gained by selling timber, but there is no doubt that they were unable to reap the benefits of the boom to the extent that the established peasant upper class in the south-west did. The purchases of peasant land are an eloquent indication of this difference. The slash-and-burn cultivation had fallen into a crisis in the middle of the century, posing great difficulties for the eastern peasants. Usually it was the heavily indebted peasants who were forced to sell their farms along with the forests. Falling into debt to merchants in the late nineteenth century seems to have been most common among the eastern peasants and the northern peasants living in the neighbouring areas (Hjelt 1893: 15-17; Niemelä 1978).

South-western Finland as a core region

In this transformation the south-west, along with the southern coast of Viipuri, developed into the core area of Finland. The position of south-western Finland is reflected in Table 2, which gives information on population and urbanization both before and after industrial take-off. Both in 1865 and in 1910, Uusimaa displays the smallest proportion of population engaged in agriculture, and the greatest percentage living in cities. (The figures are somewhat distorted by the fact that, particularly in Viipuri, the new centres were not included among the cities.) But the higher degree of urbanization in the whole of south-western Finland is much more discernible in 1910 than it was 45 years earlier. It was exactly in the three south-western provinces (and Viipuri) that urban growth and the decrease in the proportion of the agricultural population were greatest. The concentration of industry tells the same story (Figures 4 and 5).

The figures for Uusimaa are highly influenced by the growth of Helsinki. It was only in this period that Helsinki became the capital of the country, not only administratively and culturally, but also economically. During the last third of the nineteenth century, Helsinki grew rapidly, and by 1920 it was the dominant city, as may

FIGURE 6
The Finnish cities in 1815 and in 1920,
ranked in decreasing order of size of population

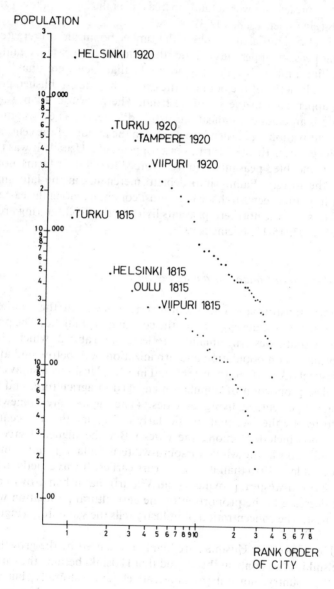

Source: Based on *Statistical Yearbook of Finland* (1922: Table 11).

FIGURE 7
Net balance of internal migration up to 1920
in Finland, according to place of birth

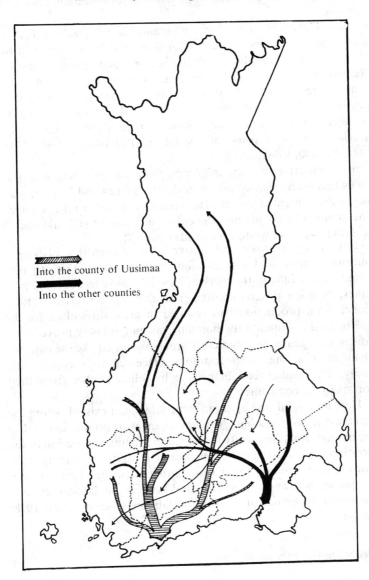

Source: Lento (1951: Appendix, Figure 8).

be seen in the distribution of city sizes in Figure 6, where the sizes are plotted by rank on logarithmic paper, as George Kingsley Zipf (1941: 11) suggests in his log-normal model for descending population size.

This change was initiated and maintained largely by the railways. The two first railways tied two great inland watercourses with Helsinki (Figure 3 above). After the opening of the railways Helsinki became the leading importing centre; the merchant fleet concentrated there, and various industries developed. Turku was deprived of its old hinterland in Häme, and eastern Finland was connected with the capital. In the process the inland cities developed at the expense of coastal cities (Jutikkala 1968: 171; 1977: 96, 100; Klinge 1975: 29).

It is indicative that the migration from the east, which in the 1800s had been oriented mainly to St Petersburg and Viipuri, was increasingly oriented towards Helsinki after the turn of the century. The majority of the nineteenth-century migrants to Helsinki came, nevertheless, from western Finland (Figure 7).

For the peasants, and above all the wealthy peasants, industrialization and urbanization created opportunities to participate more fully in the market, especially as sellers of dairy products. Both low-priced imported grain and a definite crisis in traditional grain production had resulted in great difficulties by the 1870s. In this situation the booming sawmill industry proved vital for the changeover to commercial stockraising. Unlike the east, the changeover in the south-west took place relatively rapidly and easily. The landowners financed it by selling timber from their forests to the companies.

Here an established agrarian upper class had existed before the capitalist transformation. Now the changes in production and in the market situation greatly promoted the commercialization of the agrarian upper class and the penetration of capitalist exchange relations into the countryside. Class conflict intensified both between the landowners and the crofters and between the landowners and the growing number of agricultural labourers (see Alapuro 1978: 126-8).

Declining Ostrobothnia[8]

Perhaps the most important characteristic of Ostrobothnian development in the nineteenth century is that there was no sudden capitalist upsurge comparable to that in the south and the east. The only exceptions were the more northerly towns in the region, which developed into centres of the sawmill industry. This wealthy region declined economically relative to the south: moreover, problems caused by the decline seem to have been distributed more evenly among agrarian groups than were problems owing to the increasing market involvement in the south. The decline was due first, to the breaking-off of ties with Sweden. The central areas of Sweden had been the main target for Ostrobothnian grain, which had increasingly replaced tar after the late 1700s. Sweden remained as a central trading partner during the next century, but now sales were crippled by Swedish tariffs. Second, the timber boom, which so immensely affected the south-west and the east, failed to affect Ostrobothnia because the earlier tar-burning — and shipbuilding — had severely reduced the supply of timber available to sawmills. The Ostrobothnian peasants did not have the forest incomes of the south-western peasants.

The decline resulted in the preservation of many traditional traits in the agrarian community; its structure did not change drastically during the nineteenth century. Even the very rapid increase in population throughout the century affected the social structure less than might be expected. Part of the increase was channelled into the prevailing structure: in contrast to developments elsewhere, the number of landowning peasants in Ostrobothnia increased, because it was a common practice to divide farms among the heirs of an owner. It was also more common than elsewhere for some of the children to remain on the home farm as crofters paying nominal rent. Part of the increase in population was also channelled outside the region. Emigration from Finland to the United States around the turn of the twentieth century was greatest from Ostrobothnia (Figure 8).

In addition to economic decline and the persistence of the agrarian community, the region remained isolated economically. In a sense, in the nineteenth century Ostrobothnia still 'did not belong to Finland' (Kaila 1931: 360). Characteristically, while the population increase in other regions during the late nineteenth century resulted in internal migration, in Ostrobothnia it resulted in an extensive emigration. It indicates the lasting importance of an external orientation in Ostrobothnia.

The isolation was, of course, relative. Ostrobothnia was connected to the south through railways in the 1880s, and many Ostrobothnians migrated to Helsinki in the last decades of the century. But this connection materialized very late, given the wealthy

FIGURE 8
Overseas emigration from Finland, 1870-1930,
by commune

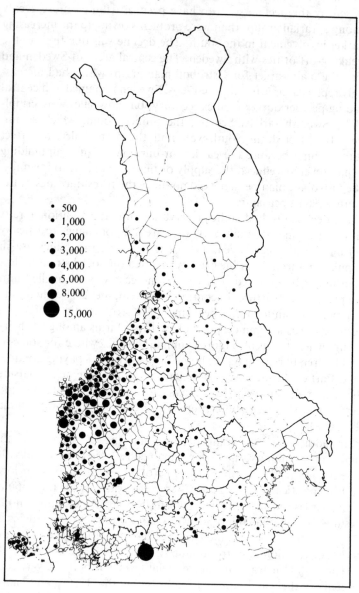

Source: Kero (1976: 25).

past of the province. Moreover, the construction of railways had considerable adverse effects. The south-north orientation of the lines broke old ties which went from the inland to the coast, and caused harm to economic life in the coastal cities (Tommila 1978: 26).

The Ostrobothnian development is in some respects reminiscent of the industrialization of the periphery as portrayed by Sidney Tarrow (1977: 23-4) in his assessment of theories of 'peripheral marginality'. In the earliest stage of industrialization, isolated factories and mines grew up wherever there were sources of raw materials. At a second stage, priority shifted to seeking concentrations of consumers and services; the industrial core developed in other regions, leaving the periphery, at least relatively, in isolation and decline. In the third stage, the isolated region was industrialized again, but this time industrialization was a dependent phenomenon.

In Ostrobothnia the early prosperity was based on the exploitation of forests — on tar-burning and shipbuilding. Later these enterprises declined, simultaneously with the consolidation of the Finnish core in the south. In the third phase, during the 1900s, Ostrobothnia had industrialized, but its industrialization was secondary to that in the south (and, after the Second World War, in Sweden). In the 1960s Ostrobothnia became a so-called development zone (see Figure 11 below). The relative decline of Ostrobothnia and its subsequent linkage with the core of Finland are crucial factors for an analysis of the nature of the religious and political mobilization in the region.

Regional division of labour and state penetration in northern Finland

Here, the connection with the Ostrobothnian cities was redefined and strengthened, and the region was linked to the integrating state with stronger ties than before. One characteristic of lasting importance remained the region's long distances and the small population: in 1900 it accounted for 42 percent of the total area of Finland but only 5 percent of the population. There arose a regional division of labour, which resembled that between Viipuri and eastern Finland. The northern Ostrobothnian cities, especially Oulu, developed into centres for the sawmill industry with the richly forested inland areas in northern Finland as sources of the raw material. This was especially so along the two rivers flowing westwards to the towns of Oulu and Kemi.

Before the timber boom the peasants in the Kainuu region were exploited by the Oulu merchants as tar producers (see Hautala 1956: 228-39), like the peasants in eastern Finland had once been exploited by the Viipuri merchants. After the growth of the sawmill industry, a large number of indebted peasants in Kainuu were also forced to sell their lands to the merchants and ultimately to the timber companies (Kyllönen 1975: 122-3). Kainuu was adjacent to eastern Finland, but because of the direction of the waterways its connections went to the Gulf of Bothnia, not the south. Certain demographic effects were also similar to those in eastern Finland. The timber companies in Oulu and elsewhere purchased peasant land, thus creating a landless population, which then migrated to the coastal centres to serve as workers for the sawmill industry (Kyllönen 1975: 131).

The ties to the coast were strengthened, but so were connections with the south of Finland. The railway reached Oulu in 1896, and it was continued further to the north after 1900: from the south-east another railway line was extended to Kajaani in 1904. The railway broke the monopoly of the Oulu merchants in that area (Hautala 1956: 247-9) and somewhat eased the miserable conditions of the agrarian population. Other factors connecting the north and other regions were the big seasonal occupations of logging and floating. After 1900 logging gave work for as many as 10,000 men every winter; and the work force was recruited not only from the north but also from other regions, especially eastern Finland.

There is one further indicator of integration with the south. The enclosures, which elsewhere had been carried through mainly in the decades around the turn of the nineteenth century, were postponed here to the late nineteenth and even to the present century. Unlike other regions, the boundary survey between state land and private peasant-owned land became a major matter of controversy. It was aggravated by the rise of forestry. There arose a bitter conflict between the state and the local peasants for whom the reform meant a violation of their rights by the state. This was especially so in the eastern areas, where 'people for a longer time than elsewhere had been allowed to live free from the societal "bonds" ' (Kyllönen 1975: 279). The state was widely regarded as an intruder in local life. Only in the northernmost areas of the region did the Lapps and others continue to raise reindeer and to trade with the centres along the Gulf of Bothnia largely as before.

In the nineeenth century, then, economic integration and state consolidation caused class structure to develop in regionally different ways. The same process redefined the role of the various regions. A knowledge of both the class structure and the specific mode of territorial integration in various regions is needed to assess subsequent political mass mobilization. South-western Finland developed into the core, and there the class conflict was heightened both between the prospering landowners and the crofters, and between the landowners and the growing number of agricultural labourers. The industrial working class was more numerous there than elsewhere. The coastal area of Viipuri was another area that developed into the core. Outside the industrial centres, the majority of the population consisted of small peasants who were dependent on the market both as producers and as workers. Eastern Finland and large areas of the north were forced into the role of a periphery. There the conflict between the wealthier landowning peasants and the landless was accompanied and partly confounded by conflicts between the landless and the companies, and between the small landholders and the companies. Ostnobothnia stagnated and retained its relative isolation from the other regions. Many traditional traits of the homogeneous agrarian community persisted in Ostrobothnia throughout the nineteenth century.

State-making and political mobilization

National integration and class integration

Political mass mobilization may be analysed in terms of two types of integration: national and class. The degree of national integration describes the recognition by the people that they are primarily members of the nation, which is territorially inclusive in scope. A society not nationally integrated may have a sectionalist movement or may be beset by civil war, as distinct from revolution or class war. Class integration describes the intensity of the class conflict (Hechter 1975: 216-7).

The question is, how did the two processes intertwine to create a specifically Finnish national identity? The rise of a national culture among the upper and middle classes was comparatively easy and trouble-free, and nationalism in Finland displayed rather strong elements of being a 'civic religion' for the territorially centralized state. Another factor relevant for national integration is the consolidation of the core-periphery structure. The question here is, what were the chances of imposing on the large population the core's view of the emerging Finnish nation? Class integration, on the other hand, was delineated in each region separately. In addition, the contemporary political upheavals in Russia had an effect on national and class integration. Political mobilization in Finland bears visible marks of the revolutionary events in Russia in 1905 and 1917.

The consolidation of variations
in party support

In the consolidation of party support national integration and class integration were intertwined in a very concrete fashion. First, the parties defined themselves to varying degrees in these terms. But more important, the mere rise of the party system had significance for national integration: when only a few parties received support in all the regions, the latter became defined in a new way as parts of a national whole. Through the parties, the various regions became linked into a single political unit, despite variations in local conflicts. The latter were defined in terms of a few nationally relevant conflict dimensions: persistent local rivalries were reformed along lines that had significance throughout the society (cf. Tilly 1964: 64). In other words, locally defined conflicts were fused on an emerging national level. From this point of view, it does not matter whether the parties were manifestly class parties or not.

Of course, a necessary precondition for the fusion of local conflicts was that they displayed a great deal of resemblance to one another in different regions. The basic similarity was due to the rise of capitalist relations throughout the country. In every region this created or heightened the class conflict between the rising bourgeoisie and the proletariat, and between the landowners and the landless. But if one hopes to view the institutionalization of party support from the perspective of state-making and nation-building, then it is highly enlightening to start from the regional differences in the initial conditions. The regional perspective may help us understand how national integration and class integration first developed and then became intertwined in the formation of the Finnish nation.

TABLE 3
Regional variations in the political mobilization of
the countryside, 1907-1932

Region	Main support for political parties	Main role in the civil war of 1918	Role in the fascist Lapua movement of 1929-32
South-western Finland	Social Democrats and the Finnish Party	Actively revolutionary	
Ostrobothnia	The Finnish Party and the Agrarian Union	Actively anti-revolutionary	Active
The county of Viipuri	The Agrarian Union	Passively anti-revolutionary	
Eastern Finland	Social Democrats and the Agrarian Union	Passive	
Northern Finland	The Agrarian Union and the Social Democrats	Passive	

In 1907 the Social Democrats and the Finnish Party gained the largest vote. Table 3 gives the different combinations of party support in the countryside for the five regions. Since 1907 the basic regional differences in party divisions have largely remained the same. However, in the 1920s Social Democrat support became divided between them and the Communists. Also, the Agrarian Union did not establish itself until the 1910s and 1920s. This happened at the expense of the Liberal Young Finnish Party, especially in eastern and northern Finland. The introduction of universal suffrage in 1906 and the first general election in 1907 were the culmination of a mobilization started by the general strike which spread from Russia to Finland in 1905.

In the south-western core both the Social Democrats and the Finnish Party received heavy support, indicating the strength of class conflict there. This combination of party support was restricted to this region only. The Social Democrats were backed mainly by the industrial workers, the landless and the crofters, with the wealthy peasant landowners and the upper classes in the cities and in the countryside voting for the Finnish Party (see Alapuro 1978: 111-16).

The Finnish Party, later the Conservatives, was the heir of the Fennoman movement, and functioned as the introducer of an agrarian and religious ideological alternative for national integration. It was in more direct conflict with the Social Democrats in the south-west than anywhere else. But the Social Democrats too had a view of national integration. While their theory was revolutionary, soon after the introduction of general suffrage, and with growing success in the general elections, their practice became rather reformist. They were opponents of the bourgeois parties, but they strongly opposed tsarist Russia as well; and they had very limited contacts with Russian revolutionaries up to 1917. The Social Democrats' view of the future of Finland was based on the overthrow of the prevailing capitalist society, but this view was diffuse and unconcrete, and was rooted in the existence of Finland as an autonomous state (Jussila 1979b: 114, 122, 135-6, 170-1, 178). In any case, the class conflict was most accentuated, in the core region, with two opposite views of national integration.

In Ostrobothnia both strong parties — the Finnish Party and the Agrarian Union — were defending the agrarian way of life. The former party gained its votes mainly from the larger farms; the latter from smaller farms. In both parties' support there was a very marked provincial spirit. The self-conscious and proudly provincial character of the political mobilization was accentuated by the strength of religious revivalism. Here the revivalist movement at the end of the nineteenth century was stronger than anywhere else, and was particularly connected with Finnish Party support.

The character of political mobilization resulted from the Ostrobothnian community structure and the role of Ostrobothnia in state-making. First, the traditional agrarian society had retained much of its strength through the nineteenth century; and second, Ostrobothnia had been in decline and had remained a somewhat separate region in the economic integration of Finland.[9]

Here religion was associated with a viable traditional society. Revivalism had its roots in that society. Typically, the clergy were central in both the traditional agrarian society and Ostrobothnian revivalism.

Also, in Ostrobothnia it was not essential to defend one's position and way of life in a situation of accentuated class conflict within the community, as was the case in the south-west. Rather, it was essential to defend the Ostrobothnian community and the way of life within Finland as a whole. Both the self-conscious provincial spirit and the moral indignation discernible in Ostrobothnian politics and revivalism may be attributed to the fact that the whole agrarian way of life that they represented was irrevocably giving way to the antagonisms and ways of life of capitalist society, which was advancing the integration of the rest of Finland.[10]

The strength of the Agrarian Union in Viipuri seems to be due to the fact that the peasants, as small producers and sellers of their labour power, were dependent on the market. This generalization is less valid in the western than in the eastern areas. In the industrial centres of the western areas the Social Democrats had several strongholds.

In eastern Finland, as in the south-west, the rural support for the Social Democrats indicates the sharpness of class conflict. In the south-western core region class conflict became manifest in the relations between agrarian groups, and Social Democrat support was accompanied by support for the Conservatives. In the eastern periphery, instead, class conflict became manifest in the relations between the timber companies and the local population: Social Democrat support was accompanied by support for the Agrarian Union with its populist ideology. There was a strong anti-capitalist feeling in both political tendencies; besides being socialist, anti-capitalism also had a peasant base.[11]

The northern party division became similar to the eastern in important respects, with the exception that the Agrarians there gained a stronger position than the Social Democrats. In this region there was, not surprisingly, distinct anti-state mobilization. During the general strike of 1905 this became manifest in violence against state officials and in their removal (Kyllönen 1975: 147-8). The anti-state tendency has been identified also in the strong revivalist movement, Laestadionism, that gained a strong foothold in the last decades of the nineteenth century both in northern Sweden and northern Finland. Politically, Laestadionism became linked mainly with the Agrarian Union (Suolinna 1977; Kyllönen 1975: 161).

Finally, turnout remained lowest here: in 1907 the turnout varied between 62 and 30 percent (in Lapland), whereas in the country as a whole it was 71 percent.

The civil war in 1918 and agrarian fascism in 1929-1932

The outbreak of the civil war — or, more exactly, of the attempt at revolution by the Social Democrats in 1918 — was due to the October Revolution in Russia, giving the Finnish civil war interesting common characteristics with the great bulk of revolutionary success in the twentieth century. Most revolutions in this century have been intimately related to world wars and decolonialization (Dunn 1977: 98). The revolution in Russia was the proximate cause of a revolutionary situation in Finland. The distinguishing characteristic of a revolutionary situation, as Charles Tilly (1978: 190-2, 200-2) says, following Leon Trotsky, is the presence of a dual power (or, in a generalized form, a multiple sovereignty). A revolutionary situation in this sense begins when a government previously under the control of a single polity becomes the object of effective, competing, mutually exclusive claims on the part of two (or more) distinct polities.

This sort of situation seems to have arisen in Finland in 1917. There existed no Finnish army, and because of the Russian Revolution the dominant classes in Finland could not invoke armed forces from Russia to defend the prevailing system in a crisis. In early 1917 the Social Democrats, having then an absolute majority in the Parliament, participated for the first time in the government of the bourgeois state. Soon after new elections, which the Social Democrats considered unlawful, and which they lost, they were pushed aside. In the autumn parliamentary activity practically collapsed. Both the bourgeois groups and the Socialists sought to obtain arms; civil guards and red guards were created. By the beginning of 1918 there existed an indisputable dual power, which essentially resulted from the incapacity of the bourgeois government to suppress the emergence of the Socialists' alternative polity. The outbreak of revolution in 1918 was only the last step in this process.

The revolution was crushed in three months, but it had (and still has) profound effects on Finnish society. Unlike several major revolutions, the abortive Finnish revolution was not preceded by the emergence of acute conflicts of interest within the dominant classes (see Moore 1972: 170-5). In this sense Finland was not 'ripe' for revolution. In Finland the existing polity was not fragmented from within: there was no break-off of a section of the dominant classes, and no coalition between established members of the polity and the challengers. The revolution was preceded not by an intensification of conflicts within the dominant classes, but, on the contrary, by an increased tightening of their ranks, owing to the fact that the revolution was triggered from the outside, not from the inside.

Because of this background the bourgeoisie in post-civil war Finland was perhaps unusually united. One telling feature is that the civil guard organization, consolidated at the side of the army in the 1920s, was a force supported by every non-socialist group. It seems a reasonable conclusion that this situation maintained the relative unity of a national bourgeois culture covering both language groups, a unity enhanced by the proximity of the Soviet Union. On the other hand, the working class became almost totally excluded from the national culture defined in bourgeois terms. Indicative is the foundation of a large number of parallel organizations for the working class in all central spheres of cultural and economic activity.

The structural conditions for revolutionary mobilization are distinct from this revolutionary situation. The structural conditions undoubtedly lay in the internal contradictions of Finnish society. While linked to the large Social Democratic support in elections (37 percent of the total vote in 1907 and as much as 47 percent in 1916), they should be analysed separately. A structural condition peculiar to Finland is that, because industrialization was based on the sawmill industry, the production units were scattered all over the countryside, and seasonal work in industry was usual. Consequently, in Finland there developed a very close connection between the industrial and agrarian proletariat, a circumstance which for Eric Wolf (1969: 292) seems an important precondition for revolutionary activity in twentieth-century peasant wars: 'It is probably not so much the growth of an industrial proletariat as such which produced revolutionary activity, as the development of an industrial work force still closely geared to life in the villages.'

In pre-1918 Finland there existed exceptionally favourable circumstances for the fusion of the industrial and the agrarian class conflict. The revolutionaries had their stronghold in the south, especially in the south-western core (see Figure 1 above). While it was the most industrialized region, it seems to have been extremely important for the revolution there that the industrial workers' readiness for collective action be fused with a corresponding readiness among the landless labourers and, to a lesser extent, the crofters (e.g. Rasila 1968: 32-65).

The Whites' stronghold in the civil war was Ostrobothnia, a region in which the peasant community had retained much of its capacity to function as a frame for collective action in a crisis. While in the core the war was an undeniable class war, here it was conceived in different terms. The Ostrobothnian peasants, rank-and-file of the 'white peasant army', considered it a struggle for the nation. For them it was a liberation war against the Russian troops still in the country: the revolutionaries were traitors to the nation. The moral indignation was directed against the disruptive class antagonism in the south, and against the increasing opposition to religion that accompanied it. 'For the revivalists, this war was a holy war'; 'one made his way to the liberation war as to a devotional meeting; the war was also a war against the devil and deniers of God' — as students have characterized the Ostrobothnian attitude (Alapuro 1978: 138-9).

The situation may be reformulated by saying that indignation was directed against capitalist integration in the rest of Finland. In a sense, for the Ostrobothnians the war was a great opportunity to demonstrate their model for national integration to the rest of Finland — an agrarian, religious and strongly nationalist one that totally denied class conflict.

In the Finnish variant of agrarian fascism — the Lapua movement — the moral indignation, springing from a threatened way of life, reappeared for the last time. The rise of the movement was closely connected with the depression. It arose in Ostrobothnia as a peasant movement against a conception of the Communist doctrine which damned everything the peasant folk held sacred. For the Ostrobothnian peasants it was a continuation of the civil war. The contrast between their way of life and that elsewhere in Finland was a very central feature in the movement. The provincial character of many of its xenophobic manifestations and demonstrations is well-known: it was as if its supporters had wanted to make clear that, when Lapua speaks, the whole of Finland must listen.

The main struggle about the nation, national identity and the state after the two Russian revolutions in 1917 was fought in the west — that is, in the south-western core and in Ostrobothnia, which, along with the south-west, had been the wealthiest and most developed Finnish-speaking region in the Swedish state. In these regions the heritage of the nineteenth-century Fennoman movement was strongest. The fragmentation of the polity in 1917-18 took place in the central regions of the state (Figure 1 above). The nation was torn in two within the very core of Finland. The Finnish revolution did not originate, nor was it fought, in the periphery; it was declared in the Helsinki headquarters of the Social Democrat Party, and it was crushed in the centres of south-western and southern Finland. After the civil war, the nineteenth century nationalist heritage fell completely and definitely to the bourgeois groups. Its most extreme expression was the fascist movement in Ostrobothnia at the turn of the 1930s.

The Finnish economy in the interface between Sweden and the Soviet Union in the late twentieth century

Regional inequalities in present-day western Europe have their roots in the territorial reorganization of the European economy in the eighteenth and nineteenth centuries. These centuries were the period of profound upheavals known as the Agrarian Revolution, the Industrial Revolution, the rise of capitalism and of the national state. This generalization is valid both for Finland's internal regional structure and for its position vis-à-vis Sweden and the Soviet Union. It is valid in the sense that the economic cores and the peripheries are still there. However, the character of the relationship between them has changed in some important respects.

Finland is still — or, in a shorter perspective, again — a kind of periphery of Sweden. The total population of Finland in the early 1970s was 4.6 million. Total net emigration from Finland since the Second World War has been approximately 300,000, and 90 percent of the emigrants have gone to Sweden. Since the abolition of passport control inside the Nordic countries in the mid-1950s, there have been no institutional barriers to emigration. During the peak years of the 1960s almost 40,000 people emigrated annually; since then the rate has slowed down. Among factors paving the way for emigration have been differences in standards of living and wage differentials (real incomes are 20-30 percent higher in Sweden), as well as cultural similarities between the two countries (Kiljunen 1979a: 298-9). Emigration has been disproportionately high among the Swedish-speaking population. At present, the number of Finnish speakers in Sweden almost equals the number of Swedish speakers in Finland. Also, because of lower wage levels, some Swedish enterprises have transferred production to Finland.

Two-thirds of the emigrants are of working age. They tend to be rather young, rather well-educated, and often skilled workers. Moreover, the emigrants usually take their families with them, and therefore they contribute very little to the direct inflow of foreign exchange in the form of remittances, as many other emigrant groups in Europe do (Kiljunen 1979a: 299).

A redefinition of the regional division of labour within Finland, causing mass migration to the south, was the central process in nineteenth-century capitalist transformation. What has taken place in recent decades is a redefinition of the regional division of labour between Finland and Sweden — in other words, across national boundaries. This development parallels a general western European development. The earlier peripheral position of the Finnish-speaking regions in Sweden has been reactivated, but in a novel form. In past centuries subordination to the Swedish state was essential; now what is essential is the capitalist division of labour across national boundaries, interfering in the production process itself and moving people from one country to another.

Typically, migration has taken place in two stages. At first the migrants move from the peripheries into the core regions of Finland, and only later do they decide to emigrate to Sweden (Figure 9). This pattern indicates that the internal Finnish core-periphery structure and an international core-periphery structure are linked to each other. This is a matter of general importance. The Finnish economy is structurally tied not only to Sweden but to the whole western European capitalist core, and the factors maintaining regional inequalities there extend their impact to the regional differences within Finland as well.

The Finnish core-periphery structure has remained essentially similar to the structure that prevailed in the late nineteenth century. However, a new kind of imbalance has developed since the Second World War. Up to the war, forestry was the crucial sector determining the development of the whole economy. It had relatively few linkages with the other sectors of the economy, and it developed like an external enclave. In addition, the wood-based industry retained a large number of the population in primary occupations. Felling and floating timber remained important subsidiary earnings for the agrarian population, helping to maintain the viability of small farms. In other words, the nature of Finnish industrialization was such that it only slowly created industrial

FIGURE 9
Internal migration in Finland and emigration from Finland to Sweden, 1970-76

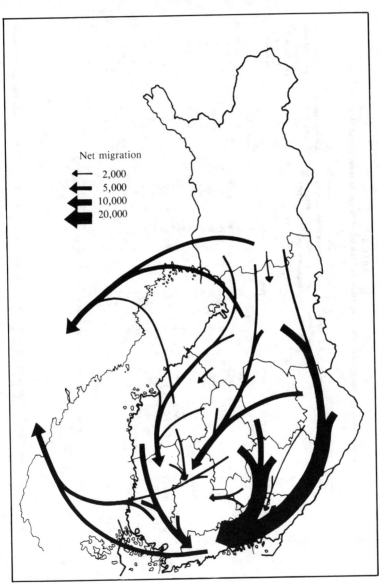

Net migration

2,000
5,000
10,000
20,000

Source: Kiljunen (1979b: 125).

TABLE 4

Population, unemployment and migration in Finland after the Second World War, by county

County	Population[1] ('000)		Growth in population, 1950-1975	% of economically active population engaged in agriculture and forestry[1]		Density of population,[2] 1975	Unemployment rate, 1977	Net in-migration (+) or net out-migration (−), 1961-73, as % of 1960 population
	1950	1975		1950	1975			
	('000)	('000)	%	%	%		%	%
Uusimaa	667.5	1,092.3	64	16	3	110.8	3.2	21.9
Turku and Pori	631.0	697.1	10	46	15	31.8	5.4	0.6
Ahvenanmaa[3]	21.7	22.3	3	56	16	15.1	2.6	6.4
Häme	521.3	659.5	27	35	11	38.4	5.7	5.3
Kymi	311.4	345.9	11	41	15	32.2	6.2	−1.3
Mikkeli	230.8	210.2	−9	63	28	12.8	8.1	−12.9
North Karelia[4]	186.7	177.1	−5	67	29	9.8	10.4	−20.1
Kuopio	267.7	250.9	−6	61	25	15.0	9.0	−12.6
Central Finland[5]	224.2	241.0	7	55	20	14.7	9.4	−8.1
Vaasa	440.6	423.6	−4	61	26	16.2	4.9	−6.3
Oulu	359.8	404.8	13	64	23	7.1	9.4	−10.5
Lapland[6]	167.1	195.8	17	56	18	2.1	11.5	−8.5
All counties	4,029.8	4,720.3	17	46	15	15.5	6.1	

1. 1950 figures are adjusted to correspond to the present division of counties.
2. Inhabitants per km² land area.
3. In Swedish, 'Åland'.
4. Pohjois-Karjala.
5. Keski-Suomi.
6. Lappi.

Sources: *Living Conditions 1950-1975* (1977: 31, 34); *Official Statistics of Finland* (VI: C 102, 11, 18-73); *Statistical Yearbook of Finland* (1975: 7); *Statistical yearbook of Finland* (1977: 47, 275).

FIGURE 10
The present division of counties in Finland

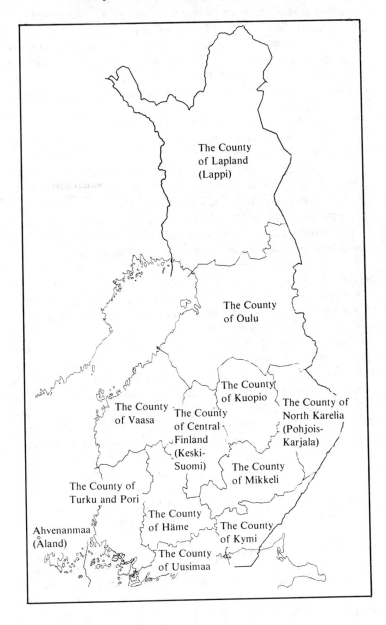

The County
of Lapland
(Lappi)

The County
of Oulu

The County
of Kuopio

The County
of Vaasa

The County
of Central
Finland
(Keski-
Suomi)

The County of
North Karelia
(Pohjois-
Karjala)

The County
of Mikkeli

The County of
Turku and Pori

Ahvenanmaa
(Åland)

The County
of Häme

The County
of Kymi

The County
of Uusimaa

employment in other sectors. Therefore, up to the Second World War the population largely stayed in the traditional rural occupations, and urbanization and internal migration did not gather new momentum (Kiljunen 1979a: 284-6; 1979b).

Only after the war, and especially in the 1960s, did migration to the south — and in many cases eventually to Sweden — strongly increase (Figure 10 and Table 4). Between 1960 and 1975 the population in the peripheral regions decreased even in absolute numbers. The increase in Table 4 for the northern counties of Oulu and Lappland (see Figure 10) is the result of colonization in the 1950s. Since the 1960s their population has declined like that in other peripheral regions. The migration was due to the diversification of the industrial structure, and the simultaneous increase of the tertiary sector, which took place mainly in the south. In addition, since the 1960s, forestry has become rapidly mechanized, causing mass unemployment among small farmers and part-time forest workers. This structural change removed hundreds of thousands of people from their traditional surroundings. In 1950 46 percent of the labour force was engaged in primary production; in 1975 only 15 percent. In the four southern counties (Ahvenanmaa excluded) the proportion of the labour force in primary production had dropped to between 3 and 15 percent in 1975 and the proportion in industry was between 25 and 38 percent.

In the peripheral eastern and northern regions, forestry and other extractive industries still prevail and with them the dualistic industrial structure. Only weak linkages exist between the externally oriented main industry and other economic sectors at the local level (see Kiljunen 1979b: 234). These regions have remained the poorest, and their population losses have been the most severe. In the southern core areas of Finland, Helsinki and its surroundings (the county of Uusimaa) have become more prominent than they were at the turn of the century (Valkonen 1980).

In the 1970s the gap between the wealthiest and the poorest region was larger than in the OECD countries on average, measured by the proportion of GNP per capita (Kiljunen 1979b: 84-5). There are no long-term studies on the development of regional inequalities, but it has been assessed that they have become more conspicuous since the take-off of industrialization (Hustich 1971: 26). The conclusions depend, of course, on the choice of indicators. Unemployment figures, to take perhaps the most central single indicator, point to an increase of differences between the

FIGURE 11
The Development Zones in Finland in 1980

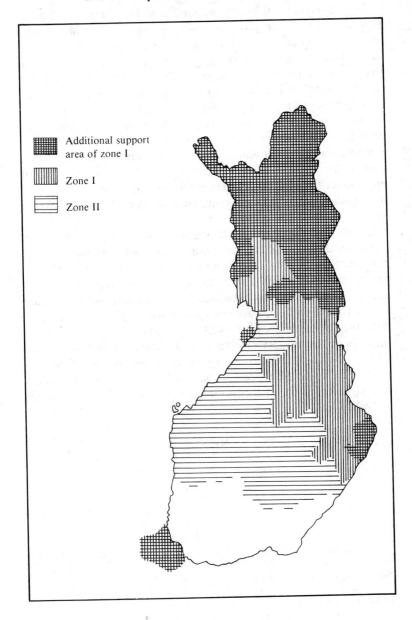

wealthy and the poor regions during recent decades (Hustich 1971: 30: Koskiaho 1979: 349-55).

An active regional policy was initiated in Finland during the 1960s. It has similarities with the development of the regional policies of several western European countries. The initial impulse came from the discovery of the regionally varying symptoms of structural change, for example, unemployment (Rinne 1978). In the 1970s regional policy became more explicitly a part of national growth policy. The so-called Development Zones were defined for the first time in 1966, and since then their boundaries have been revised several times (Figure 11).

While the dominant and most dynamic economic relations have been with western Europe, trade with the Soviet Union is of major importance for Finland. The trade relations that were broken after the October revolution were renewed at the end the Second World War, and in the 1950s the share of the Soviet Union in Finland's total trade was on average 20 percent. Since then, the long-term trend has declined; at the beginning of the 1970s the share was 12 percent. Significantly, the composition of the trade has been as favourable for Finland as during the nineteenth century. Finland exports highly processed manufactured products, while primary products, mainly fuel, have constituted four-fifths of the imports from the Soviet Union. In short, Finnish-Soviet economic relations have been highly asymmetric. In addition, Soviet trade has alleviated cyclical fluctuations in Finland's foreign trade. When the trade with western Europe has faced difficulties, Finnish industry has typically searched for new outlets in the Soviet Union (Kiljunen 1979a: 287-9).

NOTES

1. This section is reproduced, with minor revisions, from an earlier text (Alapuro 1979b: 20-8).

2. Yet another formulation of the 'diffusion' of nationalism is that of Smith (1978: 240-3). Rather than modernization or the uneven development of capitalism, he takes the centralizing reform in the ruling bureaucracy to be crucial in explaining the rise of a nationalist movement against domination.

3. As regions of 'relative overdevelopment' in eastern Europe, only Bohemia and Croatia seem comparable to Finland (see Nairn 1977: 185-7). Perhaps also the Russian partition of Poland might be cited as an example: see Kiernan (1976: 120).

4. An excellent account in English on early nationalism in Finland is Selleck (1961: see especially pp. 128-43).

5. The regional division (northern Finland excluded) is presented in more detail in Alapuro (1978: 117-25), from which this presentation has been drawn.

6. Ostrobothnia is one of the historical provinces. South-western Finland is coterminous with Uusimaa, Finland proper, Satakunta and the southern areas of Häme. Eastern Finland corresponds roughly to Savo and North Karelia, and northern Finland corresponds roughly to Lapland (see Sómme 1968: 6-7).

7. This section is based on Alapuro (1979a: 347-54).

8. This section is based on Alapuro (1978: 128-30).

9. See, in more detail, Alapuro (1978: 136-7).

10. See in more detail, Alapuro (1978: 136-7).

11. Alapuro (1978: 137). There exists a special study on national integration in the eastern region (Siisiäinen 1979).

REFERENCES

Ajo, R. (1946). 'Liikennealueiden kehittyminen Suomessa'. *Fennia*, 69:3 (Helsinki).

Alapuro, R. (1978). 'Statemaking and Political Ecology in Finland', pp. 109-43 in Z. Mlinar and H. Teune (eds), *The Social Ecology of Change*. Beverly Hills and London: Sage.

Alapuro, R. (1979a). 'Internal Colonialism and the Regional Party System in Eastern Finland'. *Ethnic and Racial Studies*, 2: 341-59.

Alapuro, R. (1979b). 'Nineteenth Century Nationalism in Finland: A Comparative Perspective'. *Scandinavian Political Studies*, 2: 19-29.

Allardt, E. and Miemois, K. J. (1979). 'Roots both in the Centre and the Periphery: The Swedish Speaking Population in Finland'. *Research Reports for Comparative Sociology*, University of Helsinki, 24.

Åström, S.-E. (1977). 'Anlagda städer och centralorts-systemet i Finland 1550-1785', pp. 134-81 in G.A. Blom (ed.), *Urbaniseringsprosessen i Norden. 2: De anlagte steder på 1600-1700 tallet*. Oslo: Universitetsforlaget.

Dunn, J. (1977). 'The Success and Failure of Modern Revolutions', pp. 83-114 and 305-18 in S. Bialer and S. Sluzar (eds), *Radicalism in the Contemporary Age*, vol. 3. Boulder, Colorado: Westview Press.

Engman, M. (1978). 'Migration from Finland to Russia during the Nineteenth Century'. *Scandinavian Journal of History*, 3: 155-77.

Gellner, E. (1964). *Thought and Change*. London: Weidenfeld & Nicolson.

Haatanen, P. (1968). *Suomen maalaisköyhälistö tutkimusten ja kaunokirjallisuuden valossa*. Porvoo and Helsinki: WSOY.

Hautala, K. (1956). 'Suomen tervakauppa 1856-1913'. *Historiallisia tutkimuksia*, 45 (Helsinki).

Hechter, M. (1975). *Internal Colonialism*. London: Routledge & Kegan Paul.

Hjelt, A. (1893). *Maakaupasta ja krediitti-oloista Itä-Suomessa*. Porvoo: WSOY.

Hobsbawm, E. (1972). 'Some Reflections on Nationalism', pp. 385-406 in T.J. Nossiter, A.H. Hanson, and S. Rokkan (eds), *Imagination and Precision in the Social Sciences*. London: Faber & Faber.

Hoffman, K. (1978). 'Höyrysahan aika'. Manuscript at the Department of Economic and Social History, University of Helsinki.

Hroch, M. (1968). 'Die Vorkämpfer der nationalen Bewegung bei den kleinen Völkern Europas'. *Acta Universitatis Carolinae Philosophica et Historica*, monograph XXIV (Prague).

Hroch, M. (1971). 'Das Erwachen kleiner Nationen als Problem der komparativen sozialgeschichtlichen Forschung', pp. 121-39 in T. Schieder (ed.), *Sozialstruktur und Organisation europäischer Nationalbewegungen.* Munich and Vienna: R. Oldenbourg.

Hustich, I. (1971). 'Om avgränsningen av utvecklingsområden i Finland', *Terra*, 83: 25-38.

Joustela, K. (1963). 'Suomen Venäjän-kauppa autonomian ajan alkupuoliskolla vv. 1809-1865'. *Historiallisia tutkimuksia*, 62 (Helsinki).

Jussila, O. (1979a). 'Kejsaren och lantdagen-maktrelationerna, särskilt ur kejsarens synvinkel'. *Historisk Tidskrift för Finland*, 64: 105-27.

Jussila, O. (1979b). 'Nationalismi ja vallankumous venäläis-suomalaisissa suhteissa 1899-1914'. *Historiallisia tutkimuksia*, 110 (Helsinki).

Jutikkala, E. (1950). 'The Economic Development of Finland Shown in Maps', pp. 159-66 in *Proceedings of the Finnish Academy of Science and Letters, 1948* (Helsinki).

Jutikkala, E. (ed.) (1959). *Atlas of Finnish History.* Porvoo and Helsinki: WSOY.

Jutikkala, E. (1963). *Bonden i Finland genom tiderna.* Helsinki: LTs Förlag.

Jutikkala, E. (1968). 'Suomen rautatieverkon synty', pp. 168-73 in E. Jutikkala (ed.), *Suomen talous- ja sosiaalihistorian kehityslinjoja.* Porvoo and Helsinki: WSOY.

Jutikkala, E. (1977). 'Finland. Städernas tillväxt och näringsstruktur', pp. 95-125 in G.A. Blom (ed.), *Urbaniseringsprosessen i Norden. 3: Industrialiseringens förste fase.* Oslo: Universitetsforlaget.

Kaila, E.E. (1931). 'Pohjanmaa ja meri 1600- ja 1700-luvuilla'. *Historiallisia tutkimuksia*, 14 (Helsinki).

Kero, R. (1976). *Suuren lännen suomalaiset.* Helsinki: Otava.

Kiernan, V. (1976). 'Nationalist Movements and Social Classes', pp. 110-33 in A.D. Smith (ed.), *Nationalist Movements.* London: Macmillan.

Kiljunen, K. (1979a). 'Finland in the International Division of Labour', pp. 279-302 in D. Seers, B. Schaffer and M.-L. Kiljunen (eds), *Underdeveloped Europe: Studies in Core-Periphery Relations.* Hassocks, Sussex: Harvester Press.

Kiljunen, K. (1979b). *80-luvun aluepolitiikan perusteet.* Helsinki: Työväen Taloudellinen Tutkimuslaitos.

Kilpi, O.K. (1913). *Suomen ammatissatoimiva väestö ja sen yhteiskunnalliset luokat vuosina 1815/1875. 1: Maaseutu.* Helsinki: Sana.

Klinge, M. (1968). *Ylioppilaskunnan historia, III.* Porvoo and Helsinki: WSOY.

Klinge, M. (1969). 'Suomen kansallisuusliikkeiden sosiaalisista suhteista'. *Historiallinen Aikakauskirja*, 67: 185-207.

Klinge, M. (1975). *Bernadotten ja Leninin välissä.* Porvoo and Helsinki: WSOY.

Klinge, M. (1977). *Blick på Finlands historia.* Helsinki: Otava.

Koralka, J. (1971). 'Social Problems in the Czech and Slovak National Movements', pp. 56-73 in *Mouvements nationaux d'indépendance et classes populaires aux XIXe et XXe siècles en occident et en orient.* Paris: Armand Colin.

Koskiaho, B. (1979). 'Regional Development, the Case of Finland', pp. 329-58 in A. Kuklinski, O. Kultalahti and B. Koskiaho (eds), *Regional Dynamics of Socioeconomic Change.* Tampere: Finnpublishers.

Kyllönen, M. (1975). 'Punaisten ilmastoalueiden synty Pohjois-Suomessa 1900-1910'. Manuscript at the Department of History, University of Jyväskylä.

Lakio, M. (1975). 'Teollisuuden kehittyminen Itä-Suomessa 1830-1940'. *Itä-Suomen Instituutti*, A:5 (Mikkeli).

Lento, R. (1951). 'Maassamuutto ja siihen vaikuttaneet tekijät Suomessa vuosina 1878-1939'. *Väestöpoliittisen tutkimuslaitoksen julkaisuja*, A:5 (Helsinki).

Living Conditions, 1950-1975 (1977). Statistical Surveys, no. 58 (Helsinki).

Molnar, M. (1971). 'Mouvements d'indépendance en Europe. Rôle de la question agraire et du niveau de culture', pp. 217-27 in *Mouvements nationaux d'indépendance et classes populaires aux XIXe et XXe siècles en occident et en orient*. Paris: Armand Colin.

Moore, B. (1972). *Reflections on the Causes of Human Misery and upon Certain Proposals to Eliminate Them*. London: Allen Lane.

Murtorinne, E. (1967). *Taistelu uskonnonvapaudesta suurlakon jälkeisinä vuosina*. Porvoo and Helsinki: WSOY.

Nairn, T. (1977). *The Break-Up of Britain*. London: New Left Books.

Niemelä, J. (1978). 'Metsäteollisuusyhtiöiden maanhankinta 1860-1917'. Mimeo.

Official Statistics of Finland, VI. Population Statistics nos. 1 and 45.

Official Statistics of Finland, VI. Population Statistics, nos. C 102, II.

Pihkala, E. (1970). 'Suomen Venäjän-kauppa vuosina 1860-1917'. *Bidrag till kännedom av Finlands natur och folk*, 113 (Helsinki).

Plakans, A. (1974). 'Peasants, Intellectuals and Nationalism in the Russian Baltic Provinces, 1820-90'. *Journal of Modern History*, 46: 445-75.

Portal, R. (1971). 'Balkans. La participation des classes populaires (masses et cadres) aux mouvements nationaux d'indépendance', pp. 93-100 in *Mouvements nationaux d'indépendance et classes populaires aux XIXe et XXe siècles en occident et en orient*. Paris: Armand Colin.

Rasila, V. (1968). *Kansalaissodan sosiaalinen tausta*. Helsinki: Tammi.

Reitala, A. (1980). 'Suomen kuvallisen henkilöitymän vaiheet'. Manuscript.

Rinne, P. (1978). 'Kehitysaluepolitiikan alkuvaiheet Suomessa'. Manuscript at the Department of Social Sciences, University of Turku.

Sarajas, A. (1962). *Viimeiset romantikot*. Porvoo and Helsinki: WSOY.

Sarmela, M. (1969). 'Reciprocity Systems of the Rural Society in the Finnish-Karelian Culture Area'. *FF Communications*, 207 (Helsinki).

Schweitzer, R. (1978). 'Autonomie und Autokratie. Die Stellung des Grossfürstentums Finnland im russischen Reich in der zweiten Hälfte des 19. Jahrhunderts (1863-1899)'. *Marburger Abhandlungen zur Geschichte und Kultur Osteuropas*, 19. Giessen: Wilhelm Schmitz Verlag.

Schybergson, P. (1973). 'Hantverk och fabriker I. Finlands konsumtionsvaruindustri 1815-1870: Helhetsutveckling'. *Bidrag till kännedom av Finlands natur och folk*, 114 (Helsinki).

Selleck, R.G. (1961). 'The Language Issue in Finnish Political Discussion 1809-1863'. Unpublished thesis, Radcliffe College, Cambridge, Mass.

Seton-Watson, H. (1977). *Nations and States*. London: Methuen.

Siisiäinen, M. (1979). 'Kansallisen kulttuurin nousu ja maaseutu. Tutkimus Pohjois-Karjalan henkisen kulttuurin organisoitumisesta vuosina 1860-1918'. *Joensuun korkeakoulu, Karjalan tutkimuslaitoksen julkaisuja*, 40 (Joensuu).

Smith, A.D. (1978). 'The Diffusion of Nationalism: Some Historical and Sociological Perspectives'. *British Journal of Sociology*, 29: 234-48.

Snellman, G.R. (1914). 'Tutkimus Suomen sahateollisuudesta'. *Suomen Virallinen Tilasto. Työtilastoa*, XVI (Helsinki).

Sömme, A. (ed.) (1968). *A Geography of Norden*. Bergen: Svenska Bokförlaget.

Suni, L.V. (1979). *Ocherk obshchestvenno-politicheskogo razvitiya Finlyandii, 50-70e gody XIX v.* Leningrad: Nauka.

164 *The politics of territorial identity*

Suolinna, K. (1977). 'Lestadiolaisuus ja agraarin väestön puolustusmekanismi', pp. 112-20 in M. Kuusi, R. Alapuro and M. Klinge (eds), *Maailmankuvan muutos tutkimuskohteena.* Helsinki: Otava.

Suomen valtionrautatiet 1862-1912 (1916), vol. II (Helsinki).

Statistical Yearbook of Finland (1907, 1916, 1917, 1920, 1922, 1929, 1931, 1975, 1977).

Tarrow, S. (1977). *Between Center and Periphery.* New Haven, Conn.: Yale University Press.

Tilly, C. (1964). *The Vendée.* Cambridge, Mass.: Harvard University Press.

Tilly, C. (1978). *From Mobilization to Revolution.* Reading, Mass.: Addison-Wesley.

Tommila, P. (1978). 'Rautatieverkko ja rautatieliikenne', pp. 15-31 in *Tietoliikenne Suomessa 1860-1939. Suomen sanomalehdistön historia-projektin julkaisuja,* 10 (Helsinki).

Valkonen, T. (1980). 'Alueelliset erot', in T. Valkonen, R. Alapuro, M. Alestalo, R. Jallinoja and T. Sandlund, *Suomalaiset.* Porvoo and Helsinki: WSOY.

Wirilander, K. (1960). *Savo kaskisavujen kautena 1721-1870.* Kuopio: Savon säätiö.

Wolf, E.R. (1969). *Peasant Wars of the Twentieth Century.* New York: Harper & Row.

Wuorinen, J.H. (1935). *Suomalaisuuden historia.* Porvoo and Helsinki: WSOY.

Zipf, G.K. (1941). *National Unity and Disunity.* Bloomington, Indiana: Principia Press.

5

Germany: From Geographical Expression to Regional Accommodation

Derek W.Urwin
University of Warwick

Historically, there has always been a disjunction between 'political' Germany and the territory occupied by German-speaking peoples. While German 'society' or 'societies' have been established on the map of Europe for centuries, a German state has existed only since 1870, and even then large German populations remained outside its boundaries. Similarly, the word 'Deutschland' was not used as an official designation until the nineteenth century, although it had been used for centuries to describe a territory that corresponds fairly closely to the political boundaries of the German state established in 1870. We shall follow this tradition here, excluding from other than peripheral consideration those concentrations of German populations in Austria, Bohemia and Switzerland.

The political boundaries of German territory have never been constant. Moreover, after 1870 territorial change went hand in hand with abrupt changes in regime. Yet despite this seeming persisting mutability — 'the jagged curve' (Spiro 1962) of German history — there has also been a considerable degree of continuity in the structure of German society and politics (cf. Urwin 1974). Any portrayal of the regional landscape of contemporary West German politics and society must take into account this historical combination of persisting patterns and dramatic change. Territory occupies an integral position in both continuity and change. Because of the historical mosaic of its political history, its late unification and the more recent boundary changes, German politics have been impregnated by marked regional differences. These in turn have contributed to the problems of and strains upon the various German regimes. By contrast, the Federal Republic today stands out in

Western Europe as a regime that has been less affected by movements mobilizing peripheries against national centres, pressing claims for cultural autonomy and demanding decentralized powers of territorial decision-making. The territorial changes involve some difficulties in nomenclature. Here we shall use the word 'Germany' eclectically to refer to all the territory integrated into the Second Reich or left under political control in 1918. The terms 'Federal Republic' and 'West Germany' will be applied interchangeably to describe the post-1949 state that arose out of the occupation zones of the three Western allies. The German Democratic Republic is omitted from the discussion, since it did not form part of the brief for this chapter.

Historical genealogy

In the late nineteenth century the German lands were not only unified politically; they also experienced an economic metamorphosis. Yet these transformations did not eliminate patterns established during the previous centuries. It is the interplay between past and present, between continuity and change, that gives rise to the particular regional character of Germany.

It is difficult to decide at what date to begin a review of regional differences. It seems most appropriate to focus upon the Middle Ages, for during these centuries there emerged a pattern of variations which in its essentials remained unchanged until at least the nineteenth century. The beginnings of this pattern were themselves determined by previous population movements and by the topography of the German territories. Together, these imposed upon the land a fundamental regional differentiation along a fairly consistent gradient running from south-west to north-east. In turn, this gradient was composed of two axes: a primarily cultural and political north – south axis, complemented by a fundamentally economic and political west – east axis.

For most of the subsequent period 'Germany' incorporated most of the North European plain, an area dominated by its river systems. The great rivers, which, apart from the Danube, flow northwards to the North and Baltic Seas, supply the key to an appreciation of early German developments. The beginnings lie with the Germanic tribes in the south-west. Under the aegis of the Roman Empire and its boundary at the Rhine, village life and tillage consolidated itself. Fortresses along the river acquired urban characteristics. Indeed, the whole of south-west Germany never entirely relinquished the urbanization and trade developments that it experienced under the Romans. The lands east of the Rhine were

very much virgin territory, the natural landscape being only intermittently disturbed by sparse settlement. Moreover, the further
north and east, the less the degree of German settlement, with the
Elbe, at a maximum, as the most eastward extent of settlement:
across the river the land was more or less wholly occupied by
Slavonic tribes.

This picture was radically transformed during the Middle Ages.
There were two major agencies of change: a broad trading system
based upon the utilization of waterways that connected nodal
cities, and a continuous colonization programme, partly planned
and partly spontaneous, which within the space of a few centuries
pushed the limits of German settlement far to the east.

The settlement of the land

The process of colonization came in two waves. One involved
expansion eastwards from the upper Rhine and lower Main to settle
what is now Bavaria and Austria. The colonization programme, initiated by the western nobility and clergy, soon reproduced an
agricultural base similar to that of the Rhinelands and a facsimile
of western patterns of social organization. Because these regions
lay athwart or close to major channels of European communication
and trade, and because they were settled early, they participated to
a considerable extent in the western economic developments of the
Middle Ages.

More significant was the colonization of northern and eastern
Germany, part of the great design to Christianize Europe. It was
Charlemagne who first pushed German power east of the Elbe and
Saale rivers. By the tenth century a string of frontier marches were,
albeit precariously, under German authority. Permanent control
and the consolidation of farming communities, however, were
attendant upon political authority and military security. These were
not guaranteed until the twelfth century. The following two hundred years was the heyday of German colonization. A firm
foothold was secured in the Baltic in 1158 with the founding of the
town of Lübeck. Political control in the north-eastern marches was
strengthened by the rule after 1134 of the Ascanian margraves in
Brandenburg. Under their aegis much of the territory between Elbe
and Oder passed under undisputed German possession. Further
east the way was opened in 1225 when local Slav rulers called upon
the Teutonic Knights for military aid. By 1308 the Order dominated
the lower Vistula basin, opening up the south-east Baltic for German settlement (Carsten 1954: Part I).

While this great exodus continued until around 1500, to all

intents and purposes it had been completed by the opening of the fourteenth century. The indigenous Slav populations were not all exterminated, expelled or placed under a servile yoke. Yet the lands remained sparsely settled. To increase their own security, status and revenue, the new noble and ecclesiastical rulers encouraged German migrants from the west. Faced by growing population pressures and feudal restrictions at home, German peasants were easily attracted by the lures of vacant land and less servile status in the east:

> The whole structure of society, as might be expected of a colonial area, was much freer and looser than it was in western Europe. It only seemed a question of time until the east would no longer be backward but would belong to the most developed parts of Europe. [Carsten 1954: 88]

Such an optimistic prognosis was not, however, to be: it was confounded by subsequent economic and political events. The immediate point is that this diffusion of German settlements was perhaps as significant as the much later expansion of the American and Russian frontiers. It altered permanently the ethnic map of Europe, delineated what later would be widely accepted as German territory, and led towards the very marked economic and political differences that later emerged between eastern and western Germany.

Transport, trade and towns

From the tenth to the sixteenth century southern Germany, especially the Rhine artery, was an essential link between the north European lowlands and the city regions of northern Italy. This dominant axis stimulated urban growth, manufacturing practices and agricultural innovation along its whole length. The Rhine Valley was one of the most developed urban regions in medieval Europe, backed by fertile soils that early led to a commercially oriented agriculture. After AD 800 the old Roman fortress of Köln rose to become the major seaport of the Rhine: by the twelfth century it largely controlled traffic along most of the river. But a host of cities was spawned along both the Rhine, especially on the west bank, and its lateral affluents, combining to serve also as nodal points along the north – south routes that led over the Alpine passes to the great cities of Italy.

In the north a trading network was to be formalized by the urban foundations of Hamburg and Lübeck, the two poles of the land isthmus that linked north-western Europe with the Baltic and avoided the Danish control of the sea route around Jutland. Under their initiative and the nominal leadership of Lübeck, the north German cities banded together in the loose confederation of the Hanseatic League (cf. Dollinger 1970). The justification of the Hansa was quite simply the need to have safe trade routes. Since political authority was weak and fragmented, this could best be achieved through a pooling of resources. At one time or another, some 200 cities were members of the Hansa. The basic condition of membership was quite simple: towns must be situated on navigable waterways, and aspiring members often sought to improve their rivers or build canals to link up with this diffused water-based trading system. The Hansa had no legal status, independent finances or a common institutional framework, while the major weapon against recalcitrant members (or opponents) was the threat of embargo. Indeed, the various cities continued to owe allegiance to a whole host of different overlords. What organization the Hansa possessed reflected its fundamental river and sea basis.[1]

By the thirteenth century the Hansa axis, from London through Bruges, Hamburg, Lübeck and Reval to Novgorod, dominated, but did not monopolize, trade in northern Europe. While several other city leagues existed, for example along the upper Rhine and in Schwaben, the Hansa remained unique, for politics and defence remained subsidiary to trade. While the fifteenth century saw the rise of the great south German banking families, the Hansa reached its zenith around 1400. Thereafter the cities entered into a decline, to become minor actors on the stage by 1500. The reasons for the decline of the Hansa are complex, but contributing factors include the rise of strong territorial states.

While the strength of the cities may have inhibited the rise of a strong political authority in western Germany itself (just as its earlier absence had encouraged inter-city co-operation), they failed to match the resources of these new political authorities. More or less simultaneously, wars, famine, plague and depression hurt the economic foundations of the sytem. The revolt of the Netherlands in 1585, for example, closed the mouth of the Scheldt, while plague and economic problems produced the phenomenon of the 'Wüstungen', the disappearance of settlement — sometimes of whole villages — from the more marginal lands (Abel 1955: 5-12). The later discovery of the Americas led eventually to a radical reshaping of the direction of Europe's major trading routes.

Paradoxically, while the north went into decline, the general expansion of trade in the sixteenth century seemed to confirm south-west Germany as the cornerstone of the Flanders – Italy trade conduit (Lopez 1952; Parry 1967), but it too eventually suffered from the American discoveries, as well as from the closing of the Danube by the Turkish advance into Europe. Finally, the great wealth of the Fuggers and other merchant bankers, tied up irrevocably with the fortunes of Charles V and his dream of a Hapsburg Europe, evaporated with the insolvency and bankruptcy of the Spanish crown in the 1550s (Wallerstein 1974: 185-7).[2] City Europe had shown a remarkable capacity for collective action, but the lack of integration showed more and more when the towns were buffeted by economic change and the appearance of a new kind of political authority.

Perhaps the most adverse consequences of this economic change were experienced by the towns of eastern Germany. All had been founded much later than their western counterparts. Because they occupied a more peripheral location in the trading system, their size and number were much more dependent upon the agricultural resources of their immediate hinterlands. In the last resort their prosperity depended upon participation in the Hanseatic axis from Bruges to Novgorod. Few had achieved more than local influence, or had begun to match the wealth and diversity of the western cities. Most already stood more directly under the control or influence of the local lords. In the sixteenth century the lords moved to cut the cities out of their middleman position, by dealing directly with the 'foreign' merchants. Economic change and decay led to the elimination of the towns as major political actors in later centuries. The strengthening of the Hohenzollern prerogative after the Thirty Years' War was the result of a pact with the Junker landlords, at the expense of the towns and peasantry: in the countryside the aristocratic grip was tightened by the consolidation of their control over a whole range of fiscal, juridical, police and conscription powers (cf. Carsten 1954: 117-48; Blum 1957; Rosenberg 1943; also Büsch 1962).

In the west urban prosperity lingered on for a while, despite the changing trade patterns. Yet even here there was economic decline in the sixteenth century (cf. Spooner 1968; Ludloff 1957; also Strauss 1966). But in contrast to their eastern counterparts, the western cities, while economically in decline or stagnant, retained a high degree of strength and vitality. By 1500 their size and number had given western Germany a polycephalous structure that not only never disappeared, but was reinforced by later political developments. Köln and Lübeck were the largest cities, but they were only slightly larger than a score or so of other cities. In the later Middle Ages some 25 cities within the Holy Roman Empire had a population of over 10,000, while their political vigour had been recognized by the grant to 85 cities of the status of free imperial cities. Polycephality meant not only a continuation of collaborative efforts such as the medieval city leagues, but also competition among cities: no single urban unit dominated much beyond its immediate hinterland. Because of this the German cities never resembled city states of the Italian type: they were not expanding political centres. The city belt was also intermingled with numerous ecclesiastical principalities. The density of cities and territorial bishoprics not only lent diversity to the region: it hindered the emergence of an expansive political centre. In the words of Anderson (1975: 250) there was 'no expanding space for aristocratic absolutism'.

The cultural heritage

The great colonization movement of the Middle Ages had two immediate consequences. It ensured the dissemination eastwards of both the German language and Christianity. The early expeditions across the Elbe were regarded by the Germans both as crusades and as quests for wealth and loot. The expansion of an agricultural and ethnic frontier went hand in hand with the advance of religion: the Church was an active sponsor of colonization, and in the process received large territories of its own. Within the space of two or three centuries the surviving Slav populations had become Christian.

It was the Reformation that divided the German lands along religious lines. The political fragmentation of the territory led to religious fragmentation, for the basic principle of the Peace of Augsburg — 'cui regio eius religio' — was designed to end religious conflict by leaving to the local prelate or ruler the choice of whether a particular region remained loyally Catholic or accepted Lutheranism. The Protestant Reformation was strongly nationalistic, yet after Augsburg any further pressing of its cause could only lead to war, and this few Protestant princes were willing to risk.[3] Religious divisions served to consolidate the barriers against territorial unification. Charles V's imperial dreams prevented him from uniting Germany: conversely, 'the Protestant Reformation which might have unified Germany against the Pope divided Germany against the Emperor' (Hurstfield 1968: 130). Indeed, there was even no likelihood of the resurrection of the Holy Roman Empire as a viable confederation of political units (Dickens

1967: 160). The schism with Rome generated a confused religious picture which heightened the distinctiveness of the various territorial units through the integration and subordination of the priesthood into the the administrative apparatus of the several states. It is a religious landscape that not only possessed strong regional contrasts, but has also endured to the present day, with significant political consequences.

By contrast, linguistic diversity, while noteworthy in the past, has rarely perhaps acquired a significant political meaning. There was in the Middle Ages no standard German culture, measured in terms of dialect or customs. Language dialects, especially the broad distinctions between high and low German, survived long after the development (mainly owing to the invention of the printing press) of a national literary German language.[4] Politically, these linguistic variations became increasingly irrelevant. The same may be said of linguistic – ethnic differences. Slav languages survived as alien islands during German colonization, and linguistic diversity in the eastern provinces endured long after the passing of the Middle Ages. The process of absorption was well established by the seventeenth and eighteenth centuries. The establishment of an elementary school system in Prussia in the late eighteenth century, which by 1820 incorporated some 85 percent of all children aged between seven and fourteen, was a powerful standardizing influence: a more rigorous Germanizing policy developed.[5] By the nineteenth century and the establishment of a German state, the German territories were largely homogeneous in ethnic – linguistic terms. The most significant minorities lay in peripheral regions pulled into the German orbit during more recent decades — in the east through the partitions of Poland, in Schleswig, and in Alsace-Lorraine.

Socio-economic developments after the Middle Ages

Events after the collapse of the Hansa and the weakening of the western cities sharpened the contrasts between east and west. The most significant changes came in agricultural structures. Agrarian crises, subsequently followed by changes in trading patterns and increased foreign competition, led to attempts by landowners in the west, buttressed by the conceptions of property inherent in the newly rediscovered Roman law, to reimpose or strengthen feudal obligations and burdens, and to enclose the common lands.[6] However, the Bauernkrieg of 1525 and numerous other agrarian uprisings exposed the limitations of the western nobility (cf. Franz 1956). The Bauernkrieg failed, but yet so did the nobility, for in due course large tracts of Germany west of the Elbe were characterized by a largely independent peasantry that, because of the proximity of so many cities and towns, were more accustomed to the concept of commercial agriculture. The south-west in particular came to be characterized by straightforward rental relationships (Weiss 1970). Landholding, judicial authority and liege rights were held by different people, squeezing further the space available for territorial political consolidation. Within the politically fragmented framework of western Germany, the presence and strength of the cities also counteracted and checked the rise of a strong nobility. Despite the disruption of the Thirty Years' War and the persisting famines and plagues of the sixteenth and seventeenth centuries, western Germany remained a more heavily populated, economically advanced and socially complex territory than the eastern regions.

The more dramatic changes occurred in the eastern provinces, and were the result of a combination of internal and external factors. The most significant changes were the eclipse of the towns and the rise of a powerful landowning aristocracy. During the Middle Ages the need to secure labour had led in the eastern provinces to a distribution of wealth and landownership that went far beyond the aristocracy. But the decline of the Hansa led to the decay of the towns, and a combination of war and disease shook the structure of the countryside. Labour was increasingly in short supply. To maintain the export of corn, upon which the prosperity of much of the region depended, landowners adopted the solution of desmesne farming, increasing labour services in order to tie the peasantry securely to the locality. With the towns in decay, the way was open for the consolidation of the aristocracy, which in Brandenburg-Prussia was quite unique in terms of its narrow internal range of wealth, its direct involvement in agricultural production, and its tendency to reside on its estates (Goodwin 1953; Carsten 1954: 148-78; Rosenberg 1943; also Nipperdey 1961: Ch. 5). Large estates were less prominent in Silesia and Saxony than in Brandenburg-Prussia and Mecklenburg, but their presence contrasted East Elbia sharply from the west. At a time when western Germany was throwing off most of the surviving vestiges of feudalism, the eastern provinces introduced a much more repressive serfdom.

In East Prussia the defeat of the Teutonic Knights in 1410 and their final abolition in 1525 helped the change in socioeconomic structures, for the Order's control of its territory had inhibited the rise of an integrated aristocracy: the Knights became landlords and joined the local nobility. Moreover, the Reformation made available huge tracts of confiscated land. The rulers who acquired these territories found that as owners and landlords their interests coincided with those of the nobility and not with those of the towns. In eastern Germany the combination of political and economic events in the sixteenth and seventeenth centuries produced an alliance between political centre and landed aristocracy that set the region in sharp contrast to the landscape of the western regions. The alliance and division of responsibility between the strong Hohenzollern bureaucracy and the Junker aristocracy has been held by many to be the fundamental cornerstone of the ascendant Prussian state (e.g. Moore 1966; Rosenberg 1958).

The locus of economic activity basically remained in the west, where commerce and industry were more highly developed, a vigorous urban life persisted, a suffocating political control was not so apparent, and where even human resources were significantly greater. By the close of the seventeenth century the population density of Brandenburg-Prussia was still only half that of the states of Württemberg or Saxony, and appreciably lower than that of less 'developed' western regions such as Hannover or Schleswig-Holstein. Yet it was from the east, from the derided 'sandbox of the Empire', that the political centralization of Germany would come, refining and changing the regional character of the German territories.

The political mosaic

It was with the Carolingian dynasty that the post-Roman integration of western Europe reached its height. From his capital at Aachen, Charlemagne also established a series of marches at the fringes of the German world: administration, such as it was, was handled at a provincial level. The effective political unity of the Holy Roman Empire disappeared after 1200, yet there was a degree of association and a feeling of common interests, buttressed by a common language and opposition to non-German states. It could be argued that these survived until the formal dissolution of the Empire in 1806.

In the thirteenth century there began a rapid and severe disintegration of territorial unity, most prominently in the south-west and along the major trade routes of the Rhine and Neckar. Ecclesiastical sees and towns became independent political entities. Many small nobles in the south-west avoided territorial incorporation by seeking and acquiring the status of 'imperial knight', owing direct allegiance to the Emperor: by the sixteenth century there were some 2,500 imperial knights. The territorial jurisdiction of any single ruler was extremely limited. Possibilities of expansion were hindered by economic developments and by the rise of stronger territorial units to the south and west. The miniscule size of most units made it impossible for them to contain major outbreaks of conflict, such as the Reformation and the Bauernkrieg (cf. Wallerstein 1974: 177; Carsten 1959: Ch. 4). Political strength, or more accurately territorial integrity, survived to a much greater degree in the more recently settled lands to the east. The major point is that territorial control was so diversified, despite the high degree of linguistic unity, that there was no congruence with concepts such as national identity: state and nation simply meant different things.

This patchwork quilt of political units, many of which possessed non-contiguous territory and even *internal* tariff and other obstacles, remained fundamentally unchanged in its kaleidoscopic nature until the close of the eighteenth century. There remained some 1,789 formally independent units. Apart from the few large units of Brandenburg-Prussia, Saxony, Hannover, Bavaria and Württemberg, the more important unified territories in the west tended to be bishoprics such as Köln, Trier and Münster.[7]

It was quite clear that this congerie of petty principalities could be united only from the outside. Unification, in fact, was the strategy of the Hapsburgs for centuries. Anderson (1975: Part II, Ch. 3) argues that Hapsburg prospects were definitively smashed by the Swedish interjection into the Thirty Years' War. But they had a further chance later. It was the failure of Joseph II's efforts in the eighteenth century to win Bavaria that finally determined the end of all prospects for a comprehensive south German state; so began that progressive exclusion of Austria from German affairs which was confirmed by the events of the Napoleonic era and the mid-nineteenth century (Bernard 1965).

In western Germany the largest secular states were Hannover, Hessen and Württemberg. While Hannover progressively expanded its territory, its northern location placed it in the shadow of Brandenburg-Prussia and it did not play an aggressively active role in German affairs. Hessen's central location was offset by disintegration through family quarrels over succession, which led to the partition of territory. Württemberg retained its territory and integrity, but remained static between the fifteenth and eighteenth centuries. Its territory was inextricably intertwined with that of free cities, smaller principalities and hundreds of imperial knights, as well as lying well within the Hapsburg and French spheres of influence. Attempts to incorporate the neighbouring principalities usually ended in humiliating failure (cf. Carsten 1959; Fauchier-Magnan 1958). Its doubling in size after 1800 was due to Napoleon: by then it was too late.

Saxony, Prussia's powerful southern neighbour, was for long an economically advanced region. The likelihood of territorial expansion was checked by a relatively weak nobility and a vigorous level of urbanization. It did acquire Lusatia in 1648, only to be dispossessed by Prussia one hundred years later. The major drive came in the early eighteenth century — not westwards, but to the east, with a Saxon – Polish union. Sweden again proved to be the destructive force. The Saxon government and nobility never recovered from their defeat during the last great Swedish incursion of 1706 (Carsten 1959: 191-204; Anderson 1975). Thereafter Saxony entered into a political, economic and population decline until the mid-nineteenth century.

Perhaps the most potent purely German opponent of Prussia was Bavaria. But if Bavaria was not an earthly paradise inhabited by animals, as described by Frederick the Great, it was no aggressively expanding state. Indeed, up to the eighteenth century its foundations seemed to be so fragile that its neighbours constantly reckoned with its demise (Bernard 1965: Ch. 1). Internally, secular power was counterbalanced by that of the Church, which by the mid-eighteenth century owned 56 percent of the land. Externally, Bavaria's role as the Catholic leader within Germany excluded it from much influence in the north without a much greater level of military prowess and economic development than it possessed (Carsten 1959: 350-2, 392-406; Anderson 1975: 253-6). In any case, Bavaria's territorial aggrandisement also came too late. Bavaria could not act as a German centralizer: it could only wage a rearguard action. It advocated a 'grossdeutsch' solution to the last, allying itself with Austria against Prussia in 1866.

It is easier now to see why the consolidation and history of these states was completely overshadowed by the dramatic rise of Brandenburg. While the latter was still a poor frontier region when assigned to the Hohenzollerns in 1411, it had achieved some degree of importance in German affairs, becoming an Electorate of the Holy Roman Empire in 1351. Its geographical location, in fact, was not insignificant, for it straddled the two important waterways of the Elbe and Oder. In 1473 the adoption of primogeniture made its territories indivisible — a marked contrast to many areas in Germany. Its vast expansion dates from the seventeenth century. Under the newly forged alliance between elector, central bureaucracy and landowning aristocracy, and the development of a military force that came to absorb some three-quarters of the state fiscal revenues, Brandenburg consolidated the eastern lowlands (Taylor 1961: 19). East Prussia was acquired in 1618, eastern Pomerania in 1648, Silesia in 1742 and West Prussia in 1772. At the same time it acquired various dispersed territories in western Germany.

Although Brandenburg-Prussia was clearly the dominant state in northern Germany, the turning point in the redrawing of the map of Germany came with the Napoleonic wars. By 1815 there remained only 39 states to form the new German Confederation. More important, Prussia received further territory in the west: now only the thin wedge of Hannover and the Hessens prevented an unbroken line of communication from the western outpost of Cleves to the furthest borders of east Prussia. At the same time, by giving up the vast bulk of its Polish possessions, Brandenburg-Prussia became an undisputed German state: in the 1860s only 13 percent of its population were Slavs (East 1966: 275). The final stage of expansion relates to diplomacy and war: Schleswig-Holstein was absorbed in 1864, and Hannover, the Electorate of Hessen, Nassau and Frankfurt in 1866. By 1870 there remained only 25 independent units, and it was these that, under Prussian pressure, yielded up independence and formed the Second Reich. The larger south German states could not agree on a viable alternative.[8] In any case, the grossdeutsch alternative to the Prussian 'kleindeutsch' solution was identified with Catholicism: certainly, it was Catholic groups who in the years before 1870 were the most articulate proponents of a Grossdeutschtum (cf. Hope 1973).

Prussian dominance dates most clearly only from the nineteenth century, for most of the surviving states were miniscule, and none could challenge the predominance of Prussia which now possessed three-fifths of the territory and population of the coming German empire (Table 1). Given the complex development of the German territories, it might seem inevitable that the core lands of Brandenburg-Prussia would emerge as the centralizing agency (cf.

TABLE 1
The states in the Second Reich: percentage of population

	%		%
Prussia	61.9	Saxony-Meiningen	0.4
Bavaria	10.6	Saxony-Coburg-Gotha	0.3
Saxony	7.4	Lübeck	0.2
Württemberg	3.8	Mecklenberg-Strelitz	0.2
Baden	3.3	Bremen	0.2
Hessen-Darmstadt	2.0	Lippe	0.2
Hamburg	1.6	Schwarzburg-Rudolstadt	0.2
Mecklenburg-Schwerin	1.0	Reuss Younger Line	0.2
Braunschweig	0.8	Schaumburg-Lippe	0.1
Oldenburg	0.7	Waldeck-Pyrmont	0.1
Saxony-Weimar-Eisenach	0.6	Schwarzburg-Sonderhausen	0.1
Anhalt	0.5	Reuss Older Line	0.1
Saxony-Altenburg	0.4		

Anderson 1975: 260). Yet there was a great deal that was fortuitous in Prussian expansion. Except at the beginning and end of its long period of growth, Prussian military feats were not unduly impressive. Indeed, territorial growth came about through two distinct processes: through inheritance and legitimate claims of succession, and as a result of peace treaties, irrespective of the military role played by Prussia.

With Prussian hegemony, Berlin grew in the nineteenth century to become the dominant German centre for the first time. Yet its claim to centrality rested primarily upon political developments: its business was political administration — as early as 1816, with a population of 198,000, almost 25 percent of its labour force was employed by the court or government. Similarly, its ability to dominate Germany in the same way that London dominated Britain or Paris France was circumscribed first by the well-established city polycephality inherited from the Middle Ages, and second by political developments after 1500. Most of the political units had experienced the phenomenon of state-building. Consequently, most had undergone some degree of centre consolidation, territorial standardization and identity formation. One consequence was the strengthening of several political capitals, to which commerce also tended to gravitate. After centuries of fragmentation, a German state had emerged, but this unity could not disguise the fact that the underlying territory was essentially polycephalous in nature.

Germany on the eve of union

Apart from the movement towards political unification, the most dramatic changes prior to unity were economic. The territory was overwhelmingly German in terms of language and ethnic identity, while the religious pattern introduced by the Reformation had persisted almost unaltered. Economically, the country was still essentially agricultural as late at 1850. The three major industrial regions, all dependent upon metal and textiles, were the ancient ones of the North Rhineland, Saxony and Silesia, but on the whole industry was essentially a rural occupation, with the cities tending more to be centres of commerce and trade.

The mid-nineteenth century was the period of economic takeoff.[9] The population of Germany grew from 24 million in 1800 to 41 million by 1871. A standard currency, the mark, was introduced in 1867. Prussian coal production rose from under 2 million tons in 1825 to 19 million by 1865. Between 1834 and 1864 exports of woollen and textile products quintupled. The trend towards political unification clearly contributed towards economic development. The latter was more clearly stimulated by the Zollverein after 1832. The shatter belt of German states had adversely affected trade, and the psychological effects of the Zollverein were perhaps as significant as its immediate economic consequences (Henderson 1959). The impetus also derived from a revolution in communications. Apart from those built by Napoleon in the west, the German roads were bad in the early nineteenth century. Between 1845 and 1870 Prussia launched a large road-building programme. More significant in pulling the various German states together was railway construction, which began in the 1830s: along with the Zollverein, it stimulated a rapid economic growth. By 1860 the rail network extended to over 11,000 kilometres, and by 1870 it was complete in its essentials (cf. Kobschätzky 1972; Stöckl 1969). The development of the railway grid (cf. Figure 1) bears eloquent witness to the polycephality of Germany. It reflects both the presence of several political centres and the demographic and economic dominance of the west. Spokes radiating outwards from the hub of Berlin are identifiable only within its proximity. They are soon lost in a welter of other routes linking the territory from both north to south and east to west.

The revolutionary effect of the railway grid is that it occurred before rapid and extensive industrialization. One consequence was that it contributed to the survival of the old cities and provincial

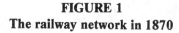

FIGURE 1
The railway network in 1870

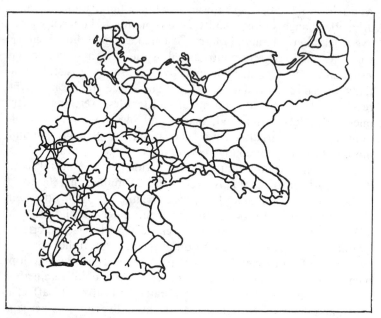

capitals. In 1870 most of the great cities were still traditional foundations such as München, Köln and Hamburg: they remained much larger than newer, more purely industrial cities such as Essen, Duisburg and Chemnitz, though the latter tended to grow at a much faster rate. A city like Köln, for example, could participate in and benefit from the industrialization of the Ruhr, even though physically removed from it. If anything, therefore, the railways and incipient industrialization increased the polycephalic nature of Germany at the same time as political events were seemingly resulting in its diminution.

   The agricultural picture remained fairly constant. Serfdom had been ended in Prussia in 1806, followed by an intensification of commercial agriculture. But the east remained characterized by large estates farmed by landless labourers. Indeed, the end of serfdom had led to further land consolidation. Some three-quarters of a million smallholdings disappeared, sold or given up as compensation by peasants to the great landlords.[10] In short, the size of landholdings decreased and the dispersion of land occupation increased fairly consistently along a gradient running towards the southwest, with the east remaining more rural and agricultural. But at the same time a new east – west distinction was beginning to emerge, with an agricultural east and the traditional Prussian aristocracy juxtaposed against a more industrial west with a new class of industrial magnates.

The establishment of a single German state in 1870 could not deny the existence of important regional variations that had accumulated over the centuries. An east – west gradient reflected economic differences in terms of urbanization, agricultural settlement and industrialization. This was crossed by the north – south axis of the religious variable. Similarly, unification and the growth of a German national identity in the nineteenth century could not remove instantaneously the accumulated weight of regional loyalties and socialization processes. This manifested itself not only in a cleavage between Prussia and the rest: an enlarged Prussia was itself a collection of disparate territorial units with different histories, identities and behavioural patterns.

Germany after 1870

Germany between the establishment of the Second Reich and the dismemberment of the unified German state in 1945 may be discussed as a whole, first because there was no fundamental deviation from or change in the patterns consolidated before 1870, and second because there was a high degree of institutional and party continuity between the Second Reich and the Weimar Republic. The regional particularism of pre-unification Germany survived well into the twentieth century. Loyalties to regions remained high, and Berlin, despite its ebullience and exceptionally rapid growth, never enjoyed the dominating central strength of some other west European capitals. Political power may have been concentrated in the Prussian capital, but administrative power was retained by the constituent states of the federation, while economic and cultural influence was widely dispersed. Moreover, substantial German populations had been left outside the boundaries of the new state. Many Germans, especially in the south, regarded the Hapsburg capital of Vienna as their 'natural' centre: the grossdeutsch option lingered on long after 1870, until Hitler attempted to turn it into reality. At least, the state had a common linguistic and ethnic structure. There were significant ethnic or linguistic minorities — Poles in West Prussia, Posen and Upper Silesia, Danes in North Schleswig, and French in Alsace-Lorraine — but the great majority of these were lost to Germany in 1918.[11] In any case, while the national minorities were occasionally salient in foreign affairs, their importance in domestic politics was more limited.

Social structure

The division of the German people between Catholicism and Pro-testantism has persisted almost unchanged since the Reformation. Germany was predominantly Protestant: Catholics formed only 33 percent of the population, declining slightly after 1918 through the loss of the Polish provinces and Alsace-Lorraine, and the separa-tion of the Saarland. The geographical distribution of the religious denominations was very uneven (cf. Figure 2). The Catholics were clustered mainly in the south and west. The north and east were overwhelmingly Protestant, extending southwards in a narrow wedge through Hessen to Württemberg. Throughout most of Ger-many the Protestant religion was Lutheran: in the south-west, however, there were significant Calvinist populations. It is this geographical distribution that has remained unchanged over the cen-turies. The territorial mix of the two religions, especially in the west, was little more than a reflection of the Peace of Augsburg and the political map of Germany as ratified at the Treaty of Westphalia.

In 1870 population distribution and density was much the same as in the Middle Ages: within the lifespan of the next generation, population growth, industrialization and urbanization altered drastically the German landscape. By 1910 the population of the Reich had risen to 68 million. Between 1830 and 1930 the aggregate population increase was 140 per cent. The most pronounced demographic trend was urban acceleration. Even given the historic multitude of German cities and towns, in 1871 64 percent of the Reich's population resided in rural communities with less than 2,000 inhabitants: within a single generation the rural sector had contracted to under 40 percent (Table 2). Urbanization was regionally selective in that it built upon past foundations: very few 'new towns' were established. The net result was, in comparative terms, a large number of densely populated cities. Urbanization was accompanied by large-scale migration not only from coun-tryside to town, but also from east to west. This strengthened the demographic predominance of the Rhinelands and the central east – west commercial belt from the lower Rhine across to Saxony (Figure 3). The north-eastern regions, including parts of Prussia,

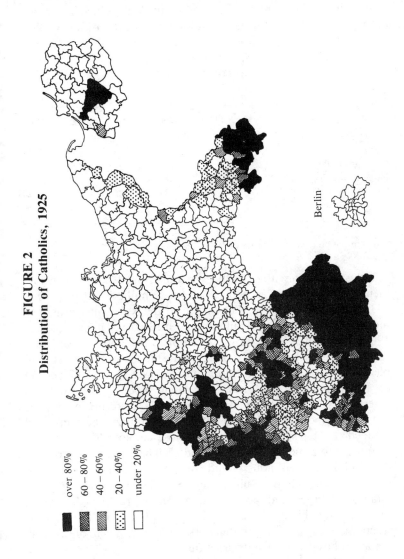

FIGURE 2
Distribution of Catholics, 1925

over 80%
60 – 80%
40 – 60%
20 – 40%
under 20%

Berlin

remained highly rural, with one-fifth or less of the population employed in industry (cf. Raupach 1967). East Prussia, where agriculture had not been able to sustain a natural population growth in the late nineteenth century, suffered particularly badly after 1918 from its physical separation from the rest of the Reich.[12]

TABLE 2
Urbanization, 1871-1939

	1871	1890	1910	1925	1939*
	%	%	%	%	%
Cities (over 100,000)	5	12	21	27	32
Large towns (20,000-100,000)	8	10	13	14	13
Small towns (2,000-20,000)	24	25	25	24	24
Rural areas (under 2,000)	64	53	40	36	30

* Within the political boundaries of 1937.

Industrialization after 1870 even outstripped earlier developments. Coal production increased from 26 million tons in 1870 to 153 million tons by 1910. The output of the Ruhr mines was even more phenomenal, from 4 million tons in 1860 to 60 million tons by 1900. The railway network spread from 18,560 kilometres in 1870 to 59,031 kilometres by 1910. By 1914 the metamorphosis was complete. Germany had become an industrial state. In 1882 42 percent of the population had been dependent upon agriculture and forestry: by 1925 this figure had declined to 23 percent. In the same year the proportion of people engaged in industry and crafts had climbed to 41 percent, and that of those employed in trade and commerce to 27 percent. The most industrialized regions were those that had dominated industry and commerce in earlier centuries — the west, especially the Ruhr region, Saxony and Silesia, and old-established cities and ports such as Hamburg, Magdeburg or Frankfurt (cf. Figure 4). Two areas of weak or non-development stand out: large tracts of the north-east and much of Bavaria remained heavily rural and agricultural.

Industrialization failed to modify radically the traditional society, and Germany, especially Prussia, remained a country characterized by the values and rigidity of a 'Ständestaat'. Dahrendorf (1967: 58) describes the Second Reich as an 'industrial feudal

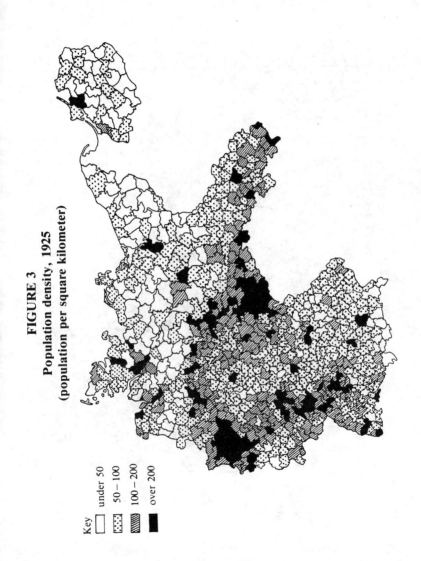

FIGURE 3
Population density, 1925
(population per square kilometer)

Key
under 50
50 – 100
100 – 200
over 200

FIGURE 4
Population dependent upon industry, 1925

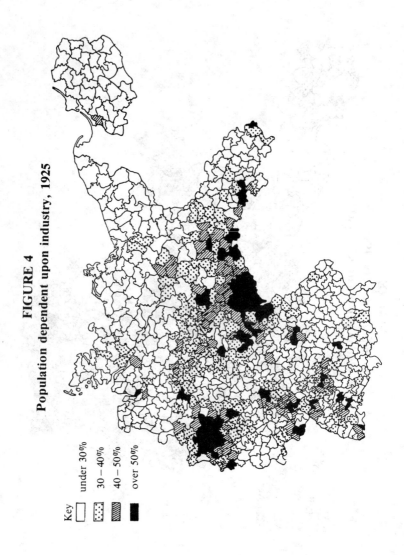

Key

☐ under 30%

▨ 30 – 40%

▨ 40 – 50%

■ over 50%

society' parallelled by an 'authoritarian welfare state'. The large agricultural estates of East Elbia were balanced by industrial units of a high magnitude and near-monopolies in production, all sponsored by the state. While it would be an error to ignore the high degree of elite co-operation that developed between, for example, the central bureaucracy and the industrial magnates, there was little space available for the kind of infrastructure that in Britain, for instance, nurtured a vibrant commercial middle class. While the fall of the monarchy in 1918 may have produced a constitutional revolution and a reduction of aristocratic influence, it did not mean the disappearance of a Ständestaat constructed around the twin aristocratic-dominated pillars of the bureaucracy and the military officer corps.

The party system

A simplified view of the complexity of the German party system between 1870 and 1933 would identify at least four main groups of parties — conservative, liberal, catholic and socialist — each appealing to and claiming to represent specific social groups and subcultures. Each 'tendance' contained, at least for part of the period, more than one political party.

There were two major conservative parties during the Second Reich, distinctive in their regional and social support. The German Conservative Party was essentially a party of the East Elbian aristocracy, a firm supporter of the monarchy, though not necessarily of the imperial government (cf. Booms 1954; also Berdahl 1972). Its conservative rival was the Reich Party. Much less particularistic than the Conservatives, it originally separated from the latter because of its more pronounced German nationalism and staunch support of Bismarck's policies. There were also several smaller conservative parties, for example the Bund der Landwirte (Agrarian Union) and the Economic Union.[13] These were primarily 'special interest' parties, tending to concentrate more on a single policy sector. There were also two important liberal groupings. Both were caught in the dilemma of giving priority to classical liberal principles or to the dream of unification. The National Liberals were similar to the Reich Party in their commitment to the national state and the economic and foreign policy of the government. The other grouping, which proved to be exceptionally prone to fragmentation, may be broadly described as Progressive. In their political principles the Progressives were similar to other nineteenth-century liberal parties in western Europe, committed to the establishment of a parliamentary democracy.[14] After 1918 there was a change of nomenclature in the conservative and liberal tendances. The National People's Party (DNVP) was basically a merger of the conservative parties (cf. Liebe 1956; Hertzman 1963). The Democratic Party (DDP) was the heir of the Progressive groups (cf. Stephan 1973), while the People's Party (DVP) was based upon the old National Liberals (Hartenstein 1962; Turner 1963). Different programmatic bases and the personal antipathy of some leaders prevented a union of the two liberal groupings.

The Catholic minority formed the Zentrum in 1871 in order to protect its interests in what was conceived to be a hostile political environment.[15] After 1918 the Bavarian wing, rejecting the Zentrum as too liberal, formed a separate party, the Bavarian People's Party (BVP), and often pursued policies quite distinct from those of the parent body. The socialist tendance was represented by the Social Democratic Party (SPD).[16] Like other Marxist-inspired parties, it claimed to be the sole representative of the working class. After 1916 it had to contest this claim with the Independent Social Democrats (USPD), which in turn divided after 1918, some supporters returning to the SPD fold with the remainder forming the nucleus of the Communist Party (KPD) (cf. Weber 1970).

The unification of the country under Prussian leadership meant that to some extent regional loyalties were translated into party divisions. But apart from the ethnic minorities, very few significant regional-specific parties emerged during the Second Reich. A greater proliferation of regional splinter parties occurred in the 1920s, when such parties, almost always with an agrarian base, appeared in several provinces. Their emergence simply confirmed the problematic nature of a national system of party oppositions; for, in a sense, all parties were regional to some extent, depending upon the spatial distribution of the economic or cultural groups supporting them. We do not have an adequate data set for a systematic cross-regional analysis of the social basis of party support in the Second Reich. Moreover, the absolute majority electoral system encouraged pre-election agreements and alliances, and positively discouraged parties from contesting constituencies they could never hope to win. Nevertheless, it will be useful to indicate the regional nature of the imperial party system before moving on to discuss the Weimar Republic.

Regional factors and the imperial party system

The electoral strength of the imperial parties was not altogether stable. However, despite the emergence of several minor parties, only two of the larger parties changed at a rate greater than 0.25 percent per annum.[17] The strong upward trend of the SPD was matched by a downward trend for the National Liberals: the Conservatives showed the least amount of change. Apart from the growth of the SPD from negligible proportions, the limited variations in the overall national trends point perhaps to the limited nature of each party's electoral support. In 1907, for example, the SPD was the only national party in that it contested all but 5 of the 397 constituencies. Only one other party, the Zentrum, contested more than one-half of the seats (265), by no means all in heavily Catholic areas. The third largest number of constituencies — only 176 — was contested by the National Liberals.

The widespread absence from competition of the parties reflects to a certain extent the alliance option offered by the electoral system (just as the nationwide spread of the SPD may reflect its lack of possible coalition partners), but it was also due to the regional nature of the parties. The SPD was very much a party of the industrial and urban areas. In addition, it proved unable to establish itself in overwhelmingly Catholic areas. In 1907 its constituency share of the vote ranged from 0.3 percent to 75.6 percent. It was, however, one of the two parties that developed a high degree of organizational self-sufficiency. The other was the Zentrum, which relied electorally upon the Catholic population. While a substantial Catholic vote was given to other parties, most notably in the ethnic areas of West Prussia, Posen and Alsace-Lorraine, virtually all the Zentrum vote came from Catholics. The number of Catholics voting for the party tended to be in constant decline after the 1880s: it has been estimated that up to 1924 the number of Catholics voting for the Zentrum ranged from a high of 86.3 percent to a low of 46.3 percent in May 1924 (Schauff 1928: 74, 128-32). Given the regional distribution of Catholics, it follows that the electoral base of the Zentrum was regionally distinctive, being clustered in the western Rhinelands, the south (especially Bavaria), and Upper Silesia.

The absence of a single cohesive or coherent middle-class grouping is striking. A stress upon status divisions was probably the major motive behind middle-class fragmentation, for none of these parties displayed any great inclination to incorporate working-class or even Catholic support. The motive of status becomes more apparent when we consider that these major parties had also to face the challenge of smaller special-interest parties. Yet the strong regional variations in the support of all these parties would suggest that status divisions do not provide the entire answer. Past socialization and acculturation experiences, as well as attitudes towards Prussia and the centre, all deriving from the polycephality of Germany, also played a part.

The most distinctive were the Conservatives, fundamentally a party of the Prussian aristocracy in the eastern provinces, and hence agrarian. The large landowners proved to be particularly successful in mobilizing their rural hinterland behind the Conservative banner (Berdahl 1972). In 1907 the Conservatives contested only 115 constituencies, the vast majority east of the Elbe. In 1912 all but 10 of their 55 elected deputies were Junker aristocrats (Schieder 1962: 121). The Conservative appeal in rural East Elbia gave it an electoral base, but one in constant decline relative to the total German electorate. Conservatives were prominent in the 1890s in sponsoring the Bund der Landwirte as a mass organization that would ally the small western farmers behind a common agrarian front. The Bund adopted an aggressive style, and was active throughout wide areas of Germany; yet it was as much an East Elbian as a western movement. Outside the Prussian heartlands the Bund in fact tapped anti-Prussian particularism, especially in Bavaria and Württemberg (cf. Puhle 1967: 37-45, 72-110, 165-212; Tirrell 1951: 331; Gerschenkron 1943: 102). After 1918 it splintered into regional movements. The Reich Party originated as a Silesian splinter from the Conservatives. It remained relatively small, and was essentially an alliance of business interests and landowners outside the old Prussian provinces: it never expanded much beyond the Rhinelands and Silesia.

The National Liberals were essentially Protestant and 'new' Prussian, remaining a loose band of notables until after 1900 (cf. Nipperdey 1961: 86ff.). The 'national' face of liberalism had first gathered force at the fringes of Germany, for example in Baden and Schleswig-Holstein (cf. Carr 1963; Gall 1968). Socially, they differed from the Conservatives in the absence of titles among their ranks. Geographically, they tended to be stronger in western Germany. In the north-west their strength rested more in urban areas; in the south their basis was more rural and agricultural. The progressive element of liberalism was more of a Prussian phenomenon, though the many mergers and splinters reflected regional as well as programmatic differences. One clear geographic basis was northern Württemberg, where the South German People's Party retained its independence until 1910. In general, the various progressive factions were more strongly urban-oriented.

Since most of the smaller parties also had regional sources of strength, it follows that the imperial party system was not a national system. Every region or province tended to have its own party spokesmen. Bavaria represents a special case, which may be deferred to a later discussion. The most outstanding regionalist parties were those of the ethnic minorities, to which may be added the interesting Guelph movement in the province of Hannover. The Polish population of the eastern provinces rallied behind their own particularist political movement. The peak of mobilization came in 1907: Polish candidates won an absolute majority on the first ballot in 19 of the 35 constituencies contested in West Prussia, Posen and Upper Silesia. Although Catholic, the Poles' co-operation with the Zentrum did not extend to a total electoral pact: in 1907 Poles and Zentrum competed against each other in 21 of these 35 constituencies.

The origins of the Guelph Party lay in the refusal of the deposed Hannoverian monarch and his supporters to accept the annexation by Prussia in 1866. Their opposition took various forms, from an illegal militia to, eventually, a political party (cf. Stehlin 1973). The Guelph Party reached its high watermark in 1881 with 38.8 percent of the Hannoverian vote. Its strength was confined to the historic parts of the old Electorate, although with industrialization its support in and around the city of Hannover declined rapidly. In essence it was a rural party, with its leadership dominated by the nobility (Franz 1957: 16). Quite similar in many ways to the Prussian Conservatives, it nevertheless collaborated closely with the Zentrum.

In Alsace and Lorraine the 1870 war had rekindled religious differences. Protestants took the lead in fanning autonomist (from France) sentiments. Yet the important political turning point was when the Catholic clergy reversed their previous opposition to the policies of Napoleon III to react against German linguistic policies, and by 1890 Berlin had to accept that Alsace and Lorraine should be treated as distinct units in linguistic matters. Overall, they represented a region that resisted integration, no matter which centre was involved (Silverman 1972). The Catholic clergy was the crucial political factor, although they did not establish a formal political party until the 1890s. While most of the deputies to the

Reichstag were priests, the clergy and party did not consider absorption by, or even formal affiliation with, the Zentrum. While parliamentary co-operation was possible, the two remained in electoral competition, especially in Lorraine: in 1907 the 7 (out of 15) constituencies with Zentrum-particularist competition included all four Lorraine seats.

Regionalism and the Weimar parties

Only two of the six 'major' parties of the Weimar Republic were survivors of the Empire: the SPD and Zentrum. Another, the KPD, was the result of an ideological split within the SPD. The other three — DVP, DNVP and DDP — were new formations, yet still possessing strong links with the past. Many small parties also successfully bridged the trauma of 1918.[18] Over all, the fragmented, regionally based party system survived the fall of empire virtually unscathed. But while the tendances may have remained operative, each political strand tended to experience further fragmentation. Between 1919 and 1928 the number of parties contesting elections separately rose steadily, reaching its peak in the 1928 election, which also saw the lowest turnout of the period. It was the non-Marxist, non-confessional parties that suffered fragmentation and, eventually, virtual annihilation. A recoalescence of splinter parties and of non-voters was achieved in the 1930s by the Nazi Party (NSDAP).

The particular variant of proportional representation introduced in the Weimar Republic encouraged parties to contest every electoral district. Since constituencies possessed large electoral magnitudes, it was also relatively easy for small and new parties to win their proportional quota of Reichstag seats (Urwin 1974). Hence, elections in the Weimar Republic offer a better opportunity than imperial elections to map the regional sources of strength of the various parties and the areas of regional disaffection, more especially as electoral data are available at the level of the 'kreis' (commune).

It would not be particularly valuable to present maps of the political strength of either the DDP or the DVP. The former proved to be electorally strong only in the first election of 1919: thereafter it was a negligible electoral force. Similarly, the DVP vote collapsed dramatically after a single strong performance in 1920. It will suffice to point out that the DDP vote persisted in strength only in the Progressive areas in Baden and Württemberg. Similarly, the DVP inherited some of the National Liberal strongholds. Its votes lay more in north-western Germany and, more generally, in urban areas. It remained very weak in the south, especially in Bavaria. In the first election of 1924, the correlation of the DVP vote with urbanization was 0.37: in 1928 it had declined to 0.30.

Figure 5 maps the median vote for the Zentrum and BVP over the four 'normal' elections between 1920 and 1928. The two did not compete against each other, except in the Palatinate. Despite the decline in the number of Catholics willing to vote for these parties, the geographical fit of their vote with the regional distribution of Catholics is impressive. The Zentrum vote clearly falters in some of the industrialized areas of the Rhinelands. More important, the BVP was not entirely successful in Catholic Bavaria, where it suffered the 'marauding' of regional and rural groups — especially the Bavarian Peasants' League, whose strength lay in Lower and Upper Bavaria. Not surprisingly, the correlation of the parties' vote with the distribution of Catholics was high: in 1924 (1) it was 0.88.

It has often been said that the Elbe – Saale line represents a significant cultural divide in German history. This boundary emerges very clearly in a mapping of the other large party with a distinctive regional base (Figure 6). The DNVP is almost entirely an East Elbian party. It inherited the core agrarian lands of the old Prussian Conservative Party, and Pomerania, rather than East Prussia, was its heartland. To the west of the Elbe, apart from strongholds in Hessen, it impressed only in Protestant Middle Franconia, a region that still stands out in the Federal Republic. Because of the strongly skewed regional nature of its support, one should not expect the DNVP vote to correlate highly with socioeconomic characteristics. In fact, its correlation with agriculture rose from 0.34 in 1924 to 0.45 in 1928.

Regional parties survived as a significant political factor in three areas: Hannover, Württemberg and Catholic Bavaria. In the first and last cases, regionalism was represented by survivors of the Empire. The Guelphs became the Hannoverian Party (DHP); in Bavaria a more specific particularism was represented by the Bavarian Peasants' League. Perhaps the most successful regional party was the Protestant Württemberg Farmers' and Winegrowers' League, which, like several rural parties in, for example, Hessen and Thuringia, arose as a fragmented survivor of the old Bund der Landwirte. These regional parties stood for rural discontent. Along

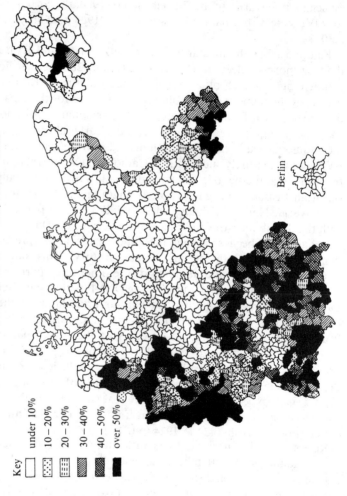

FIGURE 5
The median Zentrum/BVP vote, 1920 – 1928

Key

under 10%
10 – 20%
20 – 30%
30 – 40%
40 – 50%
over 50%

Berlin

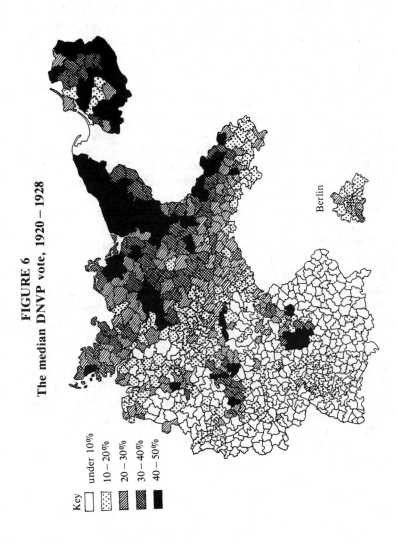

FIGURE 6
The median DNVP vote, 1920 – 1928

Key
under 10%
10 – 20%
20 – 30%
30 – 40%
40 – 50%

Berlin

with the DNVP, they represented a more or less permanent agrarian disaffection with the regime that in its origins was as much economic as political and constitutional. The persistent agricultural crisis of the 1920s might have prevented strong identification with the regime, but did not produce a strong agrarian party. Regional loyalties, the mobilization of Catholics by religious parties, western antipathy to the Conservative and DNVP parties — all hindered any form of agrarian unity. The NSDAP was the first national party able to exploit the anti-urban, anti-capitalist sentiments of the German countryside: the 1930 breakthrough came primarily in western Protestant areas of peasant smallholdings,[19] and in 1932 the agricultural east (along with most of Protestant Germany) swung overwhelmingly behind the party. Moreover, some further continuity with the past was provided by the fact that the NSDAP first rose to prominence in those Protestant western areas that in historical time, were more recent Prussian additions: that is, in areas where the conflict between nationalism and resistance to incorporation had been posed more clearly.

The KPD vote (Figure 7) is more obviously linked to industrialization. Its core areas are the industrialized regions of the Rhinelands and the Saxonies (plus the outliers of Upper Silesia and Berlin). It had a median vote of over 30 percent in only 14 kreise: in this sense, the KPD is the nearest German equivalent to, for example, the British Labour Party in the 1920s. By contrast, the SPD (Figure 8) was much more of a nationwide party, but with few areas of overwhelming strength: its median vote, 1920 – 28, rose above 40 percent only in 44 kreise, and nowhere was it above 50 percent. With the exception of the KPD in parts of the Rhineland, both were excluded more or less completely from Catholic Germany: but more generally, both were very weak in the south. By contrast, the SPD and KPD were the two dominant parties in Berlin, the political centre and symbol of Prussian dominance. Both parties in fact correlated highly with industrialization, reflecting their competition for the same social base. The correlation of the KPD vote with proportion in industry in 1924 (1) and 1928 was 0.66 and 0.73 respectively, for the SPD, 0.62 and 0.72. By contrast, the SPD correlation with urbanization was much lower, only 0.34 and 0.32.[20]

Regionalism and the institutional framework

The institutional structure of the Empire also had consequences for the regional nature of German politics. The introduction of male

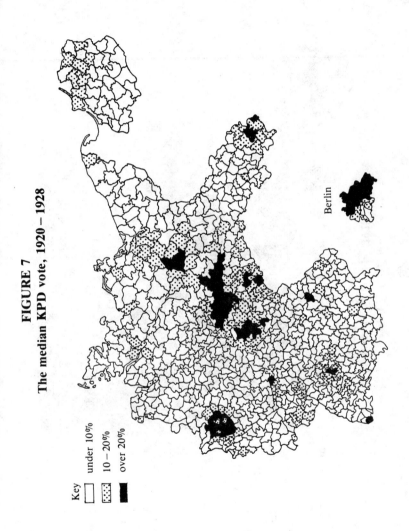

FIGURE 7

The median KPD vote, 1920 – 1928

Key

under 10%

10 – 20%

over 20%

Berlin

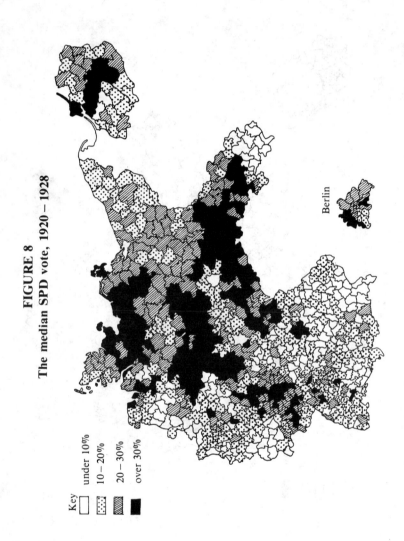

FIGURE 8
The median SPD vote, 1920 – 1928

suffrage permitted mass mobilization, but this perforce had to take place within several widely different socio-structural landscapes. Moreover, it existed alongside an authoritarian form of government.[21] The operation of the 1871 constitution revolved around the twin pivots of the emperor and chancellor. The latter, the head of government, was appointed by the emperor and had complete control of the central administration (cf. Morsey 1957). The functioning of the federation was achieved through the upper chamber, the Bundesrat, which was composed of delegates from the constituent states. Prussia, because of its size, dominated this body, with 17 of the 58 seats (extended to 61 in 1911 when Alsace-Lorraine was accorded constituent status). Since constitutional amendments could be blocked by 14 votes, Prussia had a veto. Moreover, Prussian influence was strengthened by the rule that state delegates must vote en bloc. In addition, the imperial chancellor presided over the Bundesrat, which deliberated in closed sessions.

The federal element nevertheless provided for an institutional expression of political differences between the Reich and the constituent states. An alternative outlet was offered by the dichotomy between Reich executive and lower chamber. For within the nexus of power relationships, only the chancellor was responsible in any way to the popularly elected Reichstag. But the latter could not directly appoint or remove him: its sole decisive influence was to pass or reject the budget. Since electoral victory did not necessarily lead to participation in government, the parties had neither opportunity nor incentive to moderate their views or to broaden the social spectrum of their support in the hope of gaining office. A multiplication of parties was in theory furthered by the two-ballot electoral system, since the availability of the run-off second ballot could reduce the risks of multi-party competition on the first ballot. To some extent, however, party proliferation was circumscribed by the well-pronounced regional bases of partisan politics. More important, regional imbalance was intensified by the failure to revise constituency boundaries and an increasing rural bias in the electoral system.

Nor did government policies contribute much towards the elimination or pacification of regional differences. Indeed, the cultural divide was decisively strengthened by a frontal governmental assault upon the minorities. The exterior motive for the Kulturkampf lay in the justification of protecting the rights of the Old Catholics, who rejected the doctrine of papal infallibility. But active social discrimination against the Catholic Church sprang more realistically from the desire to merge church and state, especially in education (cf. Schmidt-Volkmar 1962; also Franz 1954). It was the Poles who bore the brunt of government policy, but it affected all Catholics and all ethnic – linguistic minorities: even the Danes experienced their own, admittedly less harsh, Kulturkampf. The net result was a more intense mobilization of the affected populations: a major exception was Catholic Bavaria, where, because of the federal element, the conflict was more blurred. Poles and Alsatians collaborated with the Zentrum, while retaining formal independence. In Alsace-Lorraine it was the rise of the SPD, not religious discrimination, that stimulated the establishment of an organized party. While the Kulturkampf was, to all intents and purposes, abandoned by 1879, its cessation did little to diminish the Catholic sense of alienation. Alsace-Lorraine continued to represent a region that resisted the imposition of national standards, irrespective of the centre from which they emanated, while the Reich persisted in actively discouraging Catholic organizations (Silverman 1972: Chs 4 – 6). The imperial government also looked benignly upon efforts to dampen Polish agitation and organization in the eastern marches throughout the lifetime of the Reich. But German Catholics, too, sheltering behind a dense organizational shield, remained distinct from the rest of Germany. Despite the Zentrum's growing activism in central German affairs, there was an underlying 'silent' Kulturkampf throughout the whole of the Wilhelmine era (Ross 1976: 18-32).

As the 'official' Kulturkampf petered out, a new government assault began, this time against the SPD. For the next decade the party had to labour against a whole range of discriminatory legislation. The result was again failure, and a consolidation of SPD cohesion and organization (cf. Pack 1961; Lidtke 1966). Economic differences were further heightened by the contrast between industry and agriculture. Germany's rapid industrialization coincided with a general European crisis in agriculture. By the 1890s all German farms were affected. Caprivi's trade policies simply accentuated the gulf. Small western peasants found they had something in common with eastern Junkers. The Bund der Landwirte appeared to try to serve as an umbrella agrarian party. Outside Prussia agrarian depression even stimulated or reawoke rural particularist sentiments, most prominently in Bavaria (Thraenhardt 1973: 91).

National government action, therefore, served to reaffirm the lack of integration. The Catholics, already regionally based, were, like the SPD, a social and political periphery. The Conservatives were perhaps more concerned with agricultural problems and the retention of Prussian ascriptive values and privileges than with the problems of the Reich as a whole: they were often in conflict with government policy. It was the Reich Party and the liberals that supported unification and subsequent centralization measures most ardently. But after 1870 German liberalism was weakened and fragmented by a conflict between its desire for national unity and its belief in a democratic parliamentary framework, and by the constitutional format of the Reich. It was the more conservative National Liberals that were probably closer than any other party to the political and constitutional spirit of the Reich. In this way, the questions of the territorial extent, political structure and styles of government provided one more link in a regional disaggregation of the Reich, already formalized in the party system.

Since the new state was federal, the elections to the parliaments of the various states were perhaps as important as those to the Reichstag, and none more so than elections to the Prussian assembly, since the Reich chancellor was always head of the Prussian government. The three-tier Prussian electoral system, introduced in 1849, unreformed through to 1918 and based upon status and wealth within localities, was a system geared to the preservation of elite influence; and it was through the Prussian assembly that the Conservatives were able to wield disproportionate influence. Since some four-fifths of the electorate would elect only one-third of the deputies, there was little incentive to participate. Abstention was particularly prominent in the western regions. In 1863, for example, it reached a massive 82 percent. Even in 1912, while turnout in the Reichstag election was 84.5 percent, in the Prussian election it was only 32.8 percent.[22] Aristocratic influence was preserved also through the local administrative apparatus. Except in the recent acquisitions of the nineteenth century, there was little in the way of provincial autonomy and local self-government in Prussia, despite legislation in the 1870s. The elite basis of local government remained much the same as in earlier centuries (cf. Muncy 1944; Jacob 1963; von Unruh 1966).

The lack of franchise standardization in 'regional' elections was symptomatic of the imbalance of the whole system. The range in the degree of representativeness spanned a full gamut from the 'liberal democracies' of Baden and Württemberg to unashamedly feudal structures in the Mecklenburgs. Bismarck had attempted some concessions to particularist sentiment and decentralization:

the states retained control over the police, education, judiciary, railways and several fields of taxation. But the very existence of a central bureaucracy meant an impetus towards centralization (cf. Windell 1969). The major problem was that of dualism: the possibility of a clash between Prussia and the Empire was never institutionally resolved (cf. Born 1967). Indeed, its potential increased with the continued growth of 'national' or imperial business and the expansion of a central bureaucracy responsible only to the chancellor and, ultimately, the emperor.

In such an institutional atmosphere, it was hardly likely that suspicions of Prussian aggrandisement would be lulled. Once again, institutional structures and practices helped to freeze regional alignments. Furthermore, government in Prussia itself and in the Reich remained firmly in the hands of the traditional elite. Of the 54 chancellors and secretaries of state, for example, over one-half came from Prussian East Elbia, while some two-thirds were titled (two-fifths by descent).[23] Not only, therefore, was there a dichotomy between Prussia and the rest: there was also conflict within Prussia itself; 'anti-Prussian' groups continued to agitate both without and within the state legislature. Only a man of Bismarck's ability could find his way through the institutional labyrinth. With his departure, and the active intervention of Wilhelm II, the shortcomings of the system became more and more obvious. Institutional confusion also preserved the regional basis of the party system. Region-specific parties emerged. Conservatives and liberals split into different parties partly on institutional issues. Even the more monolithic Zentrum and SPD were not immune. The Bavarian Zentrum was persistently more particularist than its more northern brethren; and only the active intervention of the Church prevented a party split over Caprivi's trade policies in the 1890s (Thraenhardt 1973: 82). Likewise, the SPD reflected differences between the two groups that merged to form the party in 1875. Lassalle's group had been essentially Prussian and northern in origin, and was oriented more towards unification (and hence Prussian dominance). By contrast, that headed by Liebknecht and Bebel was more non-Prussian (especially Saxon), and hostile not only to Bismarck, but also to Prussia (Morgan 1965: 3). For both parties, regional tensions would be partly institutionalized after 1918.

Despite the radical transformation from imperial authoritarianism to republic, 1918 was not an outstanding watershed from a regional perspective. Although some of the problems of the past were clarified, the tensions of the overall landscape were, if anything, increased. A stronger unitary element was added in the shape of the presidency. While the new constitution clarified the problem of dualism by spelling out in detail what was and what was not the business of the central government, Weimar abandoned much of the empire's federalism. The new Reichsrat still represented the states, which retained control of justice, education, social welfare and the police; but the intention at least was that, even in these fields, the centre would be a more positive integrating and standardizing force. In particular, Berlin gained almost complete control of taxation. The electoral basis of politics was also standardized, as the multitude of franchise systems of the past were swept away.

During the confused period and political vacuum immediately after the war, regionalist sentiments gained new momentum. Yet ultimately the pendulum swung more firmly in the direction of Berlin. For the two institutional pivots of the Weimar Republic were the popularly elected Reichstag and president. The cabinet, intended to be a link between these two bodies, in practice became dependent upon both and unable to resist demands emanating from either. The presidency, which was allocated widespread emergency powers, was almost a monarchical surrogate (Bracher 1960: Part I, Ch. 2). The Reichstag, through its ability to appoint and dismiss governments, enjoyed power without responsibility. As neither had executive responsibility, there was an inbuilt potential for a political vacuum at the locus of this more centralized system.

That the vacuum was realized was due ultimately to the survival of all the regional tensions of the past. An important new cleavage was attitudes towards the new regime and state boundaries. Germany might have lost its troublesome ethnic minorities, but substantial groups remained unreconciled to the trauma of defeat, the burdens of the peace treaty and the new political and territorial shape of Germany. An extremely liberal system of proportional representation did little to prevent parties, no matter how small, from seeking legislative representation with reasonable hopes of success. The electoral system not only preserved the multi-party system: it was also a positive incentive to its further fragmentation. The important point is that new alignments could not exorcise the past completely: they merely fed into existing partisan and regional alignments.

Proposals to rationalize the internal territorial organization of the Reich were rife in the 1920s: all were abortive. The Reichstag itself set up in 1920 the Zentralstelle für Gliederung des Reichs. But apart from achieving the merging of the ruritanian Thuringian states into a new 'Land' of Thuringia, this body accomplished little before its abolition in 1929. The constitution also permitted the holding of referenda upon territorial reorganization. Predictably, this was taken up in Hannover, still seeking a distinct Land status, in 1924, but failed to win the requisite level of support, partly because the SPD, relying upon its strong support in Hannover to retain its leading position within Prussia, campaigned against the proposal.

Because of the subsequent events of the 1930s, the political and economic problems of the Weimar Republic have been a perpetual source of discussion, and do not need reiterating here. Nevertheless, they were impregnated with regional perspectives. Regionalism helped to keep the DVP, DDP and DNVP apart: the latter remained an East Elbian party.[24] The Zentrum finally lost its more conservative and agrarian Bavarian wing, while much of the Saxon SPD heartlands passed to the KPD. The quasi-federal system in fact offered parties the possibility of behaving differently at the Land and central levels. Despite their tensions at the national level, the SPD and Zentrum formed the government of Prussia throughout the lifetime of the republic. Again, despite deep hostility between the national parties, the SPD and KPD in Saxony were close to each other: 'Red' Saxony was a constant source of tension in the early years of the republic.

Some regions, like Hannover, may have agitated for an internal reorganization of the Reich. In others, peasants and small farmers may have given their support to region-specific agrarian parties. Others, however, posed challenges that threatened the territorial integrity of the state itself. The utter failure of French-inspired separatist agitation in the Catholic Rhineland in the early 1920s parallelled separatist sentiments among the landed interests of East Prussia who, hostile to an SPD government in Berlin and disgruntled by the physical separation of the province from the rest of the Reich, toyed with the notion of establishing an East German state that could evade the territorial provisions of the Treaty of Versailles (cf. Turner 1963; Hertz-Eicherode 1969). But while all regions were disaffected with Berlin for one reason or another, it was particularistic Bavaria which moved firmly into the regional limelight in the 1920s. Bavaria covered the political extremes, from an abortive attempt to set up a soviet republic in 1919 to a futile effort to revive the monarchical idea just before the Nazi

Machtergreifung. In between, it was control by the conservative
BVP that posed the problem. Right-wing groups flourished under
its protective aegis, despite edicts from Berlin, and its Reichstag
representatives regarded themselves first and foremost as
spokesmen of the Bavarian government. In particular, in extracting
a special oath of allegiance from the Bavarian units of the army,
München posed a direct challenge to the authority of the German
state (Mitchell 1965; Schwend 1954, 1960; Schoenhoven 1972;
Thraenhardt 1973).

The flawed state

Perry Anderson has written:

> The former extreme reactionary Bismarck, once the truculent champion of ultra-
> legitimism, was the first political representative of the nobility to see that this
> burgeoning force [nationalism] could be accommodated in the structure of the
> State, and that under the aegis of the two possessing classes of the Hohenzollern
> realm — Prussia junkerdom and Rhenish capital — the unification of Germany
> was possible [Anderson 1975: 276].

This is precisely the point. Through diplomacy and war, disparate
forces were harnessed to the state, whether it be regarded as Germany or a greater Prussia. How disparate became more obvious
after Bismarck's fall. East and west, themselves a kaleidoscope of
cultural and economic variations, remained worlds apart. What
made the dilemma more poignant was that the German capital
became the stronghold of the SPD and, after 1918, also of the KPD
— equally abhorrent to western capitalist and eastern landowner.

Despite elite co-operation, the new state did not, or could not,
come to terms with very different political histories, patterns of
socialization and regional loyalties. Unification led to central
attempts at standardization in both cultural and economic affairs.
The former led, for example, to the Kulturkampf. Groups that
before 1870 had been favoured by their state found themselves in
direct competition with other regions after unification. One of the
forces that kept Hannoverian particularism alive was that, after
incorporation, its small farmers had to compete with the East Elbian estates, whereas previously their interests had been protected
by the Hannoverian government (Stehlin 1973: 99).

The political activation of cultural and economic differences,
and their regional bases, were sustained through both the party

system and the institutional structures. The imperial framework in a sense ignored all these problems. The federal system permitted their full expression, while possessing very little in the way of conciliatory mechanisms. Barred from active participation in government, the parties became encapsulated, tending to retreat back into themselves, emphasizing how different they were from each other. Roth's (1963: 8, 12) comment that the attempt by the SPD to create a distinct, insulated subculture, where the party would be 'home, fatherland, and religion' — essentially an exercise in 'negative integration' — was perhaps the ideal of other parties also, but only the SPD and Zentrum approached this ideal of organizational self-sufficiency. The polarized, fragmented multi-party system survived imperial collapse virtually unscathed. Now, however, the problem was different. Weimar introduced a more centralized framework without coming to grips with the lack of integration. With the dilu-

TABLE 3
Conflict dimensions and parties in the
Second Reich*

Cleavage	SPD	ZP	Mins.	Agr.	Reg.	Prog.	N.Lib.	Reich	Cons.
Culture									
Ethnic-linguistic		(R)	R			(C)	C	C	C
Religion		R	R		(R)	(C)	C	C	(R)
Economic									
Rural-urban	(C)			R	(R)	C	C	C	R
Rural class		(R)		R	(R)				C
Urban class	R					(R)	C	C	C
Social welfare	C	C				C	R	R	R
Political/territorial									
East vs. west		R		(R)	R	C	C	C	C
Prussia vs. rest	(R)	R	R	R	R	R	C	C	C
Traditional values	R	(C)	R	R	R	R	R	R	C
Centralization	(C)	R	R		R	C	C	C	(R)

* C = centralist; R = regionalist. Brackets indicate a subsidiary and more dubious ranking.

TABLE 4
Conflict dimensions and political parties in the Weimar Republic*

Cleavages	KPD	SPD	ZP	BVP	Agr./Reg.	DDP	DVP	DNVP
Culture								
Religion		C	R	R		C	C	C
Economic								
Rural-urban	C	C		(R)	R	C	C	R
Rural class		(R)		(R)	R			C
Urban class	R	R				(R)	C	C
Social welfare	C	C	C			C	R	R
Political/								
territorial								
East-west			R	R	R			C
Centralization	(C)	C	R	R	R	C	C	(R)
Regime support	R	C	C	(R)		C	RC	R

* C = centralist; R = regionalist. Brackets indicate a subsidiary and more dubious ranking.

tion of the federal element, regional discontent could be channelled primarily only through the parties.

Both parties and institutions, therefore, combined to ossify 'the many-sided sectional, ideological, religious and economic differences in the incompletely unified nation' (Loewenberg 1966: 12). Moreover, the absence of total coincidence of cleavage lines generated a bewildering array of alliance options: at times, the various parties settled down with seemingly strange bedfellows. Conservatives and Zentrum could combine in the defence of religious orthodoxy, whatever the denomination. The Zentrum, like most religious parties, possessed strong internal economic divisions. While it may have dropped its commitment to a grossdeutsch answer to unification, it persisted in a belief in federalism, and remained a rather alien element right up to 1933. Yet after the 1890s it had few qualms about lending support to imperial governments, which saw it as the major party bulwark against the SPD, while in 1919 it joined with the SPD and DDP to establish the more centralizing Weimar constitution. The combination of SPD and

Zentrum in a national government may have been increasingly fraught with difficulties, yet the two could govern Prussia together throughout Weimar. In the presidential election of 1925, the BVP could desert the Catholic nominee of the Zentrum and advise its supporters to vote for the conservative Hindenburg. What might appear as simple opportunism arose in fact from the irresistible logic of the parties' own ideologies and cohesive bases.

The point is that the several dimensions of cleavage did not necessarily overlap. It follows that the parties could vary their positions on the various cleavage lines. Tables 3 and 4 are a crude attempt to list the salient conflict dimensions, most having a regional dimension, and the placement of the parties according to a more 'centrist' or more 'regional' stance. It becomes easier to see how varying party alliances were possible, and also the irony of the two mass parties, Zentrum and SPD, at one and the same time peripheral and central to the regimes.

The Federal Republic

The results of the first election to the Bundestag in 1949 led many observers to believe that the old multi-party system was being re-established. Fourteen parties contested the election, seemingly with social and programmatic bases similar to those of the pre-Nazi parties. But the party system of 1949 proved to be extremely short-lived. The social and political changes wrought by the Nazis, the reaction to Nazism and the geopolitical consequences of military defeat combined in a development that, in just over a decade, produced a predominantly two-party system, with by 1961 only one small party preventing the two major parties from holding all parliamentary seats.[25] In 1928 the voting strength of the leading three parties in the territory of the future Federal Republic was 54.3 percent; even in 1949, the leading three parties gained 72.1 percent of the valid votes cast: but 20 years later, two parties alone garnered 88.8 percent of the vote.

There has been a dramatic coalescence of parties since 1949: only three of those that had emerged by 1949 survived through to the 1970s as contenders for parliamentary seats. Moreover, during this period only two new groups — the Refugee Party (GB/BHE) and the National Democrats (NPD) — arose briefly as serious political contenders. The radical transformation of the party landscape is vastly greater than that of the social structure. But since extensive economic change after 1870 did not significantly alter the overall nature of the earlier party system, it would be unrealistic to suppose that economic reasons alone can account for the difference between the Weimar and Federal Republic party systems. The crucial intervening variable was the political events of the post-1945 years.

Just as political events brought about the end of the imperial and Weimar party systems, so those of the 1945-49 period of military occupation were prominent in shaping the pattern of the West German party system. Only four parties that applied for licences to operate were given permission to organize and mobilize support in all four occupation zones. Two were the old working-class parties, SPD and KPD, and two were new formations. The Free Democratic Party (FDP) — whose actual title varied regionally — was a combination of the old liberal parties, while the Christian Democratic Union (CDU) was a union of practising Catholics (mostly associated with the old Zentrum) and Protestant conservatives. In the Rhineland there did arise a new Zentrum, based not on religious particularism but in reaction to a feared CDU conservative orientation. The Zentrum co-operated with the SPD, but was never more than a tiny fragment of the once-powerful force in the Rhineland. The Bavarian wing preferred to retain a separate organization and called itself the Christian Social Union (CSU). However, the two Christian parties can be regarded as one for most purposes, and this practice is followed here: the term CDU refers to both parties unless otherwise stipulated (cf. Wieck 1953; Heidenheimer 1960; Minzel 1975). Parties that hoped to appeal specifically to refugees were banned everywhere, as were potential successor parties to the NSDAP. Other prospective parties were hindered by different policies adopted by the various Occupation authorities.[26]

After the decision was taken to form a West German state, the policies adopted earlier by the Occupation authorities were decisive in that only the original four licensed parties were in a position to contest the 1949 election throughout the whole country (Loewenberg 1968; Urwin 1974: 126-31). This fortunate position provided three of these parties with an advantage that they never lost. By contrast, the possibility of KPD consolidation and expansion was rapidly terminated by the new geopolitical cleavage between the Federal and Democratic Republics. The forced amalgamation of the SPD and KPD in the Russian zone, along with the subordination of the other parties to Communist control, resulted in the West German KPD becoming a totally negligible force long before it was legally outlawed in 1956 (cf. Kluth 1959). The other three licensed parties found themselves operating in a political vacuum. They were able to appeal to the various cultural, economic and regional groups without facing unduly severe competition, and so to assimilate most of the political tendances that existed. Through their own organizational efforts and the tendency of any working party system to produce its own inertia, they proved capable of consolidating their position, successfully resisting challenges from other parties once the new state had been established and licensing abolished. In particular, the CDU expanded to become the first democratic party capable of collecting, under one umbrella and on a non-confessional basis, most of the hitherto fragmented groups to the right of the SPD.

Social structure: its regional distribution

Pre-1933 Germany had been characterized by a complex social structure differentiated along religious, urban – rural (or industrial – agrarian) and class lines, all possessing more or less prominent regional dimensions. These remained salient features of West German social structure after 1949. In addition, the role of the Elbe-Saale watershed as a distinguishable border between east and west was given a new twist with the establishment of two distinct German states.

In the new Federal Republic the two religious confessions became more evenly balanced. In 1961 just over 50 percent was formally classified as Protestant, while Catholics constituted almost 46 percent. Most Catholics are still clustered in the south and west, while the northern provinces are overwhelmingly Protestant. The unchanging nature of the spatial distribution of the two confessions across the centuries is apparent in Figure 9, which clearly shows the heritage of the Reformation and Augsburg and the earlier fragmented political situation, especially in the south-west. There has been a slight propensity towards Catholic growth, but the increase is negligible. Pronounced geographic mobility has led to an increased mix in the industrial conurbations, but this has not significantly affected the historical balance. More important, the high level of church membership is formal rather than significant, being required by law: the procedure for opting out of formal membership is cumbersome, and has been disapproved of by society. While church attendance remains quite high, postwar

FIGURE 9
Distribution of Catholics, 1961

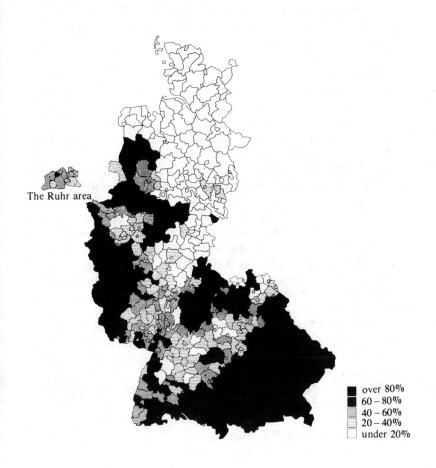

The Ruhr area

over 80%
60 – 80%
40 – 60%
20 – 40%
under 20%

secularization has contributed towards a decline in religious hostility or alienation: religious differences do not have the same degree of bitterness and suspicious hostility as in the past. Secularization has been doubly valuable in that the new religious balance between the confessions ended the status of Catholics as a minority. They were offered an opportunity to play a more positive role than that available to them between 1870 and 1933.

From the ruin of 1945 the so-called West German economic miracle has made the Federal Republic one of the dominant economic nations in the world (Wallich 1955; Arndt 1967). Economic prosperity was achieved more rapidly and was distributed among the population more equitably than before 1933, affording numerous opportunities for rapid social mobility and advancement. The bulk of the population is clustered in the western regions. The concentration of population in the north Rhine – Ruhr complex and along the Frankfurt – Stuttgart axis points to the continuing importance of the Rhine and affluents such as the Main and Neckar. The lateral axis running eastwards from the lower Rhine through Hannover is less densely populated, reflecting perhaps its severance at the new border between the two Germanies. The picture in the remainder of the country is one of rather isolated population concentrations. In particular, West Germany's two largest cities, Hamburg and München, are still very much city islands surrounded by a rural hinterland. Despite an increase in population after 1950, the rural population has declined still further: by the end of the 1960s, only one-fifth of the population lived in communities with under 2,000 inhabitants.

Two important trends emerge from a study of population changes between 1949 and 1969 (Figure 10). First, the growth of suburbs, dormitory areas and improved transport facilities has enabled a large number of rural and semi-rural residents to work in urban industrial centres. The most extensive population growth, constant over the years, has been in the immediate hinterlands of the cities. The latter, by contrast, have been static or even declining in size, despite the fact that in the early 1950s they were still growing from a decimated wartime level. München is unique in its substantial increase in size, for very few cities display even a modicum of growth. The expansion of the Bavarian capital is a reflection of the relatively greater population growth in southern Germany that marked the 1960s. Second, Figure 10 suggests a significant movement of population along an east – west axis: while the weight of the population was always in the western regions, the

FIGURE 10
Population change, 1949-1969

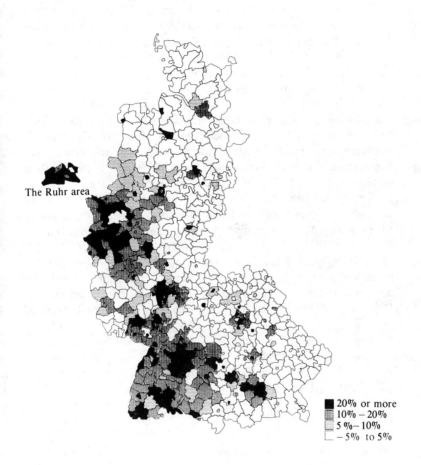

The Ruhr area

■ 20% or more
▨ 10% – 20%
▧ 5% – 10%
☐ – 5% to 5%

latter display a more substantial demographic growth than the eastern regions. The flow of population to the more resource-rich regions of the west, more marked in the 1950s, was due not only to the movement of refugees away from their original areas of settlement. The shift in the balance of population reflects the post-1945 geopolitical landscape. With few exceptions, localities ranged along the land borders with East Germany and Czechoslovakia show a very limited population growth. Some were poorer agricultural areas, relatively remote in communication terms with western centres; others were rumps of broader industrial belts, shut off from their natural markets and region across the border.

Economic growth gave West German society a character different from that of its predecessors. In 1961 the occupational structure of the Federal Republic, in percentages of the total labour force, was: self-employed middle class, 11; farmers, 9; salaried employees and officials, 30; manual workers, 48; farm labourers, 2. The pendulum, therefore, has swung even more decisively towards the secondary and tertiary sectors. Equally important is the point that West Germany before 1933 already had a much more balanced economic structure than East Germany. Not surprisingly, the geographical location of the proportion of the workforce employed in industry (Figure 11) corresponds fairly closely to population density. The North Rhine – Ruhr constellation, the Frankfurt – Stuttgart axis and the Protestant regions of Franconia are major industrial regions, as are the Saarland and the Hannover – Braunschweig axis. Each reflects a continuity with the past.

By contrast, the decline of agriculture has continued apace. Its share of the labour force declined from 22 percent in 1950 to under 10 percent by 1970. The agricultural sector is the only one that has not absorbed any part of the great population increase. Three major agricultural regions remain: the area around Oldenburg and Ostfriesland in the north-west, large tracts of Bavaria, and the Eifel region in the west. The continued contraction of agriculture has meant a reduction in the political influence of the farmer and a diminution in the political consequences of urban – rural differences; also, it has more or less removed completely the possibility of the agricultural sector generating a distinctive party of agrarian defence. On the other hand, agricultural contraction has occurred within a context of increased homogeneity. The western Occupation authorities did introduce land reforms after 1945: but these were only partially carried through.[27] In any case, they were perhaps hardly necessary, except in so far as they applied

FIGURE 11
Proportion of working population in industry, 1961

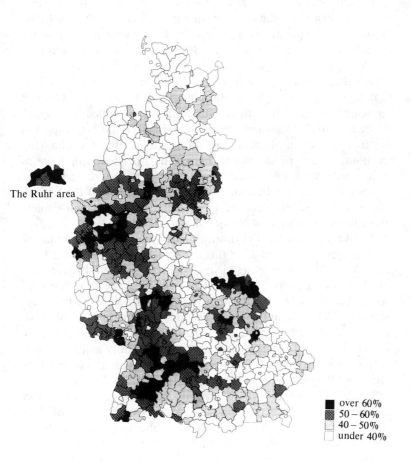

The Ruhr area

■ over 60%
▨ 50 – 60%
▢ 40 – 50%
□ under 40%

to land consolidation, still a serious problem in parts of the south-west. Large holdings of over 50 hectares are important only in Schleswig-Holstein and Niedersachsen — where, however, they constitute only a meagre proportion of landholdings. At the other extreme, dwarf holdings are common in the Saar and the south, but agriculture frequently is only a secondary source of income for their occupants (cf. Franklin 1969: Ch. 2). Most enterprises, however, are owner-operated: only 8 percent are rented by tenants. The pattern of occupation, the much narrower range in size of holdings and the style of farming have made West German agriculture a more integrated sector: this has enabled it to speak more effectively to governments, while in turn it has helped parties to treat agricultural questions more easily. But the changing economic structure has meant that political parties, if they are to enjoy electoral success, must appeal to some part of the urban and industrial sector. On the other hand, the expansion of the tertiary sector has limited the probability of electoral success coming solely from the mobilization of industrial workers.

Twelve years of Nazi authoritarianism and the vast population movements during and after the war weakened regional and local ties. The allied authorities dismantled the state of Prussia and created a new pattern of constituent 'Länder', a few of which were somewhat artificial. In the west, only Bavaria and the two city-states of Hamburg and Bremen retained a shape identifiable with that of the immediate past. The federal structure of West Germany has ensured the limited survival of regional identity and autonomy within an institutional framework that in its essentials corresponds closely to the pattern established in 1870.

There is, however, one significant difference. Before 1945 Berlin, because it was the political capital, was *the* centre of Germany. But the new geopolitical situation, the division of the Prussian capital, the unresolved constitutional status of West Berlin and its physical isolation from the Federal Republic — all have turned the city very much into a periphery of West Germany.[28] No single city has arisen to take the place of Berlin. Just as the latter was never quite as dominant as London or Paris, so the new political capital of Bonn has had no possibility of pre-empting the centrality of Berlin. The lack of any undisputed metropolitan centre to act as a political centre and a political and career magnet means that West Germany resembles more the historical pattern established during the Middle Ages. Indeed, West Berlin itself remains the largest German city. The next two in magnitude, Hamburg and München, lie at the longitudinal extremes of the state and away from the mainstream of West German life. Their demographic weight is more than offset by their geographical location, and by the host of substantial city populations in between, especially the conglomeration of cities on the lower Rhine and along the Ruhr.

The polycentric nature of the state can be confirmed by several indices. No single airport, for example, enjoys a predominance in air traffic. The largest, Frankfurt, carries only about one-quarter of all commercial flights and just over one-third of all passenger traffic. Newspaper ownership might have become more concentrated over the past three decades, but circulation is markedly regional. The comparative survey of polycephalic practices by Ahnström (1973: 65) looks at the location of headquarters of financial and industrial corporations around 1960, measuring the degree of uni- or multi-centrality in terms of the proportion of sales controlled by the city where the head office is based. In West Germany there is a clear-cut disjunction between political and economic centres. In fact, it possesses the lowest concentration coefficients of all West European states for industrial and commercial sales, bank deposits and loans and insurance income. West Germany, then, can almost be described as a structure of 'peripheries without a centre.'

At the same time the Federal Republic displays an ethnic-linguistic homogeneity even greater than that of the old Reich. The tiny Danish population in the north is the only officially recognized ethnic minority with some constitutional guarantees. Postwar economic chaos generated some agitation here, even among Germans, for border revisions, but this was dead by the early 1950s. The Danish party, the Südschleswig Wählerverband, which is exempt from the rigorous barrier clause of the electoral system, declined consistently from 1949 onwards. It had never possessed an unambiguous secessionist stance, and in the 1960s began to demote the defence of Danish culture and traditions in favour of an emphasis upon social welfare and economic development (cf. Jäckel 1959).

Quite apart from the problem of Berlin, the new territorial boundaries of 1945 had two profound effects upon West Germany. First, the establishment of a Communist regime in East Germany completed the isolation of the radical left of the Weimar Republic from the rest of society. With the two German states moving into opposing international alliances and associations, it became part of the international rather than the domestic cleavage structure. One major effect of the removal of the Communist opposition of prin-

ciple to a location external to the boundaries of the Federal
Republic was a decrease in the salience of other domestic cleavages.
A diffuse, yet intense, yearning for reunification joined with international tensions produced a desire to present the image of an
united society to the outside world.

Second, the division of the Reich produced a new social group in
the Federal Republic. Refugees and people expelled from Poland,
Czechoslovakia and, later, East Germany numbered well over 10
million. By the 1960s one citizen in five of the Federal Republic was
not a native of West Germany (cf. Lattimore 1974; Demoskopie
1964; Lemberg and Edding 1959). These refugees and expellees
were originally settled in rural areas, most especially in Schleswig-
Holstein and Niedersachsen, while the French, disclaiming
responsibility because France had not been a signatory of the
Potsdam Agreements, were extremely reluctant to permit too many
to settle in their zone. There were also geographical variations in
refugee settlement by region of origin. Those from the Sudetenland
were the largest group in most of Bavaria, North Würtemberg and
Hessen. The settlement, by Länder, of those from the ceded eastern
territories of the Reich is shown in Table 5. With little prospect of

TABLE 5
The settlement by Länder of refugees/expellees from
the Oder-Neisse territories
(in thousands)

Land	East Prussia/ Memel	Pomerania	Brandenburg	Silesia	Danzig
Schleswig-Holstein	314.9	307.1	20.1	59.7	70.1
Hamburg	36.5	26.5	3.0	21.5	8.3
Neidersachsen	423.6	264.7	47.2	722.4	49.4
Nordrhein-Westfalen	332.9	161.1	26.2	526.4	43.7
Bremen	14.1	8.5	1.2	12.6	3.4
Hessen	62.0	34.9	9.9	111.2	9.9
Rheinland-Pfalz	30.5	15.9	2.7	31.4	6.5
Baden-Württemberg	87.4	37.7	6.9	107.0	22.1
Bayern	93.2	34.8	13.8	461.1	11.4

Source: Göttingen Research Committee (1957: 23).

FIGURE 12
Distribution of refugees, 1950

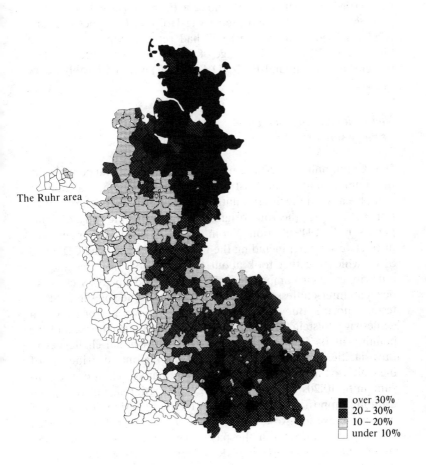

The Ruhr area

■ over 30%
■ 20 – 30%
▨ 10 – 20%
□ under 10%

employment in their original rural settlement areas, many later moved into the expanding industrial areas. Nevertheless, the basic geographical pattern of refugee settlement in 1950 (Figure 12) has remained unchanged in its broad outlines. The refugee population was settled, in relative terms, along a fairly consistent east – west gradient. In the late 1960s Schleswig-Holstein (32.1 percent) and Niedersachsen (29.6 percent) still had refugee populations much higher than the national average of 21 percent, while the Saar (3.8 percent) and Rheinland-Pfalz (11.8 percent) were noticeably below the average.

The social and regional basis of party support

If at a minimum we define a national party as one that contests all, or all but a handful, of constituencies in every region of a country, then, because of the discriminatory policies pursued by the Occupation authorities, the four original parties were the only national parties in the 1949 election. Yet in a sense all displayed a regionally distinctive support, including the fairly equally balanced CDU and SPD, which together took about 60 percent of the votes and far outstripped their competitors. The CDU lived up to its claim of being an inter-confessional party by winning some votes in the Protestant north. But apart from its impressive performance in Schleswig-Holstein, a lesser one in northern Württemberg and the inability of its CSU associate to overcome regional challenges in Bavaria, the most striking fact about the regional distribution of the CDU vote in 1949 (Figure 13) is its similarity to that of the Zentrum in the 1920s. In short, the CDU's strength relied heavily upon Catholic support, most especially in the more agricultural areas of the west and south-west. Similarly, SPD support in 1949 displayed a strong continuity with the past. The party was still not pre-eminent in the North Rhine – Ruhr complex. Its greatest area of strength was in the region to the south of Hannover, one of the most loyal SPD strongholds throughout the 1920s. However, in contrast to the CDU, which registered levels of support up to and over 80 percent in several kreise, nowhere did the SPD enjoy an undisputed hegemony. The geographical spread of the SPD vote depended, therefore, to a considerable extent upon the distribution of industry and population density, with one significant proviso. The religious factor still hindered severely the party's ability to mobilize support in Catholic Germany.

FIGURE 13
The CDU/CSU vote, 1949

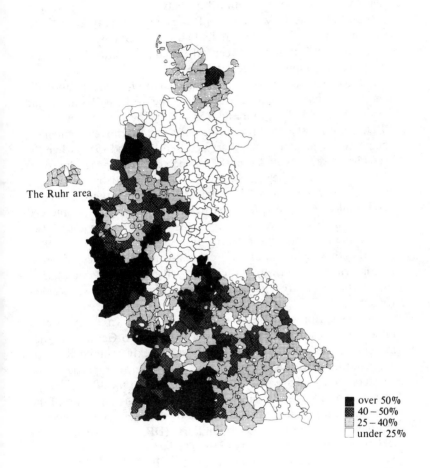

The Ruhr area

■ over 50%
▩ 40 – 50%
▒ 25 – 40%
☐ under 25%

The relationship between social structural variables and party support can be easily tested through the use of scatterplots. In 1949 there was a very strong relationship (correlation 0.60) between CDU support and percentage of Catholics by kreis. By contrast, there was no strong relationship between CDU support and agricultural areas: its weakness in Protestant Germany led to a correlation of 0.29. The SPD vote had an industrial basis, with a correlation of 0.51: that between SPD support and urbanization (measured by population density) was 0.31.

The geographical expansion of the CDU to become a nationally competitive party occurred only after the Federal Republic had been established. To a more limited extent, the same is true of the SPD. The critical period for both parties and for the realignment of the party system is the four years between 1949 and 1953, when the CDU increased its vote from 31 to 45 percent. Very generally, since 1949, CDU – SPD hegemony has been challenged by four broad types of party: region-specific parties, the FDP, a refugee party and a residual category of minor parties. Regionalist parties emerged again in the Federal Republic, if not as prolific as previously. But one of the most striking features of electoral change in the 1950s was the total elimination of regional-oriented parties, even as minor irritants, even within their own self-proclaimed boundaries: by the early 1960s regionalism as a party phenomenon was a spent force.

The Saar, for example, had developed its own party system under French administration. Upon re-entry into German politics in the mid-1950s, most Saar parties were quickly assimilated by or aligned themselves with the major West German parties. The resisters took just over 10 percent of the vote in the 1960 Landtag election, to sink the next time round to about 5 percent. Two of the three most important regional parties of the Weimar Republic re-emerged in new guises: the German Party (DP, and successor to the DHP) and the Bavarian Party (BP). The latter's major significance was its effect upon the CSU; in the late 1940s its challenge was deemed sufficiently serious to be advanced as a reason why the CSU should remain distinct from the CDU. But the BP never acquired any significant strength, even in its core region of Lower Bavaria: in Landtag elections its vote steadily declined from 17.2 percent in 1950 to only 1.3 percent in 1970. The vast increase in CDU support between 1949 and 1953 was to a large extent based upon extensive expansion in Bavaria. By the middle of the decade the BP was a spent force. The DP lingered on longer. However, its original aim — the abolition of Prussia and the recognition of Hannover — had been almost immediately recognized by the creation of the Land of Niedersachsen. The DP retained a core vote in the old Hannoverian provinces, but set out to turn itself into a national conservative party (Meyn 1965). The attempt failed, and the DP relied so heavily upon CDU benevolence that its opponents described it as 'the CDU's little sister' (Olzog 1964: 56). But not only did its attempt to turn itself into a national conservative party fail utterly: its efforts to do so distanced it somewhat from its origins and contributed to the almost complete erosion of its once impressive strength in parts of Niedersachsen.

The CDU was the only beneficiary of the demise of the BP. Niedersachsen, however, was one region where the FDP actually increased its share of the vote — at the expense of the DP. While technically a national party, the FDP has always displayed extremely marked regional variations: at the extreme, its vote in 1949 ranged from around 40 percent in a few kreise in north Hessen to under 5 percent throughout much of Bavaria. The party has, almost without exception, been confined to Protestant Germany (cf. Braunthal 1960). Four areas of relative FDP strength have persisted over time: northern Hessen (though here its vote suffered severe erosion), the Oldenburg – Ostfriesland area in the northwest, the eastern half of Rheinland – Pfalz, and northern Württemberg. These were all areas before 1933 of relative liberal strength or of resistance to incorporation into a national party system. In the latter region the FDP seems to have pre-empted the position of the third leading regional party of Weimar days, the Württemberg Farmers' and Winegrowers' League.

The possible desire to return to their homelands, and the danger of remaining an unassimilated and discontented group, has perhaps made the refugees and expellees a 'natural' target of party mobilization. This potential was certainly feared by the Allied authorities, who all refused to license such a party. Consequently, discounting the brief fling of the Economic Reconstruction in Bavaria in 1949, the first opportunity for such a party was the 1953 election. But this party, the GB/BHE, could mobilize only about one-quarter of its potential electorate. A great deal of legislation in the early 1950s had been designed to combat the plight of the refugees and assist their social integration. By the mid-1950s the refugees were already participating in the West German economic revival, and surveys indicated only a limited and declining willingness to return to their earlier homes in the east, should this become possible (cf. Kraus and Kurth 1957: 178–205; also Bluhm 1963). The GB/BHE remained an amorphous organization, unable to reconcile completely the interests of people with the diverse backgrounds of, for example, Pomerania or Silesia, and very flexible in its alliance options, although becoming increasingly conservative in the late 1950s. After a disappointing electoral performance in 1957, the GB/BHE also faded from the electoral scene (cf. Neumann 1968).

The number of minor parties has been legion: very few, however, have made much of an impact upon West German politics. We may refer only to the National Democrats (NPD), who came to prominence through winning representation in eight of the ten Länder legislatures between 1966 and 1968. The significance of the NPD was twofold. It was the first serious contender in national elections that claimed a link with, and was regarded by many as being a direct descendant of, the NSDAP.[29] Certainly, the distribution of its vote in 1969 reflected on a much more limited scale that of the NSDAP between 1930 and 1932. Second, the NPD has been the only new challenger to the two major parties to emerge since the early 1950s. Its success in Länder elections even led the CDU and SPD to consider the question of electoral reform and the introduction of a simple majority system in order to block possible NPD representation in the Bundestag. However, the party had already passed its zenith by 1969: it too subsequently collapsed.

The history of the NPD confirms the developments of the previous two decades: the growth and consolidation of a two-party system capable of resisting outside challenges. By 1969 the CDU and SPD were pre-eminent everywhere. Together, they had almost 90 percent of the vote, and their combined strength was remarkably consistent across the country: only in a handful of localities did it surpass 95 percent. The corollary is that third parties were squeezed everywhere. The few pockets of resistance where the two parties failed to collect at least 85 percent of the vote are shown in Figure 14. It can be seen that they are not distributed randomly. Four broad areas of third-party survival can be identified: Oldenburg – Ostfriesland, central and northern Hessen, the eastern Palatinate, and a broad region extending from northern Württemberg into the mid-Franconia area of Bavaria. While these are areas of relatively greater FDP strength, they are also areas where the NPD performance was above average.

The party system of West Germany today differs considerably from that of the 1920s: yet nevertheless it has a genealogical descent

FIGURE 14
Kreis where the combined CDU/SPD vote was 85% or less, 1969

The Ruhr area

from its predecessor. But while the high rate of change in the Weimar Republic reflected an unstable multi-party system in the throes of even further disintegration, the high rate of change after 1949 reflects greater stability, the disappearance of small parties and the coalescence of support behind the two largest parties. While areas of third-party concentration are still those that proved to be more immune to the larger parties before 1933, their contemporary resistance is much weaker and it is difficult to see how they can retain the potential to serve as catalysts of disintegration. Only the FDP has proved capable of surviving nationally as a continuous electoral force through to the 1970s. And even it has been caught up since the mid-1960s in a struggle to remain above the barrier clause in the electoral law.

Moreover, electoral turnout in the Federal Republic is very high, being exceeded in earlier periods only by that for the rather unique 1933 election. Nationally, the range has been from 78.5 percent in 1949 to 87.8 percent in 1957. In addition, the range in turnout across the kreise is very limited: in 1969, for example, it was from 73.3 to 94.0 percent. Yet the turnout figures for the majority of kreise were clustered closely around the national figure of 86.7 percent (see Table 6). There is a tendency for turnout to be slightly lower in the southern Länder, but the difference is not particularly significant, and is certainly much less than in the Weimar Republic. One important consequence of high and uniform turnout is that parties that wish to challenge the dominance of the CDU and SPD can do so only through a change in the direction of mobilization. The mobilization of the electorate has more or less reached saturation point. Very few untapped electoral resources remain available, and hence the possibility of sudden political change occurring through fluctuating electoral participation has been drastically reduced.

The major challenge to the established party system is more likely to come from within the two major parties. While the SPD has had its internal difficulties, the possibility is much greater in the overtly confederal CDU, most specifically with its 'sister party', the Bavarian CSU. While relationships between the two have not always been amicable, the CSU very soon distanced itself from the stance of its BVP predecessor. The decline of a BP alternative and the shock of losing control of the Land government to a grand coalition of opponents in 1954 led to a revolution in party organization and a diminution in the appeal to Catholic conservatism and Bavarian cultural particularism.[30] The confederal argument of the

TABLE 6
Kreise turnout by Land, 1969
(absolute figures)

Land	Turnout			
	Under 80%	80-85%	85-90%	over 90%
Schleswig-Holstein	—	5	16	—
Hamburg	—	—	1	—
Neidersachsen	2	13	42	18
Bremen	—	1	1	—
Nordrhein-Westfalen	—	10	64	20
Hessen	—	5	26	17
Rheinland-Pfalz	—	9	39	3
Saar	—	1	3	4
Baden-Württemberg	3	28	41	10
Bayern	4	59	118	10
Federal Republic	9 (1.6%)	131 (23.3%)	351 (62.3%)	72 (12.8%)

BVP may have been dropped, but the CSU has been careful to insist upon its separate identity. Personal ambitions of the CSU elite may have been more significant in the increased tension of the 1970s, which led to a temporary dissolution of the 'Fraktionsgemeinschaft' between CDU and CSU in 1976, for there was no reference to the necessity to protect specifically Bavarian interests or values. Since then, the threat by the CSU to compete nationally has hung like a sword of Damocles over the CDU. But this is not a new territorial threat to the party system: a national CSU would not be a peculiarly Bavarian party. Its success to date has depended upon its almost perpetual control of government in Bavaria. To be sure, a national CSU might well make severe inroads into the support of the CDU (though surveys indicate a rather limited 'secession'). But the latter, even if having to start from scratch, would not be excluded entirely from Bavaria. At least one survey has indicated that a national CSU would lose as much as a majority of its support in Bavaria if faced with competition from a Bavarian wing of the CDU (*Der Spiegel* 1976: 28). In such a situation, perhaps only the SPD would benefit. Such scenarios are, however, only speculative. But action of the kind threatened by the CSU constitutes the most probable path that disruption of the two-party system would take.

The expansion of the two major parties beyond their traditional core groups and regions has meant that they have become less ideologically and socially cohesive. The CDU cannot be characterized simply as a Catholic party: nor can the SPD be said to be nothing more than a party of workers. Catholics constitute only about 60 percent of CDU support, and regular churchgoers about 60 percent. The corollary is that the SPD is not a purely Protestant or secular party. More than one-third of SPD support comes from Catholics, and about one-quarter from regular churchgoers. Similarly, one-third of CDU support comes from workers while only 60 percent of SPD support comes from this occupational category.[31]

It is most appropriate, therefore, to characterize both as heterogeneous parties, since in comparative terms neither relies exclusively or overwhelmingly upon either religion or occupational categories for its support. Both have had to face the problem of reconciling the interests of their traditional core with those of 'new' followers, a problem that has been more acute in the SPD than in the openly confederal CDU. The changes in the party system have not abolished all differences in the social bases of support. Multivariate analyses of survey data stress the importance of organizational indicators that relate to the degree of commitment to a religion or class (Urwin 1974: 158-62). Similarly, the core areas of CDU and SPD strength tend to remain those where their Weimar predecessors had their strongholds. The CDU still gains its highest voting support in predominantly Catholic rural areas: at a minimum it has been able to win over 60 percent of the vote only in predominantly Catholic areas. Conversely, however, there have been no areas of pronounced CDU weakness since 1953. By contrast, there have been no overwhelmingly SPD areas, and the party has failed to surmount its marked weakness in the rural regions of the Catholic south and west.

These basic regional facts of political life in West Germany can be illustrated by referring to the 1961 election results and the distribution of social structure variables drawn from the 1961 census. If we correlate the SPD vote with the proportion of the industrial workforce by kreis, even allowing for the wider diffusion of industrialization and the considerable increase in SPD strength since 1949, the distribution is remarkably similar to that for 1949: in fact, the correlation was exactly the same, 0.51. The relationship between SPD vote and degree of urbanization again gave a lower correlation of 0.37.

Turning to the social bases of CDU support, we find more impressive relationships. In 1949 CDU penetration of agricultural areas had been circumscribed by the success of regionally based parties, most especially in Bavaria and Niedersachsen. By 1961 these had withered to almost nothing. Given the seeming inability of the SPD to attract a large degree of rural support, the way was open for the CDU to rally agricultural Germany behind its banner. There is a discernible relationship between agriculture and CDU support of 0.51, even when setting aside the cities. Almost without exception, the CDU occupies a majority position in the most agricultural kreise. Just as with agriculture, the simplification of the party system after 1949 led to a consolidation of the CDU position in Catholic regions: the correlation between CDU vote in 1961 and proportion of Catholics in 1961 was very high, 0.83.

These analyses confirm the continuity in German electoral politics, with the spatial distribution of social structure variables as a basis of regional variations in party support. The point can be made rather differently by raising the question of the interrelationship between economic and religious characteristics. This can most appropriately be measured by multiple regression analysis. First, from the correlation distributions, we can calculate the residuals, that is the difference between the real placement of each kreis and its location predicted by the overall pattern of the distribution. The predictive value of the models is held to be high for residuals lying between $+5$ and -5, less satisfactory for those between $+5$ and $+15$, and low where the residuals are greater than $+15$. The geographical distribution of the residual values can be plotted cartographically.

Figure 15 shows the residuals for the 1961 CDU vote predicted by the distribution of Catholics. In the far north the model tends to underestimate the CDU vote. By contrast, in the area south of München and in a broad swathe running southwards from Hannover to the Saar, it overestimates the CDU vote. The predictive power of the model is high in many of the mixed religious areas, especially in industrialized parts of North Rhine – Westphalia and much of Baden – Württemberg. The impact of urbanization upon religion can also be discerned, especially in Bavaria, where the city residuals are different from those of the surrounding country districts. The power of religion emerges even more clearly in the residuals from the 1961 SPD vote predicted by the the percentage of the workforce employed in industry (Figure 16). The predictive power of the model is lower in that the geographical areas where

FIGURE 15
Residuals for CDU vote, 1961, predicted by
proportion of Catholics

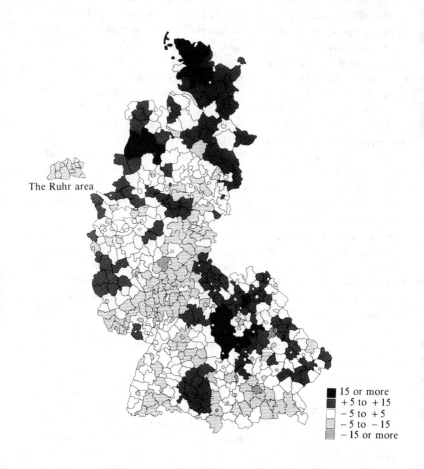

The Ruhr area

■ 15 or more
▨ +5 to +15
□ −5 to +5
▨ −5 to −15
▨ −15 or more

the residuals lie between $+5$ and -5 are smaller. But this seems to be precisely because of the religious factor. Figure 16 indicates a fairly consistent tendency for the model to underestimate the SPD vote throughout Protestant Germany, and to overestimate it in Catholic Germany.

For both parties, proportion of Catholics and proportion in industry were combined in a multiple regression analysis. The results provide a strong confirmation of the power of religion and economics in West German electoral behaviour. The correlation (multiple R) for the CDU was 0.86, with an explained variance (R^2) of 0.74. The results for the SPD were a correlation of 0.80 and an explained variance of 0.64. The residuals from these multiple regression analyses also refine the spatial distributions displayed in Figures 15 and 16. Mapping the CDU residuals removed most of the strong regional effects, although the tendencies towards under- and over-prediction persisted. Only one region of gross 'misprediction' remained: the Protestant enclave of Middle Franconia. Similarly, a revised SPD map would show a lessening of the variations in Figure 16: indeed, there were no regions of strong under-prediction. However, the multiple regression residuals outline one region of SPD over-prediction not present in Figure 16: precisely the same Protestant area of Middle Franconia.

In conclusion, therefore, aggregate analysis substantiates survey findings on the crucial importance of religion and economics for West German voting behaviour. More specifically, the multiple regression analyses show how the two have penetrated almost the whole country, leaving little room for regional variations caused by other factors. Just as the party system has been 'synthesized', so in a sense the social bases of party support have been regionally integrated. Only one deviant region stands out clearly. Middle Franconia is an area of Protestant rural conservatism: the only area west of the Elbe wholeheartedly to support the DNVP in the 1920s, it has continued to preserve its political distinctiveness. On the other hand, its weight in the West German structure is so minimal as to limit its potential for generating, by itself, any form of regional disruption.

West Germany, then, displays marked regional variations in the distribution of social structure characteristics that generate significant political consequences. Their pattern derives from developments that occurred several centuries earlier, with religion as probably the most significant factor in determining the regional shape of West German politics. The most outstanding change is the

FIGURE 16
Residuals for SPD vote, 1961, predicted by
proportion of workforce in industry

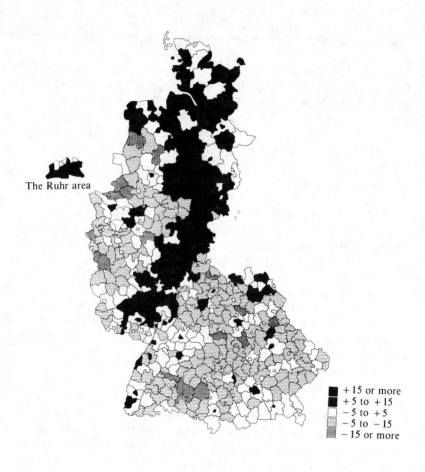

The Ruhr area

■ + 15 or more
■ + 5 to + 15
□ − 5 to + 5
▨ − 5 to − 15
▨ − 15 or more

metamorphosis in the party system: the spread of two heterogeneous parties to blanket almost all the country has blurred
regional differences, and their political prominence. Those areas
that have more successfully resisted accommodation in a catchall
party system are too few in number, limited in area and deficient in
resources to mount an effective challenge against the incorporating
parties. The CDU and SPD have proved themselves capable of
combining the role of a mass majority-oriented party with that of
the defender of subcultural interests. To that extent, regionalism is
no longer an effective factor in inter-party conflict. The obverse is
that regional strains can emerge as intra-party conflict within the
blanketing parties: the problem of the CSU has already been
discussed. A more comprehensive picture can be gained by studying
how the institutional framework has eased the path of conflict
resolution both between and within parties, and across territory.

Regionalism and institutional accommodation

The federal structure of West Germany rests upon the Länder
units, set up hastily by the Allies in their concern with decentralization and the destruction of Prussia. Despite the fact that military
and logistic requirements, rather than political links and
administrative convenience, determined the outline of the new
units, their boundaries are not too far removed from historical
realities. Some are direct descendants from the past; others are ag-
gregations of recognizable past regional entities. They are,
however, very different in their size, economy and population,
which creates obstacles to co-operative efforts in many fields. This
disparity has given rise to an ongoing debate about territorial adjustments in order to achieve a more equitable regional balance of
resources, especially since the Länder are constitutionally vested
with a considerable degree of autonomy (cf. Wells 1961; Laufer
1974).

In the early 1950s an investigatory commission explored the
possibilities of reorganization, but by then the inertia of the system
was probably too great to overcome. The Länder themselves were
opposed to reorganization, preferring to delay any changes until
after reunification. Most problems involved minor adjustments to
border regions. The area most affected has been what is perhaps
the most artificial Land of Rheinland – Pfalz: referenda in 1956
that achieved the necessary (but low) level of participation in-

dicated that the Trier and Koblenz bezirke wished to be reunited with the rest of the old Prussian Rhine province in North Rhine – Westphalia, and that the Montabaur and Rhein – Hessen areas wanted reunification with Hessen. In the old Palatinate itself, insufficient voters turned out to vote on the qustion of possible union with either Bavaria or Baden – Württemberg. Elsewhere, referenda indicated a preference for the re-establishment of Oldenburg and tiny Schaumburg – Lippe out of the Land of Niedersachsen, while there was also pressure for a separate Land of Baden. No action was taken after these referenda, and the regional discontent, if any, certainly did not affect the party and institutional balance of the country. In 1973 the, if anything, increased disparities between the states led to the reopening of the debate and the establishment of an expert commission which proposed the practicality of creating five or six more equally balanced Länder. New referenda in 1975, however, indicated the obverse: there was still some support for the disaggregation of Niedersachsen, while in Rheinland – Pfalz there was a rejection of change. Again, no further decision has been taken: the longer the delay, the more unlikely reform will become. In any case, popular prejudices seem to be tied more to historical memories and smaller units than to 'super Länder'. Furthermore, the status quo contains many vested interests, including those of the Länder themselves, and their active consent to change is very necessary.

One of the most satisfactory definitions of federalism (Riker 1974: 101) is of a state that divides governmental activities between the centre and regional units in such a way that each has the right to make the final decision in at least some fields of activity. Implicit in the definition is the understanding that the contents of the decisions allocated to lower-level units are spelled out in detail. In West Germany the federal principle is an inviolable part of the Basic Law. The central government may legislate in all fields except cultural and educational affairs, and has the right to lay down guidelines for the regulation of field administration. However, it has only a small field bureaucracy of its own, and must rely upon Länder bureaucracies for policy implementation (cf. Jacob 1963). While there is obviously some administrative overlap between federation and Land, their respective spheres of operation are constitutionally spelled out in detail. Most important is the division of revenues: some belong to the centre, some to the Länder, while others are divided between the two in a variable relationship that is adjusted periodically. Furthermore, the poorer Länder benefit from a process of revenue adjustment intended to compensate for differences in size and wealth, and to equalize the provision and standard of services across the Länder.

Opponents of the federal structure criticize the heavy duplication of bureaucratic structures; proponents defend it as a powerful barrier against over-centralization and a means of catering to regional sentiments. Certainly, the Länder have considerable autonomy even within their limited legislative competence, for the popularly elected Bundestag has authority only over national affairs. Within their constitutional confines the Länder have full control over their own affairs, and each possesses a complete executive and legislative structure. Moreover, with the exception of Schleswig-Holstein (which uses federal agencies), the Länder possess their own constitutional courts to determine issues within Länder jurisdiction. Not only do the Länder have control over education, police and mass media; they also have had the major responsibility for regional planning within their own territory. The Länder administer not only their own laws, but also federal laws on activities within their own jurisdiction, and federal laws (for example, environmental legislation) for activities that fall within the competence of the centre. The Länder therefore are the major blocks of the West German system, capable of balancing standardization with regional distinctiveness.[32] Conversely, they stand as a shield between regional sentiments and the centre, capable of blocking or resolving regionally emotive issues before the latter enter and possibly disrupt the national arena.

Some reforms occurred in 1969 with federal agencies moving to take on a larger role in planning and administration of problems that went beyond individual Länder boundaries. However, reforms can be initiated only with the consent of the Länder. Cross-Land problems are a joint responsibility, and not only that of the centre. For if the Länder are the building blocks of the system, the keystone is the upper chamber of the national legislature, the Bundesrat. The Bundesrat not only links the regions into the central decision-making process; it also serves to retain their independence. Regional power is most clearly expressed in the Bundesrat, with its virtually absolute veto on most legislation affecting Land-Bund relations, and a general suspensory veto on other legislation (cf. Neunreither 1959; Ziller 1966). Voting patterns in Länder elections may be the same as in federal elections, and the same parties may contest both: yet neither has generated absolute conformity, even though the staggered pattern of Länder elections turns each into at once a national and a regional election (cf. Culver 1966). Coalition rather than single-party government has been the rule rather than the exception in the Länder, and the expression of partisan preferences in the Bundesrat is of necessity muted by the fact that a Land delegation must vote en bloc: in addition, party coalitions have varied from those at the national level. Moreover, the alternative regional structure of representation has provided an additional channel of political career advancement, weakening the cohesiveness of party organizational coherence, as well as permitting governmental experience to accrue to those parties out of power in Bonn.

One further point distinguishes the Bundesrat from its predecessors: Land representation is not based solely upon demographic considerations. With the membership per Land fixed at a minimum of three to a maximum of five, there can be no repeat of Prussian hegemony, and it is theoretically possible for the smaller states to outvote the most populous. The Bundesrat, in short, has served to dilute both institutional and party centralization — to plug the Länder into central decision-making processes, and to ensure that federal government must take Länder perspectives into account at all stages of the decision-making process.

Federal encroachment upon Länder autonomy has been a feature of political life since 1949, but constitutional restraints have meant that this has come about through collaborative efforts rather than the abrogation of regional rights. The dominant social market philosophy might have stimulated economic growth, but it became increasingly obvious that by itself it could not eliminate or rectify regional imbalances in the system. Recognition led to the ratification by both centre and Länder of a federal regional planning programme in 1975.

The prevalent social market economy, memories of the past and the looming presence of the Communist neighbour to the east combined to demote regional planning as a desirable policy, at both national and regional level. This is not to say that planning with a regional perspective did not exist. There has been much discussion on setting up workable and functional planning regions, coordinated with West Germany's position and membership in international organizations. Some of the first regional problems, related to the huge refugee population, were tackled by legislation in the early 1950s to spread the financial burden of assimilation across the whole country. Beginning in 1953, there was also a listing of temporary redevelopment areas where social market forces alone could not eliminate general economic weakness. While the centre was constitutionally empowered to furnish regional policy guidelines, the responsibility for both legislation and execution lay with the Länder — and their response was variable.

By the 1960s, however, there was a growing recognition that the Länder shared common problems: the press of economic growth areas, and their increasing distance from poorer areas. A new emphasis was given to promoting economic growth and to replacing blanket aid with the channelling of resources through a network of central places designed to serve as growth poles. Throughout the whole period, particular attention was paid to the regions lined along the long border with East Germany and Czechoslovakia: the Zonenrandgebiet has been a major recipient of aid, though not entirely successfully. However, not until 1965, with legislation that laid down the planning responsibilities of centre, Länder and localities, do we approach the extensive implementation of a 1954 Federal Constitutional Court ruling that 'planning cannot stop at Land boundaries'. Existing programmes and areas were integrated in 1969-70 into larger regional action programmes: for virtually the first time, some of these cut across Länder boundaries. The final development was the Bundesraumordnungsprogramm of 1975 which, through establishing 38 functional planning regions, was designed to cement Land-Bund and inter-Land collaboration more firmly (cf. Hallett 1973; Blacksell 1975; Schmidt 1978; Jochimsen and Treuner 1974).

The institutional framework of West Germany is a direct descendant of past regimes, only slightly modified by the incorporation of Occupation innovations between 1945 and 1949. But in contrast to the past, the pendulum of power has swung further away from the political centre. Not only is Bonn perhaps weaker than any previous regime: the Länder themselves have less control over local activities than in the past. The focal political point in the Empire was the duality between chancellor and emperor, in the Weimar Republic that between president and Reichstag. In West Germany, however, it is that between chancellor and Bundesrat. The latter's dominant role has turned it into the effective centre of the institutional system. Regionalism, therefore, is inherent in the West German system. Regional interests enjoy a protected position at the same time as being able to influence federal policy. Regional issues can be resolved at the regional level: there is no necessity for them to disturb the national climate. By the same token, Bonn has found it difficult to co-ordinate federal policy. It is not insignificant that national direction of and responsibility for regional planning became a slogan in the Federal Republic later than in most Western European countries.

But the increasing activism of Bonn and growth of the federal bureaucracy has not entirely offset the weight of regionalism. For the federal structure itself is geographically decentralized. Polycephality and administration intermesh in the diffusion of federal functions. Karlsruhe, for example, houses the Federal Constitutional Court and a host of legal functions. The financial centre of Frankfurt has been strengthened by being the home of the Bundesbank. The Federal Tax Court is located in München, and the Federal Administrative Court in remote West Berlin. Subordinate agencies of ministries in Bonn have also been diffused. The Federal Statistical Office is in Wiesbaden, while the Unemployment Insurance Authority in Nürnberg enjoys de facto almost complete autonomy from its superior, the Ministry of Labour, on the Rhine. Hence there is no single centre of bureaucratic dominance. The federal bureaucracy is decentralized, as well as having to share functions with ten regional political capitals. Furthermore, the federal structure and the differential opportunities for governmental participation and influence that it offers have worked upon the political parties and the ambitions of politicians. Even in the centralized SPD there is not entirely congruity between party headquarters and the Land organizations.

Conclusion: a regional synthesis?

In the 1970s the interconnections of economic structures with questions of territory and identity as a factor generating political strains in the Western world were usually associated with undisputedly geographic peripheries in the state system, and most specifically with ethnic – linguistic regions and nationalist/separatist parties. On this count West Germany does not qualify for inclusion in a comparative study of this problem. Yet territory and identity have been central elements in the 'German problem' in terms of both its internal stability and its place in the inter-state system. Moreover, West Germany can be regarded as a test case in any study of geographic politics in the 1970s. It is one of the few countries where such protest movements have not appeared — indeed, where there has been hardly any disturbance of the established party system.

Until 1870 Germany was no more than a geographical expression. The geomorphological structure of German territory comprised three interlinked contexts: geopolitical, with, broadly speaking, a monocephalic east juxtaposed against a polycephalic west; geoeconomic, where a labour-intensive agricultural north-east faced a more highly commercial, economically advanced south-west; and geocultural, where religion roughly divided Germany along a north – south axis. These gradients held within a space that had no well-defined boundary to the east, nor perhaps to the west. The German territories straddled both the city belt of Europe and the more sparsely populated areas to the east. They were, moreover, contained within a European geomorphology which saw them eventually surrounded by strong centralizing states. The absence of geographical barriers to military mobility, and tensions between the expanding state centres, all helped to preserve the shatter belt of German principalities and delay unification. Just as the international climate of the sixteenth and seventeenth centuries worked against an expanding German state centre, so the military conflicts of the nineteenth century cleared the way for unification under the centre of Prussia by progressively removing from contention the alternative centre of Austria, and weakening the major opponent, France.

Prussian unification did not proceed in a vacuum. All German states had experienced consolidations of one form or another, albeit on a miniscale. Prussia itself had not fully completed the digestion of its more recent acquisitions. Germany in 1870 was as much a mosaic of political variations as it had been several centuries earlier. The new state had to bear all the fracture planes accumulated over the centuries. From being 'international' in character, they became regional differences: the empire was merely superimposed on top of them. Political life, as expressed through the party system, had to accommodate itself to both centralization pressures and the mosaic. But the parties, barred from active participation and responsibility in the centre, turned instead to emphasize and reinforce the prevailing differences. The latter, too, were preserved by economic developments; and Berlin, the symbol in the eyes of the world of Prussian values, became totally distinct in political life from old Prussia.

In this sense there was no revolution in 1918. Weimar failed to come to terms with the dilemmas of centralization, regional non-integration and the crippling geographical and symbolic weight of Prussia. Weimar's solution was to offer a more centralized state, but it was one that tended to ignore the centrifugal forces at work in the society. Tensions were compounded by the fact that the new regime was regarded by many as illegitimate, especially by those who had provided the bonds of the old Empire. The parties remained frozen in an ideological mould with regional perspectives. If they tried to break out of the mould, the usual result was further fracturing along regional planes. The Third Reich was perhaps not inevitable, but this kind of ruthless imposition of centralization — a kind of culture shock — was perhaps one of the few approaches that, at least superficially, could impose some degree of order.

Just as international events combined to enable German unification under monocephalic Prussia, so war and its aftermath served to create two German states and delineate the outline of the Federal Republic. In a sense, some of the problems of the past were resolved by separating the monocephalic Prussian core from the polycephalic west. In addition, while hopes of reunification might linger on, the progressive shrinking of German territory in the twentieth century has removed the extremes of ethnic – linguistic minorities, although the state was less vulnerable to their pressures than to those of intra-German differences.

The Federal Republic, therefore, is a country without a centre. This is reflected in a structure that is not dominated by a single unit. Federalism has swung back to the landscape of 1870, but one that lacks the massive presence of Prussia. The political centre can operate only with the active co-operation of the Länder, which have a voice in virtually all aspects of federal decision-making. Widespread autonomy with democratic restraints is the slogan in both politics and administration. What is radically different from the past is the party system — in both its structural outlines and ideological attitudes.

It is perhaps ironic that the two parties, or tendances, that survived all the twists of the jagged curve of history with some semblance of organizational cohesion and mass support were those that had been true political peripheries in the Empire — ideologically, economically and culturally remote from old Prussia. In a sense, their survival represents a peripheral take-over of the centre, one that was not possible in Weimar, which had absorbed too much of the imperial atmosphere, where strong antagonisms remained between the two and where both were exposed to the challenges of other political movements seeking to mobilize the same support. In the 1950s first the CDU, and then the SPD, spread out from their old bases to capture almost the whole of the electorate. They became mass heterogeneous parties that could no longer survive on only Catholic or working-class votes. Both are integrationist, having to balance many conflicting claims, a task that has been aided by the neighbouring presence of the Democratic Republic. The latter has provided an integrative stimulus that was not provided earlier by the 'outside' presence of Prussia in the seventeenth and eighteenth centuries. Again, in contrast to the past, it is the parties that have penetrated the institutional interstices and operate the federal structure.

The combination of party integration and institutional decentralization characterizes West Germany, working to produce a centre – region balance. Regional or peripheral protests are not only about discrimination by the centre, but also about the lack of control over their own destiny. The latter are certainly catered for by the West German institutional structure. Discrimination and the lack of resources have been further met by ongoing territorial redistribution programmes at both the federal and Land levels. Compared with Weimar, the high level of economic prosperity has enabled redistributive processes to be initiated and fairly successful. It may be that future demands will increase the strains on the system, but the latter is better suited than most for the accommodation and diffusion of strains.

The ethnic – linguistic minorities have gone. Religious tension has declined. In both comparative and historical terms, there are no great economic differences. What, then, is left? West Germany's outstanding periphery is Bavaria, which certainly has the resources — territory, economy, identity — to pose separatist demands. But in contrast to the old commitment to Grossdeutschtum and the particularism of the BVP in the 1920s, there has been very little territorial content in the contemporary political expression of Bavarian particularism. One must retain a certain degree of scepticism about strong territorial tensions in the future. First, West Germany's equivalent of a flight to the sun-belt may reduce Bavarian homogeneity. In addition, its increasing economic weight still cannot match that of the North Rhine. Finally, the CSU's strength rests in its Bavarian role and the alliance with the CDU. Discontent with the existing situation has manifested itself in dreams of becoming a nation-wide West German party rather than in claims for independence. But the CSU must weigh the possible gain of electoral expansion against the probable erosion of its core support by a competitive CDU. The arguments must remain speculative, but survey evidence, at least, suggests that the latter will more than cancel out the former.

The varying regional bases of religious and economic structures still operate to differentiate the parties in their social cohesion. But neither religion nor region activate powerful political debates. Both parties have sought the middle ground. Both, moreover, have to a considerable extent been bound to consensus by the modern geopolitical landscape, despite the furore over Brandt's Ostpolitik. The major debate has been over economic issues, and regional differences too have been over economic questions — despite, or perhaps including, the stance of the CSU. Economic differences may be successfully resolved by the adroit use of money. Economic prosperity since 1950 has meant that West Germany has had the financial resources to dampen or eliminate regional problems. Nothing is constant, and West Germany may again experience regional tensions. The latter perhaps are less likely as long as economic prosperity lasts. Even so, the latest territorial solution to the 'German problem' and the inertia of any working political system suggest that the Federal Republic has achieved a satisfactory level of regional accommodation.

NOTES

1. For administrative purposes the League came to be divided into four 'circles', each centred upon a water route. Lübeck led the Wendish circle, which extended along the Baltic as far as Stettin. Further east lay the East Baltic circle, embracing Prussia and Livonia, using the Vistula and other rivers to penetrate far inland. The remaining two circles centred around the major rivers of western Germany. The Saxon circle basically covered the drainage area of the Elbe as far inland as Leipzig, and incorporated many of the towns of Brandenburg (including Berlin), while the Rhineland circle had its centre at Köln, which served as the major Hansa link with the cities further south.

2. A further severe blow was dealt by the loss through the decline of the north Italian cities of the southern pole of the trade axis. By the 1550s southern Germany and northern Italy were desperately competing in each other's territory, to the benefit of neither.

3. The aggressive military posture of Calvinist Heidelberg did, in fact, result in just this: it was the catalyst of the Thirty Years' War (Clasen 1963).

4. The Low German areas originally coincided more or less exactly with the sphere of Hanseatic influence.

5. Cf. Bendix (1964: 216). For a brief review of the contemporary Lusation Sorbs of the Dresden and Cottbus regions of the German Democratic Republic, cf. Stephens (1976: 403-18).

6. Cf. Strauss (1972: 271-92) for an account of how concepts of Roman Law consolidated themselves in Germany.

7. In fact, three of the four western Electors of the Empire were archbishoprics (Köln, Mainz and Trier).

8. Baden and Württemberg were suspicious of Bavarian designs and demands for special privileges: cf. Simon (1969: 35).

9. Nineteenth-century economic developments are discussed in Kitchen (1978); Böhme (1966); Borchardt (1972).

10. It should also be pointed out that, at the same time, capital investment by the middle classes had led by the 1850s to some 45 percent of the Rittergüter in the East Elbian provinces passing to non-aristocratic owners (cf. Wunderlich 1961: 10; Simon 1971: 104; Gillis 1968: 113).

11. The Masurian Slavs in the Allenstein district of East Prussia were Protestant and did not regard themselves as Poles: the same could probably be said of the East Elbian Sorbs and Wends. Alsace and Lorraine are interesting cases, the former in particular being a strong German-speaking area: cf. Silverman (1972: 75).

12. Cf. the excellent study by Hertz-Eichenrode (1969).

13. The background and structure of these parties is well documented. For the Bund, cf. Tirrell (1951); Puhle (1967). For the Economic Union, cf. Gellately (1974).

14. The liberal dilemma is very well documented: cf., inter alia, Conze (1962); Eisfeld (1969); Gall (1968); Seeber (1965). For all parties cf. Nipperdey (1961); Ritter (1973).

15. Cf. the massive study of Bachem (1927-32): also Morsey (1966); Schoenhoven (1972); Windell (1954).

16. Studies of the SPD are legion. Arguably, one of the best is Roth (1963).

17. Cf. Urwin (1974: 125-6). The techniques for measuring trends and functuations are described in Rose and Urwin (1970).

18. Cf. the simplified genealogical lineage in Urwin (1974: 129).

19. Cf. the excellent ecological analysis of Schleswig-Holstein by Heberle (1945).

20. For a more rigorous methodological examination, cf. the analytical combination of geography and social structure in Urwin and Aarebrot (1981).

21. For a discussion of the constitution, cf. Becker (1958).

22. Cf. Patemann (1964); Anderson (1954). On the alienation of the western provinces, cf. also Kaiser (1963); Schierbaum (1960). Many of the other states, however, introduced universal, secret and direct suffrage: Württemberg even introduced proportional representation (e.g. cf. Simon 1969). A classification of the franchise requirements in the various states is given in Urwin (1974: 117).

23. Knight (1955). A bureaucratic career was the easiest way to acquire aristocratic status: cf. Koselleek (1967).

24. As a minor rider as to how the Elbe remained very much an internal frontier, it is interesting to note that the military, despite its antipathy to Weimar, was regionally divided over the 1920 right-wing Kapp Putsch. The military leaders in Berlin and the east aligned themselves with the coup, while those in the south and west remained loyal to the Ebert government (Pinson 1954: 408).

25. On the disruption in social and political continuity during the Third Reich, cf. Dahrendorf (1967); Schoenbaum (1966); Frye (1968).

26. On the whole the British were most liberal, while the French refused to allow any party other than the original four to operate in their zone: cf. Zink (1957); Ebsworth (1960); Willis (1962).

27. Britain and France, for example, set absolute maximum limits of 150 hectares upon individual ownership, while the Americans attempted the surrender of land on a sliding scale for farm units of over 100 hectares.

28. West Berlin is regarded as the eleventh Land of the Federal Republic — but only by West Berlin and West Germany. De facto, however, it is almost fully integrated into the Federal Republic.

29. The most thorough account of nationalist activities in West Germany is Tauber (1972): cf. also Liepelt (1967).
30. The history of the CSU is exhaustively covered in Minzel (1975).
31. Cf. the extensive analysis in Urwin (1974: 145-62).
32. Cf. the excellent Merkl (1959).

REFERENCES

Abel, W. (1955). *Die Wüstungen des Ausgehenden Mittelalters* (2nd ed.) Stuttgart: Fisher.

Ahnström, L. (1973). *Styrande och ledande verksamhet i Västeuropa: en ekonomisk-geografisk studie.* Stockholm: Almqvist & Wiksell.

Anderson, E. (1954). *The Social and Political Conflict in Prussia, 1858-1864.* Lincoln: University of Nebraska Press.

Anderson, P. (1975). *Lineages of the Absolutist State.* London: New Left Books.

Arndt, H.J. (1967). *West Germany: Politics of Non-Planning.* Syracuse: University Press.

Bachem, K. (1927-32). *Vorgeschichte, Geschichte und Politik der deutschen Zentrumspartei* (9 vols). Cologne: Bachem.

Becker, O. (1958). *Bismarcks Ringen um Deutschlands Gestaltung.* Heidelberg: Quelle & Meyer.

Bendix, R. (1964). *Nation-Building and Citizenship.* New York: John Wiley.

Berdahl, R. (1972). 'Conservative Politics and Aristocratic Landholders in Bismarckian Germany', *Journal of Modern History*, 44: 1-20.

Bernard, P.P. (1965). *Joseph II and Bavaria: Two Eighteenth-Century Attempts at German Unification.* The Hague: Nijhoff.

Blacksell, M. (1975). 'West Germany', in H.D. Clout (ed.), *Regional Development in Western Europe.* New York: John Wiley.

Bluhm, G. (1963). *Die Oder-Neisse Linie in der deutschen Aussenpolitik.* Freiburg: Rombach.

Blum, J. (1957). 'The Rise of Serfdom in Eastern Europe', *American Historical Review*, 62: 807-36.

Böhme, H. (1966). *Deutschlands Weg zu Grossmacht.* Cologne: Kiepenheuer & Witsch.

Booms, H. (1954). *Die Deutsch-Konservative Partei: Preussischer Charakter, Reichsauffassung, Nationalbegriff.* Düsseldorf: Droste.

Borchhardt, K. (1972). *Die Industrielle Revolution in Deutschland.* Munich: Piper.

Born, K.E. (1967). *Preussen und Deutschland im Kaiserreich.* Tübingen: Mohr.

Bracher, K.D. (1960). *Die Auflösung der Weimarer Republik* (4th ed.). Villingen: Ring Verlag.

Braunthal, G. (1960). 'The Free Democratic Party in West German Politics', *Western Political Quarterly*, 13: 332-48.

Büsch, O. (1962). *Militärsystem und Sozialleben im Alten Preussen 1713-1807.* Berlin: de Gruyter.

Carr, W. (1963). *Schleswig-Holstein, 1815-48: A Study in National Conflict.* Manchester: University Press.

Carsten, F.L. (1954). *The Origins of Prussia.* Oxford: Clarendon Press.

Carsten, F.L. (1959). *Princes and Parliaments in Germany.* Oxford: Clarendon Press.

Clasen, C-P. (1963). *The Palatinate in European History, 1559-1660*. Oxford: Basil Blackwell.

Conze, W. (ed.) (1962). *Staat und Gesellschaft im deutschen Vormärz 1815-1848*. Stuttgart: Klett-Cotta.

Culver, L. (1966). 'Land Elections in West German Politics', *Western Political Quarterly*, 19: 304-36.

Dahrendorf, R. (1967). *Society and Democracy in Germany*. Garden City, NY: Doubleday.

Demoskopie (1964). *Die Neubürger. Bericht über die Fluchtlinge und Heimatvertriebenen in der Bundesrepublik*. Allensbach: Institut für Demoskopie.

Der Spiegel (1976). 'Bayern gegen Strauss', 29 November 1976.

Dickens, A.G. (1967). *Martin Luther and the Reformation*. London: English Universities Press.

Dollinger, P. (1970). *The German Hansa*. London: Macmillan.

East, W.G. (1966). *An Historical Geography of Europe* (5th ed.). London: Methuen.

Ebsworth, R. (1960). *Restoring Democracy in Germany: The British Contribution*. London: Stevens.

Eisfeld, G. (1969). *Die Entstehung der liberalen Parteien in Deutschland 1858-1870*. Hannover: Verlag für Literatur und Zeitgeschehen.

Eschenberg, T. (1959). *Der Sold der Politikers*. Stuttgart: Seewald.

Fauchier-Magnan, A. (1958). *The Small German Courts of the Eighteenth Century*. London: Methuen.

Franklin, S.H. (1969). *The European Peasantry*. London: Methuen.

Franz, G. (1954). *Kulturkampf: Staat und Katholische Kirche in Mitteleuropa*. Munich: Callweg.

Franz, G. (1956). *Der deutschen Bauernkrieg*. Darmstadt: Gentner.

Franz, G. (1957). *Die politische Wahlen in Niedersachsen 1867 bis 1949* (3rd ed.). Bremen: Walter Dorn.

Frye, C. (1968). 'The Third Reich and the Second Republic: National Socialism's Impact upon German Democracy', *Western Political Quarterly*, 21: 668-80.

Gall, L. (1968). *Der Liberalismus als regierende Partei: Das Grossherzogtum Baden zwischen Restauration und Reichsgründung*. Wiesbaden: Steiner.

Gellately, R. (1974). *The Politics of Economic Despair*. London/Beverly Hills: Sage.

Gerschenkron, A. (1943). *Bread and Democracy in Germany*. Berkeley/Los Angeles: University of California Press.

Gillis, J. (1968). 'Aristocracy and Bureaucracy in Nineteenth-Century Prussia', *Past and Present*, 41: 105-15.

Goodwin, A. (1953). 'Prussia', in A. Goodwin (ed.), *The European Nobility in the Eighteenth Century*. London: A & C Black.

Göttingen Research Committee (1957). *German Eastern Territories*. Würzburg: Holzner.

Hallett, G. (1973). *The Social Economy of West Germany*. London: St Martin's Press.

Hartenstein, W. (1962). *Die Anfänge der deutschen Volkspartei 1918-1920*. Düsseldorf: Droste.

Heberle, R. (1945). *From Democracy to Nazism*. Baton Rouge: Louisiana State University Press.

Heidenheimer, A.J. (1960). *Adenauer and the CDU*. The Hague: Nijhoff.

Henderson, W.O. (1959). *The Zollverein*. London: Cass.

Hertz-Eicherode, D. (1969). *Politik und Landwirtschaft in Ostpreussen 1919-1930.* Opladen: Westdeutscher Verlag.

Hertzman, L. (1963). *DNVP: Right-Wing Opposition in the Weimar Republic, 1918-1924.* Lincoln; University of Nebraska Press.

Hope, N.M. (1973). *The Alternative to German Unification.* Wiesbaden: Steiner.

Hurstfield, J. (1968). 'Social Structure, Office-Holding and Politics, Chiefly in Western Europe', pp. 126-48 in R.B. Wernham (ed.), *The Counter-Reformation and the Price Revolution, 1559-1610,* New Cambridge Modern History, vol. III. London: Cambridge University Press.

Jäckel, E. (1959). *Die Schleswig-Frage seit 1945.* Frankfurt: Metzner.

Jacob, H. (1963). *German Administration since Bismarck.* New Haven: Yale University Press.

Jochimsen, R. and Treuner, P. (1974). 'Staatliche Planung in der Bundesrepublik', *Politik und Zeitgeschichte,* 9: 29-45.

Kaiser, R. (1963). *Die politische Strömungen in den Kreisen Bonn und Rheinbach.* Bonn: Röhrscheid.

Kitchen, M. (1978). *The Political Economy of Germany 1815-1914.* London: Croom Helm.

Kluth, H. (1959). *Die KPD in der Bundesrepublik: Ihre politische Tätigkeit und Organisation 1945-1956.* Cologne: Westdeutscher Verlag.

Knight, M. (1955). *The German Executive 1890-1933.* New York: Fertig.

Kobschätzky, H. (1972). *Streckenatlas der deutschen Eisenbahnen 1835-1892.* Düsseldorf: Alba Buchverlag.

Koselleek, R. (1967). *Preussen zwischen Reform und Revolution.* Stuttgart: Klett.

Kraus, H. and Kurth, K. (1957). *Deutschlands Ostproblem: Eine Untersüchung der Beziehungen des deutschen Volkes zu seiner östlichen Nachbarn.* Würzburg: Holzner.

Lattimore, B.G. (1974). *The Assimiliation of the German Refugees.* The Hague: Nijhoff.

Laufer, H. (1974). *Der Föderalismus in der BRD.* Stuttgart: Kohlhammer.

Lemberg, E. and Edding, F. (1959). *Die Vertriebenen in West-Deutschland* (3 vols). Kiel: Hirt.

Lidtke, V.E. (1966). *The Outlawed Party: Social Democracy in Germany, 1878-1890.* Princeton: University Press.

Liebe, W. (1956). *Die deutschnationale Volkspartei 1918-1924.* Düsseldorf: Droste.

Liepelt, K. (1967). 'Anhänger der neuen Rechtspartei: Ein Beitrag zur Diskussion über das Wählerreservoir der NPD', *Politische Vierteljahresschrift,* 8: 237-71.

Loewenberg, G. (1966). *Parliament in the German Political System.* Ithaca, NY: Cornell University Press.

Lowewenberg, G. (1968). 'The Remaking of the German Party System', *Polity,* 1: 86-113.

Lopez, R.S. (1952). 'The Trade of Medieval Europe: The South', pp. 257-354, in M.M. Postan and E.E. Rich (eds), *Trade and Industry in the Middle Ages* (Cambridge Economic History of Europe, vol. II). London: Cambridge University Press.

Ludloff, R. (1957). 'Industrial Development in 16th and 17th Century Germany', *Past and Present,* 12: 58-75.

Merkl, P.H. (1959). 'Executive-Legislative Federalism in West Germany', *American Political Science Review,* 53: 732-41.

Meyn, H. (1965). *Die Deutsche Partei.* Düsseldorf: Droste.

Minzel, A. (1975). *Die CSU.* Opladen: Westdeutscher Verlag.

Mitchell, A. (1965). *Revolution in Bavaria, 1918-1919*. Princeton: University Press.

Moore, B. (1966). *Social Origins of Dictatorship and Democracy*. Boston: Beacon Press.

Morgan, R. (1965). *The German Social Democrats and the First International, 1864-1872*. Cambridge: University Press.

Morsey, R. (1957). *Die obere Reichsverwaltung unter Bismarck 1867-1890*. Münster: Aschendorff.

Morsey, R. (1966). *Die deutsche Zentrumspartei 1917-1933*. Düsseldorf: Droste.

Muncy, L.W. (1944). *The Junker in the Prussian Administration under William II*. Providence, RI: Brown University Press.

Neumann, F. (1968). *Der Block der Heimatvertriebenen und Entrechten 1950-1960: Ein Beitrag zur Geschichte und Strukur einer Interessenpartei*. Meisenheim: Hain.

Neunreither, K. (1959). *Der Bundesrat zwischen Politik und Verwaltung*. Heidelberg: Quelle & Meyer.

Nipperdey, T. (1961). *Die Organisation der deutschen Parteien vor 1918*. Düsseldorf: Droste.

Olzog, G. (1964). *Die politische Parteien*. Munich: Olzog.

Pack, W. (1961). *Das parlamentarische Ringen um das Sozialistengesetz Bismarcks 1878-1890*. Düsseldorf: Droste.

Parry, J.H. (1967). 'Transport and Trade Routes', pp. 155-219 in E.E. Rich and C.H. Wilson (eds), *The Economy of Expanding Europe in the 16th and 17th Centuries* (Cambridge Economic History of Europe, vol. IV). London: Cambridge University Press.

Patemann, R. (1964). *Der Kampf um die preussische Wahlreform im ersten Weltkrieg*. Düsseldorf: Droste.

Pinson, K.S. (1954). *Modern Germany*. New York: Macmillan.

Puhle, H-J. (1967). *Agrarische Interessenpolitik und preussischer Konservatismus*. Hanover: Verlag für Literatur & Zeitgeschehen.

Raupach, H. (1967). 'Der interregionale Wohlfahrtsausgleich als Problem der Politik des Deutschen Reiches', pp. 13-34 in W. Conze and H. Raupach (eds), *Die Staats- und Wirtschaftskrise des Deutschen Reiches 1929-33*. Stuttgart: Klett.

Riker, W.H. (1974). 'Federalism', in F.I. Greenstein and N.W. Polsby (eds), *Handbook of Political Science*, vol. 5. Reading, Mass: Addison-Wesley.

Ritter, G.A. (ed.) (1973). *Deutsche Parteien vor 1918*. Cologne: Kiepenheuer & Witsch.

Rose, R. and Urwin, D.W. (1970). 'Persistence and Change in Western Party Systems since 1945', *Political Studies*, 18: 287-319.

Rosenberg, H. (1943). 'The Rise of the Junkers in Brandenburg-Prussia, 1410-1653', *American Historical Review*, 49: 1-22, 228-42.

Rosenberg, H. (1958). *Bureaucracy, Aristocracy and Autocracy: The Prussian Experience 1660-1815*. Cambridge, Mass.: Harvard University Press.

Ross, J. (1976). *Beleaguered Tower: The Dilemma of Political Catholicism in Wilhelmine Germany*. South Bend, Ind.: University of Notre Dame Press.

Roth, G. (1963). *The Social Democrats in Imperial Germany*. Totowa: Bedminster Press.

Schauff, J. (1928). *Die deutschen Katholiken und die Zentrumspartei*. Cologne: Bachem.

Schieder, T. (1962). *The State and Society in Our Times*. London: Nelson.

Schierbaum, H. (1960). *Die politische Wahlen in den Eifel und Moselkreisendes Regierungsbezirkes Trier, 1849-1867*. Düsseldorf: Droste.

Schmidt, M.G. (1978). 'Party Control and Public Policies: The Case of the West German States'. Paper presented at the ECPR Workshops, Grenoble.

Schmidt-Volkmar, E. (1962). *Der Kulturkampf in Deutschland 1871-1890*. Göttingen: Musterschmidt.

Schoenbaum, D. (1966). *Hitler's Social Revolution*. Garden City, NY: Doubleday.

Schoenhoven, K. (1972). *Die Bayerische Volkspartei 1924-1932*. Düsseldorf: Droste.

Schwend, K. (1954). *Bayern zwischen Monarchie und Diktatur*. Munich: Richard Pflaum.

Schwendt, K. (1960). 'Die Bayerische Volkspartei', pp. 457-519 in E. Matthias and R. Morsey (eds), *Das Ende der Parteien*. Düsseldorf: Droste.

Seeber, G. (1965). *Zwischen Bebel und Bismarck. Zur Geschichte des Linksliberalismus in Deutschland, 1871-1893*. Berlin: Akademie Verlag.

Silverman, G.P. (1972). *Reluctant Union: Alsace-Lorraine and Imperial Germany 1871-1918*. University Park: Pennsylvania State University Press.

Simon, K. (1969). *Die württembergischen Demokraten*. Stuttgart: Kohlhammer.

Simon, W.M. (1968). *Germany in the Age of Bismarck*. London: Allen & Unwin.

Simon, W.M. (1971). *The Failure of the Prussian Reform Movement 1807-1819*. New York: Fertig.

Spiro, H. (1962). 'The German Political System', in S. H. Beer and A.B. Ulam (eds), *Patterns of Government* (2nd ed.). New York: Random House.

Spooner, F.C. (1968). 'The Economy of Europe, 1559-1609', pp. 14-93 in R.B. Wernham (ed.), *The Counter-Reformation and the Price of Revolution, 1559-1610* (New Cambridge Modern History, vol. III). London: Cambridge University Press.

Stehlin, S.A. (1973). *Bismarck and the Guelph Problem, 1866-1890*. The Hague: Nijhoff.

Stephan, W. (1973). *Aufstieg und Verfall des Linksliberalismus. Geschichte der DDP*. Göttingen: Vandenhoek & Ruprecht.

Stephens, M. (1976). *Linguistic Minorities in Western Europe*. Llandysul: Gower Press.

Stöckl, F. (1969). *Die Eisenbahnen in Deutschland — Vom Adler Zum TEE*. Vienna: Bohmann.

Strauss, G. (1966). *Nuremberg in the Sixteenth Century*. New York: John Wiley.

Strauss, G. (ed.) (1972). *Pre-Reformation Germany*. London: Macmillan.

Tauber, K. (1972). *Beyond Eagle and Swastika: German Nationalism since 1945* (2 vols). Middletown: Wesleyan University Press.

Taylor, A.J.P. (1961). *The Course of German History*. London: Hamilton.

Thraenhardt, D. (1973). *Wahlen und politische Strukturen in Bayern 1848-1953*. Düsseldorf: Droste.

Tirrell, S.R. (1951). *German Agrarian Politics After Bismarck's Fall*. New York: Columbia University Press.

Turner, H.A. (1963). *Stresemann and the Politics of the Weimar Republic*. Princeton: University Press.

Urwin, D.W. (1974). 'Germany: Continuity and Change in Electoral Politics', pp. 107-70 in R. Rose (ed.), *Electoral Behavior*. New York: Free Press.

Urwin, D.W. and Aarebrot, F.H. (1981). 'The Socio-Geographic Correlates of Left Voting in Weimar Germany, 1924-1932', in P. Torsvik (ed.), *Mobilization, Centre-Periphery Structures and Nation-Building*. Oslo: Universitetsforlaget.

von Unruh, G.C. (1966). *Der Landrat*. Cologne: Kiepenheuer & Witsch.

Wallerstein, I. (1974). *The Modern World System*. London: Academic Press.

Wallich, H. (1955). *Mainsprings of the German Revival.* New Haven: Yale University Press.

Weber, H. (1970). *Die Wandlung des deutschen Kommunismen.* Frankfurt: Europäische Verlagsanstalt.

Weiss, E. (1970). 'Ergebnisse eines Vergleichs der grundherrschaftlichen Strukturen Deutschlands und Frankreichs vom 13. bis zum Ausgang des 18. Jahrhundert', *Vierteljahresschrift für Sozial- und Wirtschaftsgeschichte,* 1: 1-14.

Wells, R.H. (1961). *The States in West German Federalism.* New York: Bookman.

Wieck, H.C. (1953). *Die Entstehung der CDU und die Wiedergründung des Zentrums.* Düsseldorf: Droste.

Willis, F.R. (1962). *The French in Germany 1945-1949.* Stanford: University Press.

Windell, G. (1954). *The Catholics and German Unity, 1866-1871.* Minneapolis: University of Minnesota Press.

Windell, G. (1969). 'The Bismarckian Empire as a Federal State 1866-1880: A Chronicle of Failure', *Central European History,* 2: 291-311.

Wunderlich, F. (1961). *Farm Labor in Germany.* Princeton: University Press.

Ziller, G. (1966). *Der Bundesrat.* Frankfurt: Athenäum.

Zink, H. (1957). *The United States in Germany 1944-55.* Princeton: University Press.

6

Regional Imbalances and Centre – Periphery Relationships in Belgium

André P. Frognier, Michel Quevit
and Marie Stenbock
Catholic University of Louvain

Many of those who have analysed the problem of coexistence facing the two linguistic communities in Belgium have tended to focus attention on the linguistic issue. However, there are other conflicts, stimulated by economic, social and even denominational issues; and these have added to the complexity of the regional problem, making any political solution even more difficult. The intention of this paper is to present a chronological overview of those factors that progressively came to structure the linguistic, economic, social and denominational issues at stake in the relationships between Flanders, Wallonia and Brussels. More specifically, it will emphasize the consequent disparities between the regions and the political mutations of the structure of the state generated by such imbalances.

From a theoretical point of view, the complexity of the Belgian case can be easily demonstrated by employing the conceptual framework of centre versus periphery. Since the creation of the sovereign state in 1830, centre – periphery relationships in Belgium have consistently changed over the years to include very different kinds of situations and issues. Indeed, reviews of the existence of a centre – periphery dimension need not be limited to an analysis of the relationships between the different decision-making levels of the state — for example, between the central decision-making bodies and the local communities. Such a relationship is characterized much more by the territorial distribution of material and human resources, the localization of which is largely determined by those economic and political power relationships that characterize any society. In Belgium these relationships, and hence their territorial consequences, have undergone important modifications since 1830.

As an introduction, we shall briefly trace the great mutations in Belgian society since national independence in 1830, viewed from the point of view of a centre – periphery relationship.[1] The independence of Belgium gave birth to a highly centralized and French-speaking state in which Brussels was chosen as the administrative capital. The independence movement resulted essentially from an alliance between different groups within a French-speaking bourgeoisie: aristocracy, gentry, provincial bourgeoisie and an emergent industrial bourgeoisie in the growing urban centres. These different groups gave rise to two political formations and/or spiritual families: the Catholics and the liberals. Various reasons led to the choice of Brussels as the capital of the new state. Traditionally, Brussels had been the 'city of the court', under the rule of various foreign occupying forces, and the administrative centre from which everything was controlled by them. Later, thanks to the development of trade and industry, Brussels also became an expanding financial centre. Thus, with the creation of the Banque de la Société Générale in 1822 and the Banque de Belgique in 1837, Brussels, in addition to being a centre of political decision-making, became the financial centre par excellence of Belgium in the nineteenth century and the first half of the twentieth. However, the national economic boom took place in the French-speaking Walloon region of the country, and more specifically in what was called the industrial furrow of the provinces of Liège and Hainaut. Because of its large reserves of coal, the 'Walloon furrow' experienced a rapid industrialization, especially in the steel and iron sector. By the middle of the nineteenth century this region had become one of the most industrialized areas of Europe. Its development was further characterized by an economic concentration in the main industrial sectors, under the impulse of several large financial groups located in the capital. This association between the capital of Brussels and the industrial furrow gave Belgium a strong economic position in Europe and even in the world.

From that time onwards, however, the linguistic issues loomed large in the background and gave rise to many quarrels between the two language communities of Belgium. Since the Belgian state was created by French-speaking notables, French was chosen as the official language of administration, law and military affairs, as well as politics: it was taught in the schools both in Flanders and Wallonia. In contrast to Wallonia and its economic expansion, Flanders was mainly agrarian and rural. It did have some urban centres, but these were built around traditional and familial industries such as textiles and food, which were soon to suffer from an economic crisis. Yet at the end of the Middle Ages, and especially during the sixteenth century, the Flemish region had been particularly prosperous. Flemish cities such as Gent and Bruges had grown to such an extent that they could be considered northern equivalents of the great Italian cities. After the Treaty of Munster in 1648, however, these cities fell into a decline. Flanders was separated from the present-day Netherlands, and in the Spanish Empire of Philip II the province was kept apart and treated as a remote periphery. As a result of long neglect and isolation, by 1830 Flanders had practically no elites that were integrated culturally with the people and identified with the Flemish community: the majority of the elite — aristocrats, the bourgeoisie and local notables — were gallicized, more especially after invasion and occupation by French revolutionary and Napoleonic forces. The Flemish language itself had fragmented into a multitude of local dialects.

At the birth of the Belgian state the centre was thus French-speaking, politically and economically as well as culturally. From a cultural point of view, the French-speaking inhabitants of Flanders, of Brussels (situated within Flemish territory but with a French-speaking minority since the tenth century) and of Wallonia constituted the centre. Economically, the centre comprised a nucleus of financial groups which backed and directed economic development in accordance with a national 'politique'. Investment was primarily in the Walloon industrial furrow, but also was in certain regions of Flanders: for example, the expansion of the port of Antwerp, Gent's cotton industry and, at the beginning of the twentieth century, the coal industry in the Flemish province of Limburg all benefited from this investment policy.

Brussels is sometimes considered, somewhat mistakenly, as the real 'centre' of a unitarian Belgium. But in fact, it was only one element of a centre, which both culturally and economically extended well beyond it. At the most, it could be considered the 'functional seat' of the centre. That is why it increasingly became a city with a French-speaking majority, even though it lies in Flemish territory.

A little more than one century later, the landscape of Belgium has changed drastically. Culturally, Flanders has, to a considerable extent, acquired autonomy. Flemish has become an official language of the state, on equal footing with French. Constitutionally, the Flemish region is even regarded as being unilingually Dutch-speaking, although there remain French minorities within Flemish territory that consider themselves to be victims of recent language policy, as for example in the Fourons (or Voers), an enclave of the Flemish province of Limburg within the French province of Liège. For the Flemish, Brussels is the only remaining part of their territory where Dutch is in a minority and in peril. The city of Brussels, though in Flemish territory (about 10 kilometres from the linguistic border), is officially bilingual. However, French is spoken by at least 80 percent of the city's population and it seems to be still increasing. Moreover, the French-speaking inhabitants of Brussels tend to leave the city and settle in the outskirts, where they form outposts of gallicization. Economically the contemporary contrast between Flanders and Wallonia is even sharper. Walloon industry, based essentially on coal and steel, has suffered from a serious and ongoing economic crisis, especially since the 1960s. By contrast, Flanders has experienced a remarkable economic boom, built largely upon the expansion of the port of Antwerp and the petrochemical and electronics industries. Nowadays, it is quite clear that Flemish industry prevails in Belgium, and that it has reversed all the trends of the last century, and even of the interwar years.

In short, the cultural 'peripheralization' of Flanders seems to have ended, with the almost full recovery by the Flemish culture of its total autonomy. Economically, one could say that Flanders has become the centre in Belgium and Wallonia a declining and peripheral region, where government subsidies seem to have had little effect. Brussels itself has lost prestige at the same time. Attacked by the Flemish movement, the city has shrunk into 19 communes (what francophones call the 'Yoke'), and only a few 'facilities' have been granted to the francophones of six peripheral communes (for example, the ability to use French in dealings with the administration and judiciary).

Under such conditions, strong reactions from the French-speaking side were not surprising. These are essentially of two kinds. In Brussels the protest is mainly linguistic, and results from what is felt by the majority of the population to be linguistic persecution. In Wallonia the protest movements are based more on the need to counter economic and social decline. Like the Flemish, they demand greater regional autonomy. Nevertheless, most French-speakers have maintained their commitment to unitarism, although it is declining. Paradoxically, some elements in the Flemish movement, aware that the Flemish now control numerous essential levers of the political and economic life of the state, seem more and more to accept the necessity of maintaining a certain unitarism, as long as it does not endanger the regional gains that have been achieved.

This has been only a brief outline of the developments of the centre – periphery relationship in Belgium. Within a century, certain tendencies have been reversed, mainly in the economic field. This inversion of fortunes could only give rise to persistent conflicts in which everybody may be accused of having been or of being dominant. Thus the growth effects of the twentieth century have not served as homogenizing influences in a state characterized by regional imbalances. On the contrary, this growth has always occurred at the expense of one of the two halves of the country. Belgium's difficulty in finding a solution to this problem is therefore surprising for a country that is often called a 'consociational democracy', and which has found a way to pacify its ethical, and to a certain extent its social, conflicts. Will it one day find a formula for community pacification?

To illustrate this problem, we shall look especially at the non-linguistic aspects of the community conflicts in Belgium, in order to highlight some basic reasons for the present difficulties. Hence this chapter is devoted more to the evolution of regional imbalances from purely economic, social and religious points of view. Examining these imbalances will illustrate more precisely the regional inversion phenomenon mentioned earlier. The data on which the analysis is based are drawn from two sources: the Aiken – Quevit file on 30 electoral districts (to which some additions have been made), and the Quevit file on regional policy.[2] In the analysis, data will be presented at the broader regional level as well as at that of the nine provinces of Belgium.

The linguistic problem

Ever since the large-scale Germanic migration into western Europe after the decline of the Roman Empire, Belgium has straddled the linguistic border separating the German and Romance language worlds. Indeed, this border represents the most stable division that Belgium has ever known. The linguistic border is shown in Figure

FIGURE 1
The linguistic border in Belgium

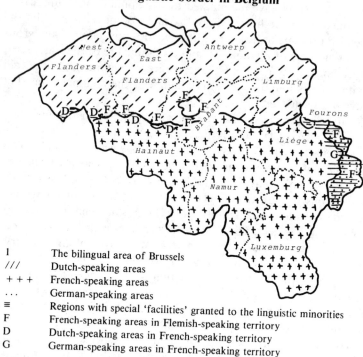

1	The bilingual area of Brussels
///	Dutch-speaking areas
+ + +	French-speaking areas
. . .	German-speaking areas
≡	Regions with special 'facilities' granted to the linguistic minorities
F	French-speaking areas in Flemish-speaking territory
D	Dutch-speaking areas in French-speaking territory
G	German-speaking areas in French-speaking territory

1, a map of Belgium's nine provinces. It separates the Flemings of the north, who speak Dutch, and the French-speaking Walloons of the south. Belgium also has a small German community of 60,000 people, on the eastern fringes of Wallonia.[3]

The line drawn in Figure 1 represents the linguistic border as it was legally recognized for the first time in 1962. It divides the province of Brabant, with Brussels, the capital and primarily a French-speaking city, lying within Dutch-speaking territory some 10 kilometres north of the linguistic border. In the Constitution, Brussels is considered a bilingual territory, but other concentrated areas of linguistic minorities can also be found on either side of the linguistic divide, especially towards its western end: the minority language group in these areas has been granted certain exemptions from the general principles of the linguistic legislation of recent years.

The Flemings have for long been more numerous than the Walloons, and their numerical superiority has been increasing steadily since the middle of the nineteenth century. Table 1 shows the evolution of the Belgian population by region from 1890 to 1970. The increase of the Flemish population in comparison to that of Wallonia is very clear. The difference in birth rates dates back roughly to 1855. Until then, the regional birth rates had been more or less equal. Afterwards, the Flemish rate increased while that of Wallonia declined rather rapidly. It has been argued by some students of Belgian history that it is cultural factors rather than economic differences between the two regions that can explain this

TABLE 1
Population changes in Belgium, by region, 1890-1970

| | Percentage changes | | | | |
	1890	1920	1947	1961	1970
	%	%	%	%	%
Flanders	50.3	51.5	54.2	55.7	65.1
Wallonia	41.5	38.2	34.6	33.1	32.7
Brussels	8.2	10.2	11.2	11.1	11.1
(Total population)	(6,069,321)				(9,650,944)

Source: National Institute of Statistics.

strong variation in demographic developments. Indeed, the birth rate in Wallonia during the last century was very similar to that of France, and according to some the practical application of Malthusian concepts is the explanation (cf. Becquet 1972: 90-2). As a result, Wallonia's population is older (14.6 percent over 65 in 1970) than the Flemish (13.2 percent over 65 in 1970).

Table 2 shows the distribution of languages in each of the three regions of Flanders, Wallonia and Brussels. The most important point is that Brussels has a French-speaking majority. Moreover, this majority has rapidly increased in size: between 1920 and 1947 it increased by some 10 percent. Since the 1950s this increase seems to have been even more pronounced, but the lack of a linguistic survey since 1947 prevents us from estimating current language use with any reasonable accuracy. Therefore, while there seems to be a Flemish majority in the country, there is a French-speaking majority in Brussels (with more than 10 percent of the Belgian

population). This is one of the main reasons for the complexity of
the Belgian community problem. There is also a small but meaningful French-speaking minority in Flanders (5.4 percent in 1947):
its presence, too, is essential for understanding the whole of the
linguistic problem in Belgium. For centuries the Flemish elite was
gallicized, and the linguistic division within Flanders thus corresponded also to a social division. As a result, the Flemish movement, when it appeared, was opposed both to its own elites and to
the French-speaking unitary state. Nowadays, compelled by the
success of the Flemish movement to take stock of itself, this elite
seems to be assimilating itself very rapidly to its immediate
linguistic environment. At least, that is what can be deduced from
current trends, though in the absence of a more recent linguistic
census any confirmation is impossible.

The linguistic problem that Brussels poses has been, and is, one
of the major stumbling blocks to all linguistic pacification projects
in Belgium. Although Brussels has a legal bilingual status, French is
the language in general use. Moreover, the expansion of the city into the surrounding countryside, together with the other classical urban phenomena, has meant that many francophones have settled in

TABLE 2
Language use by region, 1920-1947*

	Dutch	French	German
	%	%	%
Flanders			
1920	94.9	5.0	0.0
1947	94.0	5.4	0.2
Wallonia			
1920	2.4	94.5	3.1
1947	2.1	95.1	2.4
Brussels			
1920	37.2	62.6	0.1
1947	24.6	72.7	0.4
Belgium			
1920	52.7	46.0	1.2
1947	54.8	43.7	1.0

* The last linguistic survey in Belgium was carried out in 1947.

Source: National Institute of Statistics.

FIGURE 2
The proportion of French-speakers in Brussels and the surrounding communes, 1971

* Notre-Dame-au-Bois, Zuun, and Beauval are parts of communes.

Source: Martens (1976: 464).

The commentary accompanying the map reads: 'this document is an extract from the publication called *Le pays de Bruxelles* (Brussels country), no. 118, February 1970. Even though we have our doubts about the spirit in which this newspaper is written, we thought that it would be interesting to reproduce this map of the Brussels region which, according to its author, M. Fernand Rigot, has been established on the basis of the following documents: (1) identity cards; (2) tax declarations; (3) driving licences; (4) television licences; (5) militiamen's linguistic choice; (6) report to the Institute of Technology (investigation made by Mr Vander-Eycken, then VUB professor); (7) birth, marriage and death certificates; (8) the movements in the attendance of school children from the peripheral communes to the French-speaking schools of the localities with such schools. According to its authors, the map in question was drawn after also taking into account answers to parliamentary questions and "after polls, estimates made by French-speaking personalities of the communes around Brussels". As we can see, this map gives percentages for the distribution of francophones settled in the Brussels agglomeration and the neighbouring communes.'

the Flemish peripheral communes where they now constitute important French-speaking minorities.

Figure 2 shows the map of Brussels and its peripheral communes. Some of these outlying communes have been granted special linguistic 'facilities': the francophones who live there may use French in some administrative and judicial matters. The map also provides an estimate of the percentage of francophones. The data come from politically involved French-speaking circles and seem to be more stringent for Brussels and the 'facilities' communes than for the others. The map was drawn in 1971, and is based upon the employment of several indicators of language usage.[4] If the figures are reasonably accurate, then Figure 2 clearly indicates not only the overwhelmingly French nature of Brussels itself, but also the strongly bilingual nature of many of the surrounding communes. The problem, then, consists of finding, for both Brussels and the latter, a status that would make the city an acceptable capital to the Flemings, since it is on Flemish territory. At the same time, that status ideally should endanger neither the language nor the employment of Brussels' French-speaking majority (as employment may depend on language) living in the city or in the periphery; nor should it jeopardize the economic development of the area.

TABLE 3
The number of steam engines by province, 1830-1850

	1830	1850
Antwerp*	9	46
West Flanders*	2	73
East Flanders*	77	273
Limburg*	4	14
Hainaut**	118	822
Liège**	171	524
Namur**	11	99
Luxembourg**	0	3
Brabant***	36	173

   * Flemish provinces.
  ** Walloon provinces.
*** Brabant (includes Brussels and is divided by the linguistic border).

Source: Neuville (1976a: 20).

The economic situation

Industry based upon capitalist concentration first developed in Belgium in the middle of the eighteenth century (cf. Neuville 1976a: Ch. 1). The real boom, however, occurred in the wake of independence in 1830. Statistics are few, and even fewer at the regional level than at the national. Table 3 takes the increase in the number of steam engines as an indicator of industrial growth. It shows quite clearly that there was a marked variation between the provinces in terms of their economic development. Industrial development was localized mainly in the Walloon provinces of Liège and Hainaut, or more specifically in the 'industrial furrow' between the Sambre and Meuse. Flanders was essentially rural except for a few traditional and declining industries around Gent. Walloon industrial expansion was due mainly to the arrival and settlement of English industrialists such as Cockerill, who brought with them the technology needed to transform iron ore into finished metal products.

TABLE 4
Development of the primary, secondary and tertiary
sectors by region, 1920-1970

	Primary	Secondary	Tertiary
	%	%	%
1920			
Flanders	24.2	46.0	29.9
Wallonia	15.2	57.8	27.0
Brussels	9.3	47.9	42.8
1947			
Flanders	15.4	48.3	36.1
Wallonia	12.0	54.1	33.9
Brussels	5.4	40.1	54.5
1961			
Flanders	8.5	49.9	41.6
Wallonia	8.6	49.6	41.8
Brussels	2.8	37.6	59.6
1970			
Flanders	5.2	47.3	46.8
Wallonia	5.4	44.1	49.5
Brussels	0.2	26.4	62.6

Source: National Institute of Statistics.

What has changed since the nineteenth century? The statistics on the evolution of the number of industrial workers within the active working population are significant enough. They reveal one of the fundamental trends of regional economic development in Belgium: the reversal of territorial patterns of industrialization. During the twentieth century, and especially since the 1960s, Flanders has caught up with Wallonia, ultimately even to lead the way in industrialization. Table 4 shows this regional evolution by economic sector. Industrialization took place, as usual, at the expense of the primary sector, especially in Flanders, which in 1890 had 29 percent of its economically active population in the primary sector and 31 percent in the secondary. At that time only 16 percent of the Walloon population were employed in the primary sector. In 1947, 1961 and 1971 the primary sector had more or less the same size in both regions; the tertiary sector too has developed more or less equally in the three regions, especially since the end of the Second World War. In Wallonia, however, the increasing size of the tertiary sector seems to have been on a par with a transfer of population from secondary to tertiary activities: this does not seem to have been the case in Flanders. Flanders possesses a more balanced economic structure and thus is more resistant to the mutations that have been provoked by current problems of structural economic change. Indeed, an over-developed tertiary sector, together with an ageing industrial sector, are proof of the greater vulnerability of the Walloon region.

Figure 3 shows the development of the secondary sector in each province from 1919 to 1971. From 1947 onwards, the Walloon decline is obvious, and manifests itself especially between 1954 and 1971 when the Flemish curves 'meet' with those for Hainaut and Liège. The improvement of Flanders is indicated quite clearly by the way in which the Flemish provinces have reversed the historical pattern. Whereas Hainaut and Liège were dominant in 1919, by 1971 Hainaut had been outstripped by Limburg and East Flanders, and equalled by West Flanders. Liège was outstripped by all four Flemish provinces. Limburg's spectacular growth is due to the discovery of an important coal basin at the beginning of this century and to the development of economical and technologically advanced methods of exploitation.

The evolution of Flemish industrialization is also evident in worker – employer ratios, which may be taken as indicating the rate of industrial concentration. Table 5, which compares the rates of industrial concentration in 1920 and 1961, clearly shows the in-

FIGURE 3

Development of the secondary sector by province, 1919-1971

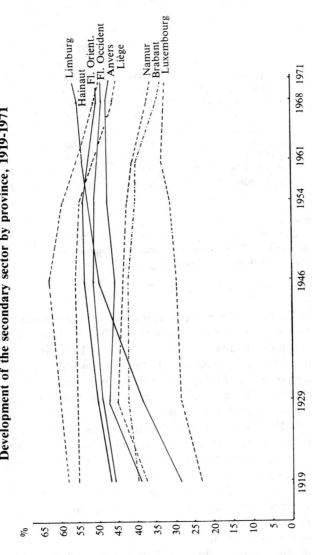

Note: The percentages are calculated from the total active population. The Walloon provinces are represented by – – –, the Flemish by ——, and the Brabant (which includes Brussels) by ——.

Source: Aiken-Quevit data.

TABLE 5
The extent of industrial concentration by province, 1920 and 1961

	1920****	1961****
Antwerp*	7.87	8.98
West Flanders*	6.24	7.53
East Flanders*	7.03	8.9
Limburg*	4.13	11.65
Hainaut**	11.06	11.76
Liège**	10.47	8.60
Namur**	5.85	6.76
Luxembourg**	2.52	4.70
Brabant***	6.08	5.54

* Flemish provinces.
** Walloon provinces.
*** Brabant includes Brussels and is divided by the linguistic border.
**** The rate is: % workers/% employers.

Source: Quevit (1978: 49, 85).

crease in Flemish industrial concentration. In 1920 Hainaut and Liège were most obviously in the lead. In 1961 Hainaut remained firm but its figure had been matched by that for Limburg. Concentration has definitely decreased in Liège, which was outstripped by the Flemish provinces of Antwerp and East Flanders.

The effect of this inversion of the economic trends of the two main regions should also be registered in data on the gross domestic product (GDP). Figure 4 shows that this indeed is the case. Only the trends for the Flemish provinces are rising ones. From 1965 onwards, the Flemish provinces progressed at an even higher rhythm. The collapse of Brabant's GDP and that of Brussels could be an indication of the capital's loss of momentum, owing perhaps to the linguistic tensions aroused by its territorial location and linguistic structure, and to the refusal by certain influential Flemish groups to recognize it as a capital.

The catching up of Wallonia by Flanders does not presuppose a greater homogeneity of industrial structure or in the structure of economic ownership. During the last century, Walloon expansion was characterized by both a concentration of certain specialized sectors (for example, electronics, metallurgy) and the relative

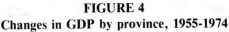

FIGURE 4
Changes in GDP by province, 1955-1974

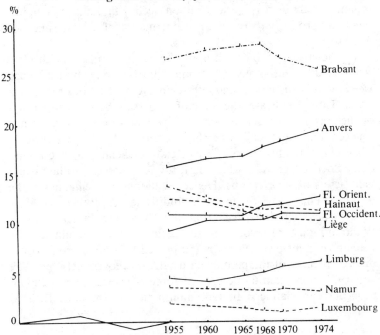

Note: It is the percentage of the country's GDP (GDP at standard prices) that is shown for each province.   *Source:* Aiken-Quevit data.

disappearance of small-and medium-sized firms. On the contrary, in Flanders the balance between the production sectors has been better, with an attempt to achieve concentration without destroying the network of small- and medium-sized firms. On the other hand, the Walloon industrial structure became more out of date and did not take advantage, if one can say such a thing, of the destruction caused by the Second World War in order to set up more modern and more profitable enterprises. The more recent Flemish structure did not have this problem, except for a few cases of antiquated industry in the Gent region. Finally, Walloon expansion occurred during the coal age, whereas Flemish expansion took place partly during the oil era, for which Flanders had the incredible advantage of the port of Antwerp, and partly during the 'Second Industrial Revolution' (petrochemicals, nuclear power, electronics). The obsolete industrial structure of Wallonia implied, for financial groups, the need for a rationalization of investment policy and even, in the case of the Société Générale, a reorientation of investments in sectors such as banking, finance and insurance. That intention only increased the Walloon decline by reducing employment and productivity (on industrial structure, cf. Biname and Jacquemin 1973).

The location of ownership of the larger firms also displays regional differentiations which may clarify the relationship between regional disparities and inter-employer disparities or even tensions. Table 6 indicates the share of ownership by different groups of the permanent assets of the 93 largest Belgian industrial firms. The economic forces of ownership are different in the two regions. Wallonia is mainly controlled by Belgian financial groups, especially the Société Générale, which itself totals 49 percent of the permanent assets of such groups. Hence, Wallonia has remained at the development stage of national industrial capitalism, mainly French-speaking, inherited from the 'founding fathers' of the Walloon industrial revolution. In Flanders there is domination by foreign or joint-owned groups (60 percent): Belgian groups total only 27 percent of the permanent assets. Moreover, between 1959 and 1973 more than 63 percent of foreign investment in Belgium took place in Flanders. In 1968 multinational companies realized

TABLE 6
The location of ownership in the 93 largest Belgian industrial firms, 1969

	Firms located in Wallonia	Firms located in Flanders	Multiregional or multinational firms
	%	%	%
Independent/ familial groups	10.5	12.9	69.2
Belgian groups (of which,	56.2	27.0	17.8
Société Générale)	(49.0)	(24.2)	(17.8)
Foreign groups	26.4	45.3	10.4
Joint-owned (Belgian-foreign) groups	5.9	14.6	2.6

Source: Quevit (1978: 117).

one-third of the industrial turnover and employed 18 percent of the industrial labour force. But 70 percent of the new jobs were created in Flanders, especially in the Antwerp – Gent – Brussels triangle. In particular, the importance of the multinationals in Flanders enabled the region to rank before Wallonia in the size of investment relative to technology. (On the problem of industrial property, cf. de Vroey and Carton 1970.)

The social situation

It is usually recognized that poverty was widespread in eighteenth- and nineteenth-century Belgium. In 1828 an official report stated that there were 14.5 paupers for every 100 inhabitants. In 1928 the relief rate was 1 inhabitant out of 6.9, as opposed to 7 in 1839 and 6.2 in 1846 (cf. Neuville 1976b). Is there a regional differentiation in this respect? It is difficult to answer this question accurately. According to what has been written, the extreme poverty of the Walloon workers in the 'industrial furrow' was similar to the conditions of agricultural workers of Flanders. Nevertheless, it is normally recognized that a certain 'Flemish crisis' existed, especially around 1839 – 46, which was a combination of a crisis in the flax industry and a succession of very bad harvests, mainly potatoes. The poverty of that time was probably the greatest Flanders had ever known, and it led to a migration of labour towards Wallonia, as well as favouring gallicization.

The few statistics available for this period relate to the number of people who received assistance. Figure 5 shows this percentage by province for 1828, 1839, and 1846.[5] It indicates that in 1828 the Flemish provinces did not know greater poverty than Wallonia — on the contrary, Liège and Hainaut were two of the three provinces with the highest levels of assistance. But between 1830 and 1846 the trend was towards a decrease in the rate for all provinces except Antwerp. This decrease is particularly noticeable for the two Flanders and Limburg.

In any attempt to measure poverty in greater depth, it is difficult to be more precise. However, an examination of school attendance clearly indicates the inferiority of Flanders. Table 7 shows the situation of the population of primary school age in 1832. All the Walloon provinces and Brabant are clearly in the lead. This gap seems to have persisted. In order to follow the evolution of school attendance over time, we examined the statistics on literacy bet-

TABLE 7
Primary school attendance as a percentage of the population by province, 1832

	One schoolboy in primary education per N inhabitants
Antwerp*	17.8
West Flanders*	17.4
East Flanders*	19.1
Limburg	18.9
Hainaut**	9.9
Liège**	13.4
Namur**	7.9
Luxembourg**	7.2
Brabant***	15.8

  * Flemish provinces.
 ** Walloon provinces.
*** Brabant includes Brussels and is divided by the linguistic border.

Source: Bequet (1972: 38).

FIGURE 5
Percentage of people on public assistance, by province, 1828-1839 and 1846

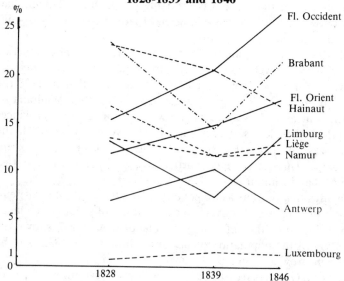

Source: Neuville (1976b: 63).

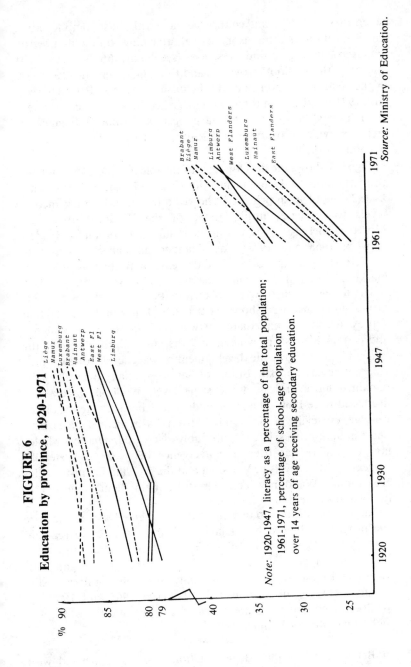

FIGURE 6

Education by province, 1920-1971

Note: 1920-1947, literacy as a percentage of the total population; 1961-1971, percentage of school-age population over 14 years of age receiving secondary education.

Source: Ministry of Education.

ween 1880 and 1947, and contrasted these with those compiled after the institution of compulsory school attendance on the number of children entering secondary school (see Figure 6). Between 1880 and 1947 the Walloon provinces and Brabant systematically have a higher education level than the Flemish provinces. Between 1961 and 1971 the situation is more even. The educational backwardness of Flanders thus seems to have been a serious imbalance in Belgium, which proved very difficult to remedy, at least before the Second World War.

Figure 7 shows the development of taxable incomes, for land and property until 1961, and of individuals in 1971 and 1975. As early as 1929 the Flemish provinces showed a tendency for their income levels to increase, unlike those of the Walloon provinces. Moreover, here too there was an economic acceleration in Flanders between 1968 and 1971: East Flanders in particular came to outstrip Hainaut and Liège. The decrease of Brabant is especially marked, and parallels strongly the decline in its share of the GDP.

The development of wages (which make up one of the elements of individual incomes) shows particularly clearly both the gap that still existed in the immediate postwar years between the Walloon and certain Flemish provinces, and the later improvement of the latter. Figure 8 shows the development of wage levels for each province. Whereas wages in Brabant (that is, especially Brussels) have remained more or less at the same level, Walloon wages have decreased since 1960. The two Flanders have been at a strong disadvantage, especially before 1960. Antwerp has remained stationary, but Limburg wages have appreciably decreased. On the whole, however, there has been a convergence between the generally decreasing wage levels in Wallonia and the growing Flemish levels, even though Walloon wage levels still remain slightly higher (cf. Leroy and Furnemont 1973).

An analysis of unemployment curves also enables us to measure the progress achieved by Flanders. Figure 9 shows the regional development from 1947 to 1971. From 1947 to 1965 unemployment rates were in general higher in Flanders, though the gap with Wallonia began to narrow, especially after the significant difference between the two in 1954. Between 1965 and 1968 the Walloon unemployment rate increased sharply, and by 1968 it was higher than that of Flanders.

Religion

In Belgium, a religious differentiation is often associated with linguistic differentiation. A mainly agricultural and always

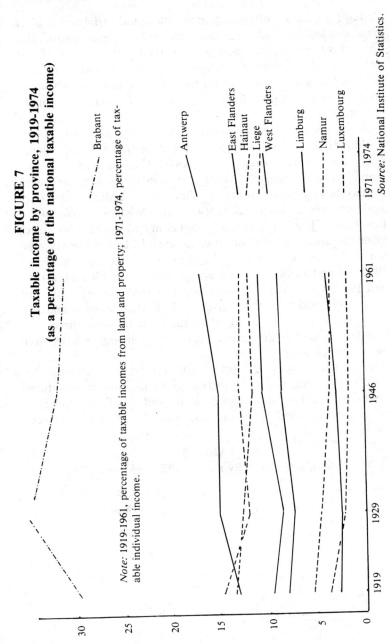

FIGURE 7

Taxable income by province, 1919-1974

(as a percentage of the national taxable income)

Note: 1919-1961, percentage of taxable incomes from land and property; 1971-1974, percentage of taxable individual income.

Source: National Institute of Statistics.

Catholic Flanders is often compared with an industrial and secular, even anticlerical, Wallonia. There are very few quantitative data about this problem, and those that exist relate only to the 1960s and later years. Apart from these data, some political confirmation of the link between regional differences in economics and religion is given by the fact that the Christian Party has always been dominant in Flanders.

Historians are ready to recognize the particular importance of the lower clergy in the Flemish emancipation movement (the higher clergy was French-speaking). Hill (1974: 33) speaks of them as a 'combination of religious and linguistic activism': by defending Dutch, the priests were aware that they were defending Catholicism against a French nation guilty of the anti-Christian excesses of the Revolution. The link between Catholicism and the Flemish movement has always been close, though several Flemish leaders were not practising Christians. The thorough researches made in 1950 – 51 and again in 1967 on Sunday church attendance (cf. Collard 1952; Voye 1973) are summarized in Table 8. These data are essentially based on statistics given by the Church about attendance at Sunday mass. They clearly show the Catholic predominance in Flanders, as well as the general decrease in religious practice over the whole country.

Nevertheless, we should not generalize the situation within the two principal regions, since there may be meaningful variations within each. That this is so is confirmed by Table 9, which reproduces 1967 data on attendance at Sunday mass by province.

TABLE 8
Church attendance by region, 1950/1951 and 1967

	Churchgoers
1950/51	%
Flanders	60.2
Wallonia	40.7
Brussels	36.9
1967	
Flanders	48.3
Wallonia	31.1
Brussels	22.6

Sources: 1950/51, Collard (1952); 1967, Voye (1973: 37).

FIGURE 8

Wage levels by province, 1948-1970

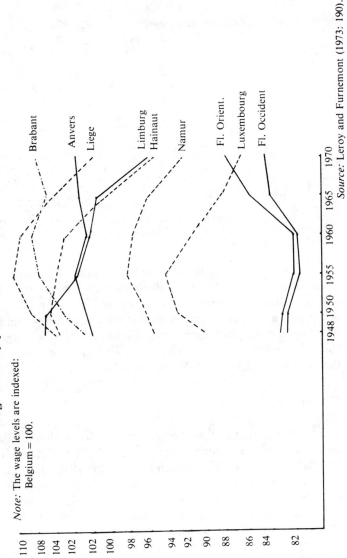

Note: The wage levels are indexed:
Belgium = 100.

Source: Leroy and Furnemont (1973: 190).

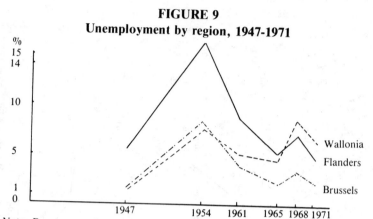

FIGURE 9
Unemployment by region, 1947-1971

Note: Percentage of unemployed and short-time workers out of the total of people who could have unemployment difficulties.
Source: Aiken-Quevit data.

While religious practice tends to be higher in most of the Flemish provinces than in most of the Walloon districts, the more urbanized and industrialized province of Antwerp is essentially non-practising, as is East Flanders. In Wallonia it is the provinces of Namur and especially Luxembourg that stand out in their degree of church attendance: in other words, religious practice is stronger in the two non-industrialized Walloon provinces.

Conclusion

The main intention of this chapter was to offer, within the limitations of available information, a more precise picture of the regional problem in Belgium and especially of the regional imbalances that characterize the country. The intensity that the conflicts based on this problem have taken in a developed and industrialized country makes one think (cf. Frognier et al. 1974) about the warning that Deutsch (1963) outlined in his classic study of social mobilization and political development. Social and economic change leads to a social mobilization which tends to favour integrative tendencies in systems where there is cultural homogeneity, whereas in systems that are not culturally homogeneous such change can produce a hardening of existing disparities. The two main Belgian regions, Flanders and Wallonia, have never flourished at the same time. Wallonia's industrialization coincided with Flemish stagnation, and the later decisive progress

TABLE 9
Church attendance by province, 1967

	Churchgoers
	%
Antwerp*	40.9
West Flanders*	54.6
East Flanders*	46.7
Limburg*	61.4
Hainaut**	21.8
Liège**	31.7
Namur**	43.7
Luxembourg**	62.1
Brabant***	31.3

* Flemish provinces.
** Walloon provinces.
*** Brabant includes Brussels and is divided by the linguistic border.

Source: Voye (1973: 34).

of Flanders coincided with Wallonia's decline. Therefore, the process of economic growth has always led to the prosperity of only one region, and not to the prosperity of Belgium as a whole. This situation is different from that in a country like Switzerland, which is in many ways similar to Belgium but which seems long ago to have settled most of its linguistic problems. The economic growth of French-speaking Switzerland and German-speaking Switzerland occurred simultaneously, with development poles on both sides of the linguistic frontier. The inequality that has marked the development of the Belgian regions does not parallel the Swiss experience, where economic development and benefits were more harmoniously distributed.

The economic and social indicators presented here show the inequalities inherent in Belgium's growth, especially the fact that during the course of the twentieth century Flanders has caught up with — and in some cases even progressed beyond — Wallonia. However, the data on education levels and wages seem to show that in some fields the Flemish advance was or is slower than in others. Could these drawbacks explain the persistence of Flemish nationalism in a period of regional prosperity? Other indicators would be necessary to establish whether this is so: such a hypothesis cannot be confirmed or rejected by the available data.

Regional differences are not only linguistic or socioeconomic. Indeed, there is a clear religious differentiation between Flemings and Walloons. In fact, Catholicism had always been one of the bases of the Flemish movement in its confrontation with a secular Wallonia. Since the last century the lower clergy, especially in rural areas, has played an important role in the Flemish movement. Although the conflicts about religion have lost some of their intensity, this is one further differentiation between Flanders and Wallonia, and can lead to very different attitudes and thoughts. The existence of different social and economic problems between the two linguistic communities influences any attempted solution of the community issue: the linguistic aspect is only a facet of a broader problem. Economic change has interfered with cultural and political mechanisms, to such an extent that the divorce between the two communities has in many ways increased, so hindering and delaying the possibility of their coexistence.

Since the relationships between centre and periphery have reached the situation where Wallonia is economically and politically dependent upon Flanders, the credibility of the central government lies in its ability to face up to the problem of regional disparities at the same time as attempting to establish an institutional structure that can accommodate the two communities. The Belgian situation shows the theoretical problems inherent in the concept of centre versus periphery. The relationships between the latter are not limited to a simple territorial configuration of ecological variables; they also imply studies at different levels of analysis:

1. the competition of different fractions of the bourgeoisie in economic activity;

2. the power relationships and class alliances formed by different fractions of the bourgeoisie with other social forces (for example, workers, farmers or craftsmen), as well as the ideological factors emerging from them;

3. the political strategies worked out by these alliances, and the power relationships that they create within the operation of the apparatus of the state.

Such studies go beyond the aim of this chapter, but the information gathered and classified here may contribute by offering a more solid empirical basis for the analysis of economic, political and cultural mechanisms which structure the relationships between centre and periphery in Belgium.

NOTES

1. For a general historical overview, see Luykx (1973). Economic history is more specifically the theme of Baudhuin (1944; 1958). For a history of Brussels, see Martens (1976).

2. The files are available at the Belgian Archives for Social Sciences (BASS), Catholic University of Louvain.

3. Data on the German community are incomplete; therefore the problem of this small minority, a legacy of the 1918 postwar settlement, will not be studied here.

4. Such evaluations are necessary because of the lack of a linguistic census since 1947. The abandonment of census data on language represents an earlier Flemish success. The political motive behind the demand was to avoid having precise data concerning the gallicization of the communes around Brussels. An estimate based on electoral results is impossible because the traditional parties have always presented joint electoral lists in the bilingual communes.

5. These statistics may be ambiguous. They must be interpreted as follows: the greater the number of people on assistance out of the total population by province, the more sources there were of poverty.

REFERENCES

Baudhuin, F. (1944). *Histoire Economique de la Belgique, 1914-1918* (2 vols). Brussels: Bruylandt.

Baudhuin, F. (1958). *Histoire Economique de la Belgique, 1945-1956*. Brussels: Bruylandt.

Becquet, C-F. (1972). *Le Différend Wallon-Flamand*. Couillet: Ed. Institut Jules Destrée.

Biname, J-P. and Jacquemin, A. (1973). 'Structures industrielles des régions Belges et grandes entreprises', *Recherches Economiques de Louvain*, no. 4.

Collard, F. (1952). 'Commentaire de la carte de la pratique dominicale en Belgique'. *Lumen Vitae*, no. 7: 644-52.

Deutsch, K. (1963). 'Social Mobilization and Political Development', in H. Eckstein and D. Apter (eds), *Comparative Politics: A Reader*. New York: Free Press.

De Vroey, M. and Carton, A. (1970). 'La Propriété et la part de marché des principales entreprises industrielles en Belgique', *Courrier Hebdomadaire du Centre de Recherche et d'Information Socio-Politiques* (CRISP), nos. 495, 509, 510.

Frognier, A-P., McHale, V. and Paranzino, D. (1974). *Vote, Clivages Socio-Politiques et Développement Régional en Belgique*. Brussels and Paris: Vander.

Hill, T.K. (1974). 'Belgium: Political Change in a Segmented Society', in R. Rose (ed.), *Electoral Behavior*. New York: Free Press.

Leroy, R. and Furnemont, A. (1973). 'L'Évolution des disparités régionales des salariés, Belgique (1950-1970)', *Recherches Economiques de Louvain*, June: 173-97.

Luykx, T. (1973). *Politieke geschiedenis van België*. Brussels and Amsterdam: Elsevier.

Martens, M. (ed.) (1976). *Histoire de Bruxelles*. Brussels: Privat et Ed. Universitaires.

Neuville, J. (1976a). *L'Evolution des relations industrielles en Belgique*. Brussels: Ed. Vie Ouvrière.

Neuville, J. (1976b). *La Condition ouvrière au 19 ième siècle*. Brussels: Ed. Vie Ouvrière.

Quevit, M. (1978). *Les Causes du déclin Wallon*. Brussels: Ed. Vie Ouvrière.

Voye, L. (1973). *Sociologie du geste religieux*. Brussels: Ed. Vie Ouvrière.

7

Nationalism, Religion and the Social Bases of Conflict in the Swiss Jura
David B. Campbell

Whenever the problems of multinational states are discussed, Switzerland, almost invariably, is invoked as the model of cultural coexistence. However, in the course of building their country, the Swiss have fought five civil wars, with continuing challenges posed by tensions among the various linguistic and religious groups.

The most serious threat in recent times to the integrity of the Swiss nation-state is the Jura question. For over a century and a half the Jura has been a French-speaking appendage to the German-speaking canton of Bern. For the past 30 years a separatist group, invoking linguistic, cultural and national distinctiveness, has claimed the right of the region to secede from Bern and establish an independent canton within the Swiss confederation. The separatist cause acquired a mass following, and the canton of Jura is now a reality. Still, far from all of the residents of the Jura supported independence. And the region has been divided — quite literally — as a result: the north, predominantly agricultural and Catholic, is now the twenty-third Swiss canton; the south, for the most part industrialized and Protestant, remains part of Bern. What was the nature of the conflict in the Jura? Who supported the claims for independence? And why?

The case history

Prior to 1815 the Jura, then known as the Evêché de Bâle, was an autonomous state within the Holy Roman Empire.[1] Although the sovereign, the Prince-Bishop, and the northern part of the country were Catholic, the south had been Protestant since the Reformation. In addition, the southern districts of the Evêché enjoyed a considerable degree of independence from the Prince-Bishop and had had links with the Swiss canton of Bern for many years (Bessire 1976: 83).

After the Napoleonic wars, the Congress of Vienna awarded the Evêché to Bern as compensation for that canton's loss of territories in Vaud and Aargau. With this act, the Jurassians became a French-speaking minority, representing only 15 percent of the population, in an overwhelmingly German canton.[2] Since then there have been five major periods of conflict in the Jura (see Campbell 1978). Two are of special interest for an understanding of the present situation: the period of religious unrest known as the 'Kulturkampf', and the time of 'pan-Germanism' and the First World War.

Although the religious split in the Jura dates from the sixteenth century, it did not become a major dividing force until much later. From the Reformation until the middle of the nineteenth century, the two religious groups lived quite separately: the religious frontier was clear (Bessire 1976: 306). But with industrialization this changed. Between 1850 and 1870 the number of Protestants living in the Catholic Jura increased by 285 percent. In the Protestant regions the number of Catholics increased almost as dramatically.

The breakdown of the religious frontier coincided with the battle between church and state that became known as the 'Kulturkampf' (see Bessire 1976: 304ff; Amweg 1974: 81ff.). In response to the publication of the *Syllabus of Errors* and the promulgation of the doctrine of papal infallibility, the government of Bern deposed the Bishop of Basel, the spiritual leader of the Jurassian Catholics. In place of the Roman Catholic Church, the Bernese recognized the 'Old Catholic' church, and all priests who did not cease communication with the Roman Catholic hierarchy were expelled. Later, a law was passed that required all priests to be elected by their parishes; parishes that did not conform were suppressed. Although the federal government intervened and declared the grossest offences of the Bernese to be unconstitutional, severe restrictions on the practice of the Roman faith were maintained; and it was not until 1935 that their last vestiges disappeared altogether (Bessire 1976: 338).

The wounds of the religious dispute continue to fester in the Jura. There has been resistance by both Protestants and Catholics to teachers of the other confession entering their districts. And each commune, but particularly the smaller ones, keeps careful note of residents of the other faith. The most recent example of the continuing importance of the religious question was the 1973 vote on the suppression of the constitutional articles banning the Jesuit order in Switzerland.[3] In the Catholic portions of the Jura the move was endorsed overwhelmingly; in the Protestant districts it was vigorously opposed.

Industrialization not only brought an intermingling of Catholics and Protestants. The watch-making factories in the south attracted large numbers of German-speaking immigrants, usually Bernese, to the Jura 'méridional'. In all, the proportion of German-speakers in the population rose from 3.2 percent in 1870 to 18.7 percent in 1920. These immigrants were also largely of the Protestant faith. Settling in the south, they found a receptive religious, if not linguistic, environment.

A number of German schools were established in response to this immigration. These were viewed with a jaded eye by most French-speaking Jurassians, whose misgivings increased sharply with the 'pan-Germanism' efforts of the German Empire. When it was revealed that a number of German-speaking schools had received financial aid from groups in Germany, the French-speaking population reacted strongly to this perceived cultural threat. The French-speakers insisted that the German language schools be closed and, in accordance with the Swiss principle of territoriality in language matters, the Bernese authorities upheld these demands (Bessire 1976: 334).

But shortly afterwards the First World War erupted, and linguistic differences became very salient throughout Switzerland. Whereas the German-Swiss tended to sympathize with the Kaiser, French-speakers supported the Allied cause. In the Jura the tension of the war, coupled with the effects of the schools problem, spawned a short-lived separatist movement. Although the movement was highly romantic in flavour and never acquired a mass following, it was important in that it raised directly the question of the linguistic and cultural distinctiveness of the Jurassian population. It was also significant in that it united Jurassians of both the north and the south in a stand against 'Germanization' of their land. The apparent unanimity was shattered, however, when a Protestant pastor recalled the religious differences between the north and the south and, calling upon southerners to remember their long historic ties with Bern, resisted the idea of an independent Jura (Bessire 1976: 336). Shortly afterwards, the separatist cause died out completely.

The social and economic
context of the Jura question

The current conflict also has its immediate origin in the linguistic differences between the Jura and Bern. But the social and economic context of the present dispute contrasts sharply with previous ones. Prior to the First World War the Jura as a whole was an economically privileged area. It was highly industrialized and, owing to the German occupation of Alsace, an important transportation centre as well. But after 1920 the situation began to change. Since then the Jurassian economy has been in decline and the region has been pushed increasingly to the margins of the Swiss system.[4] This decline has been characterized by two principal changes. First, the industrial sector lost its diversified character and became increasingly centred around watch-making. Second, whereas the Jurassian economy had been controlled by the indigenous population during the early phases of industrialization, it increasingly passed into the hands of non-Jurassian industrialists. One of the first important factors that contributed to the turnabout of the region's economy was a change in the status of the railways. When Alsace was returned to France after the First World War, Basel became the favoured rail centre for the transport of French goods. As a result, the rail line through the Jura drastically declined in importance.

Jurassian industry has become dominated more and more by the watch-making sector. In 1950 watch-making already accounted for the employment of over 50 percent of the active population of the region. This dependence on a single industry, particularly one that is export-oriented, made the Jura extremely vulnerable to shifts in the world economy. During the depression of the 1930s the Jura was struck particularly hard. In some communes unemployment rose to 30 percent of the population. Today much of Switzerland's still relatively small unemployment is to be found in the Jura.

Concentration of industry has also been growing within the watch-making sector. Whereas the industry had once been dominated by small workshops spread over the entire region, this is no longer the case. Between 1955 and 1965, for example, the number of watch-making enterprises declined by 232, largely through the decline of the small factory (Prongué 1974: 288). And with the concentration of industry has come a loss of local control. But this tendency is true for the whole of Jurassian industry. The number of independent businessmen in the Jura has fallen steadily.

Between 1950 and 1970 the proportion of independents in the industrial sector declined by 5.7 percent. As external decision-making centres have increased in importance, the Jura has tended to lose control of its own development.

Changes in the primary sector have been no less important. In general, agricultural enterprise has been declining, particularly in the north. Whereas 15 percent of the population had been employed in agriculture in the northern Jura in 1920, only 5 percent was so employed in 1970. In the south the figures were 9 percent in 1920 and 4 percent in 1970. The independent farmer has also decreased in numbers: between 1939 and 1950 the number of farming establishments declined by 90 percent. Prongué (1974: 286) reports that this decline particularly involves small farmers. In 1969, for example, 52 percent of the total number of farms exceeded 10 hectares in size, and between 1950 and 1970 the number of independent farmers decreased by 5 percent. On the basis of a typological analysis of communes between 1940 and 1960, however, Bassand (1975: 147-8) has demonstrated that in the north the number of rural communes increased by 12 percent. This finding, coupled with the information on farms and farmers themselves, suggests that the concentration of industry and the decline of the independent farmer go hand in hand.

But the growth of the industrial sector at the expense of the agricultural sector is a universal phenomenon. Why then should it be particularly important in the Jura? The salient point is not simply that the agricultural sector has been overtaken by an industrial imperative, but that developments in the industrial sector have coupled the move from the farm and to industry with a move from a Jurassian-controlled farming sector to a non-Jurassian-controlled industrial sector. And this has been particularly pronounced in the north, where the decline of agriculture has been both more recent and more precipitous. There, industrialization occurred late. As a result, the industrialization process has been capital-intensive, and this necessary capital has come, for the most part, from outside the Jura.[5] This is reflected in the considerably greater decline of independents in the secondary sector in the north.

Between 1920 and 1970 the population of the Jura increased, on average, by only 0.25 percent per year: during the 1920s and 1930s the population actually declined. This period marks the first time that population growth in the Jura lagged behind that of the rest of Bern, and it has had important political consequences, for since 1920 the size of the Jurassian delegation to the Bernese Grand Conseil has fallen consistently. In order to maintain a steady proportion of legislative seats, the Jura had not only to increase its population, but to do so at least at the same rate as the rest of the canton.

In language, there has been an overall decline in both the absolute number and the proportion of the population that is German-speaking. But this decline has not occurred at the same pace in each district. In Porrentruy and Courtelary the proportion of German-speakers continued to climb until 1930, and in Franches-Montagnes until 1941. The overall decline can be accounted for by the fact that there are few German schools left in the Jura. The children of immigrants, therefore, are taught French. But a decline in the number of German-speakers should not be confused with an increase in the number of Jurassians. After 1920 the proportion of the population that were citizens of their commune of residence (see note 7) continued to decline, falling to 26 percent by 1941. More recently, the figure has risen slightly.

In the Catholic areas, the proportion of the population that is Protestant has remained unchanged at 15 percent. In the south, however, the proportion of Catholics has increased from 28 to 38 percent. This shift would seem to indicate continued migration from north to south. The decline in the agricultural population of the north adds further weight to this suggestion. In general, then, the conflict that has been unfolding in the Jura since 1947 has taken place in an environment of economic decline and demographic stagnation.

The unfolding of the Jura conflict

The catalyst of the current conflict was the 'Affaire Moeckli'. In September 1947, Georges Moeckli, a Jurassian and a member of the Bernese executive council, was refused an important ministry in the cantonal government when German-speaking members of the legislative assembly argued that the position was far too important to be entrusted to a French-speaker. A wave of protest swept the Jura, and many became convinced that only by completely separating itself from Bern could the Jura protect its own interests. These separatists formed an organization, the 'Rassemblement jurassien' (RJ), which has led the campaign for a canton of Jura. The RJ based its claim on the linguistic and cultural distinctiveness of the Jurassians.[6] But not all residents of the Jura have supported the creation of a canton of Jura. The anti-separatists organized a group known as the 'Union des patriotes jurassiens' (now the 'Force démocratique') to oppose the creation of an independent canton.

The goal of the RJ is to 'free the Jurassian people from Bernese domination' (*Statuts du Rassemblement Jurassien*, Article 1). And its principal weapon has been the referendum. In 1957 the RJ launched an initiative calling for a consultative vote that would allow Jurassians to decide whether they wished to form a new canton. Voting on this initiative took place in 1959. Although the Canton of Bern rejected the notion overwhelmingly, the vote in the Jura itself was close: 52 percent of the electorate rejected the idea, 48 percent were in favour. But opposition to and support for the initiative followed a clear north-south split. In the north 68.7 percent of the electors cast ballots in favour of a new canton, whereas in the south 70.1 percent voted against the move.

In 1974 a second referendum was held. Unlike the 1959 vote, this one was limited to the Jura itself. And unlike the 1959 vote, it would yield definite results. The mechanisms of the process allowed that any district that expressed an opinion contrary to the results of the Jura as a whole could, upon petition, hold a second referendum to decide whether it wished to form a new canton or not. Finally, if there was to be a split, any commune on the border could petition for a third referendum on the subject. And three referenda there were.

In the first, 52 percent of the electors in the Jura voted in favour of leaving Bern. As a result, a new canton seemed assured. But in the south 66.5 percent of the electorate voted against such a move. In the subsequent referendum in the south 73 percent of those who voted opted to remain within Bern. The third referendum shifted a number of communes back and forth between north and south. The Jura was divided. The northern districts were to become the twenty-third canton in the Swiss confederation: the southern districts remained a part of Bern. Clearly, if one can account for the split in the Jura, one will have a better understanding of the nature of the conflict in the region.

Since the most obvious difference between north and south is religious, many writers have argued that the Jura conflict is, in reality, religious. Meyer (1968) claims that:

> the religious factor is a 'sleeper' in the separatist movement. Although religious peace and toleration prevails today on the surface the sharp conflicts of the past have left subterranean residues of resentment and antagonism which are concealed by the cultural-linguistic grievance proclaimed by the RJ [Meyer, 1968: 738].

Less sweepingly, Bassand (1975: 142) has argued that 'in the anti-separatist region, religion is the factor which best accounts for separation whereas in the separatist region the linguistic factor is decisive'. Gasser (1978) notes that the border separating the new canton from that part of the Jura remaining in Bern is exactly the line of demarcation between the Catholic and Protestant districts, and Prongué (1976: 372) suggests that, although the Jura question is not essentially a religious one, the memory of the Kulturkampf must surely play a role. But the rhetoric of the separatists never touches on the religious issue, nor indeed on any issue that seems to divide the north and the south except for the presence of a large number of immigrants in the south. And, as Table 1 shows, it is not only religion that divides the separatist and anti-separatist regions of the Jura: the north and south differ also in terms of the percentage of the population who originate from the German-speaking part of Bern, the percentage of German-speakers in the population, and the percentage of the population employed in the agricultural sector of the economy.[7]

Most of the research that has stressed the importance of religion has been based upon the observation that the correlation ($r = 0.95$) between the proportion of the population that is Catholic (or Protestant) and the percentage of votes for a new canton has been significantly higher than that ($r = 0.86$) between the proportion of the population that is French-speaking and the percentage of 'yes'

TABLE 1
Voting results in the Jura, 1959 and 1974

Administrative district	% voting in favour of a new canton		% Catholic	% French	% origin Canton Bern excluding Jura	% in agriculture
	1959	1974	1970	1970	1970	1970
Jura North						
Delémont	72	76	82	77	17	8
Franches Montagnes	76	76	85	86	13	20
Porrentruy	66	66	85	86	14	11
Jura South						
Courtelary	24	23	31	69	41	8
La Neuveville	34	35	23	65	34	9
Moutier	34	42	46	73	37	7

votes. Yet barely 10 percent of the total population have German as a mother tongue. And since the separatists stress that their claims are cultural in the broadest sense, it is surprising that other factors have not been taken into account. For example, the correlation between the proportion of the population whose origin is a commune in German-speaking Bern and the percentage voting 'yes' in the 1974 referendum is − 0.96. Still, there remain good reasons for imagining that the religious element might be important in the current conflict. And there appears to be justification for claiming that the 'national' element is crucial. But ecological analyses can carry us only so far. To proceed further we require evidence of a different sort. The data upon which the following investigations are based are drawn from a random sample of the Jurassian electorate. This survey was conducted, by mail, in the summer of 1978. In all, there were 508 respondents to the questionnaire.

Religion, nationalism and autonomy

As a first step towards understanding the bases of conflict in the Jura, respondents were asked to rank in importance four factors frequently mentioned as reasons for the division of the region: the religious differences between north and south; the economic differences between the two areas; the presence of non-Jurassians in the south; and the historic ties that have existed between the south and the Canton of Bern. As Table 2 clearly shows, the presence of non-Jurassians in the south and the religious difference are regarded as being the most important factors dividing the Jura. The non-Jurassian factor was selected by 44 percent of the respondents as being most important. The religious factor was placed first by 29 percent. And if first and second placings are considered together, more than 65 percent of the respondents ranked non-Jurassians in the south and religious differences in these categories. By contrast, less than half of the respondents ranked the historic ties between the south and Bern in either of the first two places, and less than one-third of the sample ranked economic factors first or second. When respondents were asked to state other factors that they considered important, 34 percent of those replying claimed that 'differences in the mentality' of the two regions were crucial, and 17 percent claimed that linguistic differences played an important role.

TABLE 2
Position assigned to selected factors as causes for the division in the Jura

	1st	2nd	3rd	4th	Overall rank
	%	%	%	%	%
Religious differences	29	36	16	19	2
Economic differences	13	17	33	36	4
Non-Jurassians in south	44	22	18	16	1
Ties between south and Bern	16	24	31	29	3

The apparent importance of religious, national, cultural and linguistic differences is highlighted still further when the rankings assigned to these factors are compared with the way in which the respondents voted in the 1974 referendum. Table 3 presents the relationship between vote and the importance assigned to religion; Table 4 the relationship between vote and the ranking of the presence of non-Jurassians as a factor. In both cases the relationship is strong, but whereas the former is strong and negative, the latter is strong and positive. Similarly, whereas 44 percent of those who opposed the formation of a canton of Jura claimed that there were cultural differences between northern and southern residents, only 30 percent of those who favoured a new canton mentioned this

TABLE 3
Rank assigned to religious differences, by vote in 1974 referendum

	1st		2nd		3rd		4th		Total
Yes	41	17 (50)	74	40 (117)	77	19 (55)	85	24 (72)	67 (294)
Abst.	7	39 (9)	4	30 (7)	3	9 (2)	6	22 (5)	5 (23)
No	52	54 (64)	21	28 (33)	20	12 (14)	9	7 (8)	27 (119)

Gamma = −0.51
$r = -0.33$

TABLE 4
Rank assigned to the presence of non-Jurassians in the south, by vote in the 1974 referendum

	1st	2nd	3rd	4th	Total
Yes	95 ^{64}(195)	61 ^{19}(58)	43 ^{10}(31)	27 ^{6}(19)	68(303)
Abst.	3 ^{26}(6)	8 ^{35}(8)	4 ^{13}(3)	8 ^{26}(6)	5(23)
No	2 ^{3}(4)	30 ^{25}(29)	53 ^{32}(38)	65 ^{40}(46)	26(117)

Gamma = 0.78
r = 0.58

factor. And although only 13 percent of those voting against the new canton mentioned language, 20 percent of those favouring this option claimed that linguistic differences were important. Clearly, the perception of the factors that have divided the Jura is related to whether or not individuals favoured the creation of a new canton.

Although those favouring a canton of Jura stressed the importance of non-Jurassians in the south as the key factor dividing the region, and while those opposing a new canton stressed the importance of the religious factor, these two elements are not entirely independent. The period of industrialization brought many immigrants to the Jura, particularly to the south. These immigrants, mostly from communes in German-speaking Bern, were overwhelmingly Protestant. Hence, one might suspect that religious dif-

TABLE 5
Origin by religion

	Jurassian	Non-Jurassian	Total
Catholic	78 ^{79}(212)	30 ^{21}(55)	59
Protestant	22 ^{32}(61)	70 ^{67}(127)	41
Total	60	40	(455)

Gamma = 0.78
r = 0.48

ferences and differences of origin are two sides of the came coin. The non-Jurassians, two-thirds of whom live in the south, tend very strongly to be Protestant (Table 5). Similarly, over three-quarters of the Jurassians are Catholic, and almost 80 percent of the Catholics in the sample are of Jurassian origin. The relationship between origin and religion is strong and positive. However, in order to sort out the various effects of the several factors discussed, it is necessary to relate them directly to the vote in the 1974 referendum.

The strong positive correlation between religion and the vote observed at the aggregate level is also evident at the individual level (Table 6). Three-quarters of the votes favouring the creation of a new canton were cast by Catholics, and 77 percent of the 'no' votes by Protestants. But, whereas Catholics were overwhelming in their support of the separatist cause, no such general accord existed among Protestants. Barely one-half of the Protestants in the sample opposed a canton of Jura, and over 40 percent supported it. This suggests that, rather than being a straightforward sectarian split, it is a division within the Protestant ranks that accounts for the religious elements of the current conflict. Although opposition to a canton of Jura comes principally from Protestants, the latter are not incontestably anti-separatist.

If the relationship between vote and religion is examined separately in the north and the south, another interesting point emerges. In the north, where Catholics are in the majority, 43 percent of the Protestants opposed the independence of the Jura and 45 percent supported it. But in the south, 53 percent of the Protestants voted 'no' to a new canton and only 40 percent supported

TABLE 6
Vote by religion, 1974

	Yes		Abst.		No		Total
Catholic	74	86 (229)	45	4 (10)	22	10 (28)	59 (267)
Protestant	25	41 (78)	56	7 (13)	78	51 (97)	41 (188)
Total	67	(307)	5	(23)	27	(125)	(455)

Gamma = 0.77

$r = 0.45$

the move. Among Catholics, 19 percent of those living in the south, where they form a distinct minority, opposed a new canton, whereas only 7 percent in the north were against leaving Bern. Being in a minority position vis-à-vis the other religious group tended to moderate the propensity of Catholics to support or Protestants to oppose the independence option.

Table 7 presents the relationship between origin and the vote in the 1974 referendum. This relationship, like that between religion and the vote, is positive, although not so strong. And the same general patterns are discernible. Over 70 percent of the 'yes' vote comes from persons of Jurassian origin, whereas nearly two-thirds of the 'no' vote comes from non-Jurassians. Similarly, whereas almost 80 percent of the Jurassians voted in favour of forming a canton of Jura, the distribution among non-Jurassians was less clear-cut: 49 percent report having voted 'yes', and 44 percent 'no'. The effect that being in a majority or minority religious position had on the propensity to support or oppose the separatist option is also apparent in the case of origin. Almost 90 percent of Jurassians living in the north, where they form a majority, favoured independence from Bern. In the south, however, only 63 percent of Jurassians voted for a new canton. Similarly, in the north only 28 percent of non-Jurassians cast 'no' votes, whereas 52 percent of the non-Jurassians in the south voted 'no.

But there is a strong, albeit non-causal, relationship between origin and religion. Hence, does origin affect the vote independently of its covariation with religion? Table 8 presents the relationship between origin and the vote for each of the religious groups. Among Catholics there is no relationship between origin

TABLE 7
Vote by origin

	Yes		Abst.		No		Total
Jurassian	71	80 (222)	50	4 (12)	35	16 (45)	60 (279)
Non-Jurassian	29	49 (92)	50	6 (12)	65	44 (82)	40 (186)
Total	67	(314)	5	(24)	27	(127)	(465)

Gamma = 0.58
r = 0.32

and support for an independent Jura. Among Protestants, however, origin continued to play a role, albeit an attenuated one. Whereas 54 percent of the Protestant Jurassians voted 'yes', 41 percent voted 'no'. On the other hand, only 35 percent of the Protestant non-Jurassians supported independence and 57 percent opposed it. Among Catholics, origin seems not to have influenced their vote, but among Protestants those of Jurassian origin were more likely to favour the creation of a canton of Jura. Clearly, some of the division within the Protestant group can be accounted for by the split between Jurassians and non-Jurassians.

Language is the most obvious factor separating the Jura, taken as a whole, from the Canton of Bern. But it is less certain what effect this has had on the conflict within the Jura. The difficulty stems from the very low percentage of persons of German mother tongue who live in the region. Since there are now very few German language schools in the Jura, children of immigrants from German Switzerland have been assimilated fairly rapidly into a French-speaking milieu. Although there is a strong positive relationship

TABLE 8
Vote by origin, controlling for religion

	Yes		Abst.		No		Total	
Among Protestants								
Jurassian	42	54 (33)	23	5 (3)	26	41 (25)	32 (61)	
Non-Jurassian	58	35 (45)	77	8 (10)	74	57 (72)	68 (127)	
Total	41	(78)	7	(13)	52	(97)	(188)	

Gamma = 0.32
  r = 0.17

	Yes		Abst.		No		Total	
Among Catholics								
Jurassian	80	87 (184)	80	4 (8)	71	9 (20)	79 (212)	
Non-Jurassian	20	82 (45)	20	4 (2)	29	14 (8)	21 (55)	
Total	86	(229)	4	(10)	10	(28)	(267)	

The relationship is not significant.

between mother tongue and the vote (gamma = 0.66), German-speakers constitute only 8 percent of the sample.

In order to examine the linguistic element of the present situation, a subjective measure has been adopted. If language is a salient element in the conflict in the Jura, there should be a relationship between the importance that an individual attaches to his language and his desire for independence from Bern. In order to examine this possibility, respondents were asked to rank themselves on a nine-point scale ranging from least important to most important, when asked the following question: 'Among all of the things that you consider important to you, how much importance would you assign to your being part of a French community?' Table 9 displays the relationship between the (now trichotomized) measure of linguistic salience and the vote in the 1974 referendum. Once again, the 10 percent of those who claimed that being part of a French community was highly important to them opposed the separatist position: 85 percent were in favour. Among those who ranked being part of a French community low in importance, 56 percent opposed the creation of a new canton and only 38 percent supported it.

TABLE 9
Vote by the salience of the French community

	Yes		Abst.		No		Total
High salience	71	85 (209)	58	4 (11)	22	10 (25)	57 (245)
Medium salience	17	54 (50)	19	4 (4)	33	41 (38)	21 (92)
Low salience	12	38 (35)	24	5 (5)	45	56 (51)	21 (91)
Total	69	(294)	5	(20)	27	(114)	(428)

Gamma = 0.66
$r = 0.47$ (based upon uncollapsed nine-point scale).

On the basis of the evidence presented so far, we might speculate on the possibility of a tension between the salience ascribed to linguistic community and that to religious beliefs. In view of the history of religious disputes in the Jura, it should not be surprising to find that, among Protestants, those who consider their religious views to be more important than attachment to a linguistic community would be inclined to oppose the idea of a new canton. Since

Protestants would be in the minority in an independent Jura, there would be less inclination to oppose the separatist claims among Protestants who regard their religious beliefs as being less important than their attachment to the French community. Since the Catholic population of the Jura differs from Bern on both religious and linguistic grounds, their tendency to vote for independence based on linguistic group membership would not be diminished by the importance of their religion.

In order to examine these possibilities, respondents were asked to rank themselves, again on a nine-point scale, when asked: 'Among all of the things that you consider important to you, how much importance would you assign to your religion?' Scores on this question were then subtracted from those on the question concerning the importance of language, and the sample was divided into three groups: those for whom language was more important, those for whom religion was more important, and those who scored equally on both. Although there is no relationship between this constructed factor and voting preference among Catholics, Table 10 indicates that there is a fairly strong and positive relationship among Protestants. As expected, some of the division within the Protestant group as regards support for the new canton can be accounted for by the relative salience attached to linguistic and religious group affiliations.

TABLE 10
Vote by salience of religion versus salience of French community among Protestants

	Yes		Abst.		No		Total
French community	62		5		32		45
more important	64	(48)	33	(4)	29	(25)	(77)
Both ranked	34		14		52		17
equally	13	(10)	33	(4)	52	(15)	(29)
Religion more	26		6		68		38
important	23	(17)	33	(4)	53	(45)	(66)
Total	46	(75)	7	(12)	49	(85)	(172)

Gamma = 0.51
r = 0.34

It remains unclear whether the parties to the conflict are defined along religious, linguistic or national lines, or whether national differentiation is based upon linguistic or religious factors. In order to examine these questions, it is necessary to probe further into the subjective elements of the problem. Before there is nationalism, there must be identification with the nation. If people support the creation of a canton of Jura for nationalist reasons, there should be a strong relationship between the extent to which people identify with the 'Jurassian nation' and the tendency to vote 'yes' in the referendum. If, on the other hand, there is not a nationalist element to the conflict, then there is no reason to suppose that such a relationship would exist. In order to examine this, respondents were asked to state the extent to which they considered themselves to be Jurassian or Swiss. Their answers were cross-tabulated with the reported vote in the 1974 referendum. The strong, positive relationship between identification and voting preference (Table 11) suggests a clear nationalist element to the Jurassian conflict. Everyone who claimed to be solely Jurassian supported the creation of a new canton. And 95 percent of those who identified primarily with the Jura favoured the separatist option. At the other extreme, 59 percent of those who did not identify at all with the Jura opposed leaving Bern, while 53 percent of those who identified only secondarily with the Jura were adverse to the new canton.

TABLE 11
Vote by identification with the Jura

	Yes		Abst.		No		Total
Jurassian	100		0		0		3
only	5	(15)	0	(0)	0	(0)	(15)
Jurassian	95		1		4		29
first	40	(126)	4	(1)	5	(6)	(133)
Equally Jurassian	71		8		21		34
and Swiss	36	(111)	50	(12)	27	(33)	(156)
	43		5		53		24
Swiss first	15	(47)	21	(5)	47	(58)	(110)
	27		14		59		10
Swiss only	4	(12)	25	(6)	21	(26)	(44)
Total	68	(311)	5	(24)	27	(123)	(458)

Gamma = 0.72
    r = 0.48

Although it now appears that there is a strong nationalist base to the conflict in the Jura, the role of religion remains unclear. And if the conflict in the Jura is based upon national differences, but these in turn are based upon religious differences, there is little to be gained and much to be lost by focusing only on the nationalist element. Table 12 presents the relationship between religion and national identification. The strong positive relationship suggests that religion does affect, in part, the tendency to perceive oneself as being Jurassian or non-Jurassian. But whereas an overwhelming majority of Catholics consider themselves to be at least equally Jurassian and Swiss, only a bare majority of Protestants claim to be Swiss first and only. Once again, Catholic near unanimity is contrasted with a strong division in the Protestant ranks. And because of this division, it is impossible to argue that national differences simply mirror religious differences. Although religious persuasion may inhibit some Protestants from identifying with the Jurassian nation, it is not the case that Protestantism excludes an individual from identifying with the Jura.

But if nationalist sentiments form a basis for the conflict, and if religion is related to, but does not serve as a defining characteristic of, this nationalism, are religious factors nevertheless important in determining the structure of the conflict? That is, do religious differences serve as a basis of conflict in the Jura independently of any co-variation with national identification? And if so, which basis of

TABLE 12
Identification with the Jura by religion

	Catholic	Protestant	Total
Jurassian only	5 ^{87}(13)	1 ^{13} (2)	3 (15)
Jurassian first	39 ^{80}(106)	13 ^{20}(26)	28 (132)
Equally Jurassian and Swiss	36 ^{61}(98)	32 ^{39}(63)	34 (161)
Swiss first	17 ^{40}(45)	34 ^{60}(67)	24 (112)
Swiss only	3 ^{18}(19)	20 ^{82}(40)	10 (49)
Total	58 (271)	42 (198)	(469)

Gamma = 0.59
r = 0.41

division — nationalist or religious — has more impact on the current situation? In order to answer these questions, the vote in the 1974 referendum was regressed on to both religion and national identification. The beta weight (path coefficient) for national identification was 0.36 and for religion 0.32: that is, the Jurassian conflict is affected by both national and religious divisions, independently of their shared variance. However, the independent direct effects of national identification are greater than those of religion.

But why do people identify with the Jura? Religious affiliation does not provide an adequate basis for identification, or rather, for inhibiting identification with the Jura. The same is true of origin. Despite a moderate relationship between origin and identification, 47 percent of those of non-Jurassian origin consider themselves to be at least equally Jurassian and Swiss. Of the objective factors, only language seems important, for no German-speakers consider themselves to be either solely, or even primarily, Jurassian. But again, the extremely small number of germanophones in the population must be noted. Objectively, there is very little that can be said about who is a Jurassian and who is not. Only minimal criteria can be established: namely, it is necessary, but not sufficient, to be a French-speaking resident of the Jura.[8]

This evidence accords with the view of such writers as Ernest Renan (1970), who have argued that objective factors do not serve as adequate bases for distinguishing between nations, and that subjective factors are crucial. Similarly, Weber (1978: 921-6) has argued that national differences revolve around distinctions in systems of meaning and interpretation. That which binds individuals together in a nation is not the sharing of this or that objective factor, but the sharing of a system of thought that allows for mutual understanding (Weber 1978: 390). From this perspective, a person is a Jurassian because he shares a 'meaning framework' with other Jurassians. When respondents volunteered reasons for the division of the Jura, the factor most often cited was a 'difference in mentality' between residents of the north and south.

The reason we tend to focus on objective factors as 'defining' criteria is, in most cases, that there is a probabilistic relationship between possession of a given objective characteristic and sharing in the system of thought of a given group. In the Jura, however, such relationships are not at all clear. And this is the source of confusion, for it makes it very difficult to make practical distinctions, on an individual basis, between Jurassians and non-Jurassians.

The social bases of Jurassian nationalism

Although Jurassians may oppose nationalist claims for an independent canton for religious reasons, who supports these claims? What, if any, are the social bases of Jurassian nationalism? The attempt to assess the social bases of nationalist sentiment in the Jura is impeded by two major obstacles. First, the inability to make a priori distinctions between Jurassians and non-Jurassians threatens to mute, or exaggerate, any relationships between socio-structural variables and support for independence. Further, because most non-Jurassians live in the south, and because the south is heavily industrial whereas the north is largely agricultural, it is impossible to assume that distributions along any given variable will be similar for both Jurassians and non-Jurassians. In order to control for this as much as possible, distributions will be examined in the Jura as a whole and in the north and the south individually. If the trends are consistent, confidence that real relationships exist will be increased. The second obstacle is the success of the separatist movement itself. The independence cause in the Jura is now mature. The Rassemblement jurassien has succeeded in mobilizing virtually all of its potential supporters. If those who oppose an independent Jura on religious grounds are excluded, almost 80 percent of the population supports the creation of a new canton. And if it were possible to distinguish between Jurassians and non-Jurassians, this figure, undoubtedly, would be higher.

But not all Jurassians favour autonomy with the same intensity. And if there are social bases to Jurassian nationalism, they should be evident among the staunchest advocates of independence. Since the RJ has led the struggle for an independent Jura throughout the course of the current conflict, its members must be counted as the strongest supporters of autonomy. Therefore, the social bases of Jurassian nationalism, if any, should be apparent in its membership. The following analyses examine the social bases of the membership of the RJ along five variables that tap potentially important dimensions of the social structure: age, subjective social class, education, income, and occupation.[9] Neither age nor subjective social class appeared to be related to support for Jurassian independence. On each of the other variables, however, interesting patterns do emerge.

Table 13 presents the pattern of support for the RJ that is associated with education. In the Jura as a whole, a disproportionate number of RJ members are drawn from among those with

TABLE 13
Support for the RJ among educational groups

	Primary %	Secondary %	Prof./technical %	University %	Ecole Normale %	Other %	Total N %
Whole Jura							
Activists	31	10	49	5	3	1	(166)
Others	25	10	53	7	3	1	(314)
Total	27	10	52	6	3	1	(480)
Jura North							
Activists	32	9	47	6	3	2	(108)
Others	33	12	49	4	2	1	(138)
Total	33	11	48	5	2	2	(246)
Jura South							
Activists	29	10	53	3	3	0	(58)
Others	19	9	57	10	4	1	(176)
Total	21	9	56	8	4	1	(234)

only a primary education. When the north and the south are considered separately, however, it becomes apparent that it is only in the southern regions that this tendency occurs. Although those with a primary education represent only 19 percent of the non-activist population, 29 percent of the support for the RJ comes from this group. And whereas 10 percent of the non-activists in the south have a university education, only 3 percent of the RJ membership has attended university.

Table 14 shows that, in general, the RJ is under-represented among those from the higher income groups. As was the case with education, this is particularly true in the south. Whereas over 16 percent of the non-activists are found in the upper two income brackets, fewer than 7 percent of the members of the RJ come from either of these groups. And although slightly under-represented among the lowest group, the RJ is markedly over-represented in the next to lowest group. In the north the same pattern prevails, though it is less sharp in its distinctions. There, 6 percent of the membership in the RJ is drawn from among the highest income groups, whereas 12 percent of the non-activists fall in these categories. Also, in the north, there is a slight over-representation in the middle income groups.

TABLE 14
Support for the RJ among income groups

| | Income group | | | | | |
	1	2	3	4	5	Total N
	%	%	%	%	%	
Whole Jura						
Activists	17	56	22	3	3	(162)
Others	17	46	23	9	5	(306)
Total	17	49	22	7	4	(468)
Jura North						
Activists	19	55	20	4	2	(104)
Others	19	53	16	9	3	(138)
Total	19	54	16	7	2	(242)
Jura South						
Activists	12	57	24	2	5	(58)
Others	15	40	29	10	6	(168)
Total	14	45	27	8	6	(226)

TABLE 15
Support for the RJ among occupational groups*

	Farmers	Independents in business and commerce	Upper managers	Middle managers	Clerical/ sales	Workers	Total
	%	%	%	%	%	%	
Whole Jura							
Activists	9	14	6	12	17	26	(167)
Others	9	11	13	12	18	22	(324)
Total	9	12	10	12	17	23	(491)
Jura North							
Activists	10	14	4	11	19	22	(108)
Others	11	10	11	7	17	22	(144)
Total	11	12	9	9	18	22	(252)
Jura South							
Activists	7	15	5	14	12	34	(59)
Others	6	11	12	14	16	19	(200)
Total	6	12	11	14	15	23	(259)

* Percentages do not sum to 100 because the 'other' occupational category has been omitted from the table.

The distribution of support for the RJ among various occupational categories is presented in Table 15. In the Jura as a whole, there is disproportionate support for the RJ among independents in business and commerce and among workers. There is also a marked under-representation among upper management. This pattern is repeated in the south, where 15 percent of the RJ support comes from independents and over one-third from workers. In the north the RJ continues to be over-represented among independents and under-represented in the upper management group: however, there is no over-representation of workers.

Among the groups that give disproportionate support to the RJ, an examination of the social backgrounds of the supporters and opponents of Jurassian independence is revealing. Of the workers who supported a new canton, nearly one-third come from farm backgrounds. This compares with only 15 percent among those workers opposed to a new canton. In general 29 percent of those who opposed it had farm backgrounds. Clearly, individuals who come from farm backgrounds and who are now working in an industrial setting tend to give disproportionate support to the independence cause. Among those independents in business and commerce who supported an independent Jura, 34 percent come from independent business backgrounds, whereas only one-quarter of those who opposed an independent Jura comes from this group: 42 percent of those independents who oppose a new canton come from a worker background. Although there is an overall tendency for independents to support Jurassian nationalism, it is considerably less pronounced among those from worker or farm backgrounds. Finally, Table 16 presents the pattern of educational attainment associated with support for the RJ among workers and independents in business and commerce. The clear indication is that, among those groups that tend to give disproportionate support to the RJ, it is the less well-educated who are especially supportive. Only university-educated workers, a very small group indeed, counter this trend.

Support for Jurassian nationalism appears to be stronger among the lower-income, less well-educated segments of society, although there appears to be no tendency for support to be concentrated in any particular age group, or, for the most part, in any particular set of class identifiers. The clearest patterns of support are associated with occupational groups. A disproportionate number of RJ members come from independent businessmen and workers. And the RJ is significantly under-represented among upper-level

TABLE 16

Level of education for activists and non-activists among independents in business and commerce

	Primary A	NA	Secondary A	NA	Prof./technical A	NA	University A	NA	Ecole Normale A	NA	Other A	NA	Total N A	NA
	%	%	%	%	%	%	%	%	%	%	%	%		
Independents in business and commerce	12	6	21	14	67	75	0	6	0	0	0	0	(24)	(36)
Workers	43	39	4	6	50	56	2	0	0	0	0	0	(44)	(70)

managers. Nevertheless, the individual propensity of members of these groups to support an independent Jura can be enhanced or weakened depending upon their social background. And the interaction of education and occupation has a pronounced effect on the propensity to support the independence movement. But if this is the picture of support for Jurassian independence, how is it to be explained? Why do the patterns of support appear as they do?

Towards explanation

In order to understand why certain groups are particularly strong supporters of Jurassian nationalism, it is necessary to recall the context in which the phenomenon has been unfolding.[10] The general picture over the past 50 years portrays a region that, increasingly, is less able to survive as an independent social unit: political dependence has been compounded by an erosion of economic and social self-sufficiency. And with this has come a corresponding loss in the power, prestige and wealth of the Jura. This point was presented most forcefully during the Moeckli Affair: there were some government positions that were simply too important to be entrusted to Jurassians.

The falling fortunes of the Jura have been keenly felt by certain sectors of society. In particular, small independent businessmen and independent farmers have borne the brunt of the Jurassian decline. And, if the Jura as a whole is losing power, prestige and wealth, independent businessmen and farmers are threatened with this loss, not only in the larger context, but also in terms of their immediate social relationships. In this light it is not surprising that independent businessmen are particularly ardent supporters of the independence movement. The changes wrought in the Jura threatened their personal social position. Further, these threats appear to emanate from persons and institutions that are based in an entirely different cultural system. As such, they are seen as challenges, not simply to one's personal position, but to the Jurassian culture as a whole. And in this context, nationalist claims become important vehicles for redress.

The plausibility of this interpretation is strengthened by the patterns of support for Jurassian independence that appear within the occupational group of independent businessmen. The less well educated members of this group are likely to be the most susceptible to change. Not only is it likely that their enterprises are small, familial operations and, as such, especially vulnerable to the centralizing tendency of modern industry; it is less likely that they would be able to obtain significant positions in larger enterprises. The strength of the interpretation is enhanced also by the countercases. Despite the overall tendency for independent businessmen to support the separatist movement, those from worker or farm backgrounds were much less inclined to do so than one would generally expect. However, these individuals have become independent businessmen despite the general decline of this group. As such, their individual experiences have contradicted the general pattern in the Jura. It would not be surprising to suppose that they would perceive potential threats to the Jurassian society as being far less severe.

But what about farmers? The decline of the independent businessman has been matched by a decline of the independent farmer, and goes hand in hand with the concentration of industry in the Jura. Although farmers themselves do not seem to give disproportionate support to the RJ, workers from farm backgrounds do.[11] Workers who come from farming backgrounds find direct personal evidence of the declining fortunes of the Jura. And the loss of personal independence coincides with the movement to spheres of activity that are controlled by persons from different cultural systems. Again, nationalist claims serve as a vehicle for redress.

Although an independent Jura will be able to do little, if anything, to reverse the general trend of increasing dependence upon other parts of Switzerland, it does provide Jurassians with a way of affecting the rate, and perhaps the direction, of change within their own environment. More important, it keeps one sphere of Jurassian life free, if only formally, from the increasing presence of other cultures. But how does one account for the success of the separatists in the Jura? In general, one would expect that any group that was becoming less able to survive as an independent social unit would be less able, over the long run, to press for independence from the larger society. In this regard two factors are crucial: territory, and the context of the Swiss political system.

Territory is a political resource. And when social groups are territorially based, they have access to this resource. The territorial basis of the Jurassians allowed for the possibility of independence as a solution to the problems that had arisen between the French and Germans in Bern. It is the possession of territory, too, that allowed the religious opponents of the Jura to remain a part of

Bern. And it is the lack of the territorial resource that will likely frustrate the claims of southern Jurassian nationalists in the years ahead.

But just as important as territory for the success of Jurassian nationalism is the Swiss political context. Had separation from Bern entailed separation from Switzerland the Jurassian nationalists could never have succeeded in establishing an independent Jura: the costs to the region would have been far too severe. However, the Swiss federal system allowed for the Jura to become independent from Bern while remaining an integral part of Switzerland. As a result, the Canton of Jura is not economically dissociated from those territories upon which the region had become increasingly dependent. Only conditions such as these, which allowed for Jurassian independence while not demanding a level of self-sufficiency beyond Jurassian capabilities, permitted the success of the Jurassian nationalists.

NOTES

1. For a fuller account of the history of the Jura, see Bessire (1976); Amweg (1974); Rossel (1914). All of these histories were originally published before 1945. For a history of the recent period see Prongué (1974; 1976).

2. For an account of the reactions within the Jura to the decision, see Ruffieux and Prongué (1972).

3. The Jesuit Order was outlawed in Switzerland after the Sonderbund War of 1848. In general, this civil war pitted Catholic cantons against Protestant ones. When the latter proved victorious, the newly written Swiss constitution forbade Jesuits from establishing themselves in Switzerland. The 1973 vote was a federal referendum on whether or not this ban should be removed. It was: see Martin (1971: Ch. 10).

4. Recently, there has been an economic upturn in the Jura, particularly in the northern portion. This has not, however, really affected the relationship of the region to the larger national economy.

5. For a brilliant treatment of the social effects of late industrialization, see Gerschenkron (1952).

6. See in particular the writings of Roland Béguilin, the undisputed leader of the Rassemblement jurassien (for example, Béguilin 1963; 1972).

7. 'Origin' is an awkward concept for non-Swiss to grasp, but it is very important in Switzerland. Every Swiss is a citizen of Switzerland, of one of the cantons, and of one of the country's communes. The commune of which one is a citizen is one's 'commune d'origine'. This is not necessarily the commune of one's birth. Rather, origin is passed from the father to his children. Although a woman assumes the spouse's origin upon marriage, it is very difficult for a male to change his origin. In order to do so, he must be accepted by the 'bourgeoisie' of the new commune. Further, every Swiss is aware of his origin. It is included on his passport as well as in the papers that everyone must deposit with the 'contrôle d'habitants' in his commune of residence.

8. Even this is too stringent, for there are many Jurassians who actively identify with the Jura but who, for one reason or another, live outside the region.

9. For a more detailed examination of the social bases of the leadership of the Rassemblement jurassien, see Campbell (1979).

10. For a fuller description of the theoretical bases of the following explanation, see Campbell (1979).

11. One likely reason for this is the very high proportion of farmers in the Jura who are of non-Jurassian origin.

REFERENCES

Amweg, G. (1974). *Histoire populaire du Jura bernois*. Porrentruy: Editions Jurassiennes.

Bassand, M. (1975). 'The Jura Problem', *Journal of Peace Research*, 12: 139-50.

Béguilin, R. (1963). *Le Jura des Jurassiens*. Lausanne: Cahiers de la renaissance vaudoise.

Béguilin, R. (1972). *Un Faux Témoin, La Suisse*. Paris: Editions du Monde.

Bessire, P.O. (1976). *Histoire du Jura Bernois*. Moutier: Editions de la Prévoté.

Campbell, D.B. (1978). 'Troubles in the Jura, 1815-1977'. Paper presented at the ECPR Workshop on Nationalism, Glasgow.

Campbell, D.B. (1979) 'Some Phenomenological and Structural Bases of Jurassian Nationalism: The Rassemblement jurassien'. Paper presented at the ECPR Workshops, Brussels.

Gasser, A. (1978). *Berne et la Jura*. Bern: Imprimerie Fédérative.

Gerschenkron, A. (1952). 'Economic Backwardness in Historical Perspective', pp. 3-29 in B. Hoselitz (ed.), *The Progress of Underdeveloped Areas*. Chicago: University Press.

Martin, W. (1971). *Switzerland: From Roman Times to the Present*. London: Elek.

Meyer, K. (1968). 'The Jura Problem: Ethnic Conflict in Switzerland', *Social Research*, 35: 707-41.

Prongué, B. (1974). *Histoire populaire du Jura, 1943 à 1974*. Porrentruy: Editions Jurassiennes.

Prongué, B. (1976). 'Postface: de 1936 à nos jours', in P. Bessire, *Histoire du Jura Bernois*. Moutier: Editions de la Prévoté.

Renan, E. (1970). 'What is a Nation? pp. 61-83 in *Poetry of the Celtic Races and Other Studies*. London: Kenikat Press.

Rossel, V. (1914). *Histoire du Jura bernois*. Geneva: Editions Atar.

Ruffieux, R. and Prongué, B. (1972). *Les Pétitions du Jura au Canton de Bern durant le XIX siècle*. Fribourg: Editions Université Fribourg.

Weber, M. (1978). *Economy and Society*. Berkeley and Los Angeles: University of California Press.

8

Regionalism and Autonomy in Alsace since 1918

Solange Gras
University of Strasbourg II

Land of two cultures yet German-speaking, Alsace has suffered since 1871 from the vicissitudes of Franco-German relationships. Torn between these two countries, which have been hostile more often than friends, its attitude has often been misunderstood. Alsace has never ceased to question itself about its identity. Even today, because of its particular history, the sentiment of regional identity is perhaps more difficult to pin down in Alsace than elsewhere (cf. Hoffet 1951; Philipps 1978).

The region, which has persistently displayed a certain particularism, seems nevertheless to be well integrated into the French state, and more placid than Corsica or Brittany. For over a decade the appeals of those who have rushed to the defence of an Alsatian identity that they deem to be threatened have scarcely disturbed the region's tranquillity: a marked contrast to the autonomist agitation of the interwar years. There is no longer an Alsatian party committed to opposing centralization and assimilation: political participation and debate are dominated by French parties and organizations. Immersed in a 'gentle somnolence' (*Rot un wiss*, January 1980), Alsace's distinctiveness is confined largely to cultural matters: its political expression is minimal.

Between 1871 and 1918 Alsace was transformed and modernized. Despite a sentimental attachment to France, it more and more risked being assimilated by German culture. The fear of germanization led it to fall back upon itself, its past and its traditions: Alsatians opted for the security of the 'petite patrie', whose virtues were praised (cf. Fleurant 1907). In 1918 euphoria over the return to the 'mother country' was rapidly replaced by disillusionment. France and Alsace no longer understood each other. Between the secular republican state and the deeply religious region there was a cultural

divorce and political resentment. The efforts of French governments to gallicize the province were unreasonable and clumsy, and
Alsace was a major victim of French anti-German sentiment. Faced
with an oppressive centralization, Alsace looked back with longing
at the 1911 Constitution and the autonomy it had given the province. France condemned the province, but without going beyond
accusations of accepting German financial support and of pro-
German sentiment.

The coming to power of Hitler and the Second World War led to
a rapprochement with France and a rejection of the German
aspects of its culture, the burden of which was now much heavier.
Often suspect in the eyes of many Frenchmen, Alsatians became
even more bent on gallicization in order to demonstrate their
loyalty: this inevitably led to a crisis of identity, which has been
gradually eased by economic prosperity, positive government attitudes and the new opportunities offered by European co-
operation. The new situation is radically different from that between the wars. After 1918 the loss of the German market and the
glacis policy of the French centre were catastrophic for the
province's economy. After the Second World War, because of
more liberal government policies and a high rate of foreign investment, Alsace developed more rapidly than most French regions. In
1970 the average household income was higher than the overall
French average. Despite recent economic problems, Alsace is still
an advantaged region.

Distrusted after 1918, Alsace has regained a privileged place in
France. Presidents and ministers continually make conspicuous
visits to the province, accompanied by subsidies and the tricolour:
historic antagonisms are to be laid at rest. European co-operation
has restored Alsace's pride in its two cultures and its language.
Revitalized within the Rhine corridor by the new Franco-German
reconciliation, it has acquired importance as a crossroads while
Strasbourg plays the role of a European capital. Its bilingualism
and the Germanic aspects of its culture, once the symbols of guilt,
have become precious assets. Alsace has changed its 'complexes'
(Philipps 1978: 175), exchanging inferiority for superiority.
Undoubtedly, the change in fortunes was helped by the changed
nature of Europe, but this cannot provide answers to all the questions that Alsatians pose about their identity. There have emerged
other assessments, which have attempted to weigh up the crushing
dominance of Paris, the power of foreign capital, the threat of
French cultural hegemony upon the regional culture and language,

the effect of industrial pollution, and the impact of new techniques of communication. Linguistic and ecological conflicts have been the major consequences of this new self-consciousness, though their influence upon the whole population is less certain.

The options between France and Germany

The history of Alsace between the wars was marked especially by a rapid autonomist upsurge which sought roots in the German period as much as in the ferment of ideas spawned by the war. The Alsatian paradox was that, just as the German regime appeared to have become accepted, its legitimacy collapsed with the defeat of the imperial army, while by contrast France, greeted with a sincere wave of enthusiasm in November 1918, rapidly provoked malaise, discontent and an autonomist backlash.

By the end of the nineteenth century, toleration between Alsace and Germany, with the decline of protest and direct action indicating acceptance of the Prussian regime, had been aided by greater economic prosperity (Dollinger 1979), although agriculture, despite state aid, lagged well behind industry. The province's wellbeing was assisted in 1902 by a softening of the regime, which abolished the dictatorial paragraph that governed German occupation and control. The parties began to argue that Alsace was identical to any other German state, and aspired for autonomy within the federation. The 1911 Constitution, which established in Strasbourg an assembly with legislative and budgetary powers, satisfied the claims to some extent. However, the emperor retained the right of veto, and could nominate one-half of the new Landrat's representatives, which diminished that body's independence. Despite this, the 1911 election was a shock for those nostalgic for a French Alsace. The Nationalbund received only 2.1 percent of the vote on the first ballot in Lower Alsace and 7.7 percent in Upper Alsace, despite a programme that hinged upon the demand for even greater autonomy, respect for the Alsatian 'individuality' and instruction in French at all educational levels. The results were an indication of the progress made in the germanization of Alsace. But the deteriorating international climate came to disturb the relative harmony established between Alsace and the Empire. Friction between the military and civilians, and the derision that arose over the Saverne incident in 1913, poisoned relationships to the point of provoking the resignation of the Statthalter and several of his senior officials. Such incidents were the first signs of a growing alienation from Germany that later was aggravated by mobilization, the financial cost of the war and suspicions that the military were taking over political control. Above all, the likelihood of German defeat revived the possibility of a return to France, for which previously there had seemed little hope. The offer by the new parliamentary government in Berlin of autonomy for Alsace came too late.

In the autumn of 1918, in a situation fundamentally modified by the German defeat, both public opinion and political activists seemed to be polarized between the supporters of a temporarily 'neutral' Alsace and those of a French Alsace. In October the region was swept by a wave that favoured self-determination by plebiscite. The protagonists of a plebiscite hoped to gain time from remaining uncommitted, which they felt could only benefit their cause. However, they were divided into two radically opposed groups: one preferring Germany and opting for an independent Alsace, the other choosing France. The former included not only German immigrants, but also several Alsatian groups. Several writers were apprehensive of losing their publics, and some would effectively choose to live in Germany or Switzerland after 1918. This choice, according to René Schickelé, should be considered as treason (*Catalogue* 1978: 459). Many employers were also hesitant and fearful of breaking the economic links with Germany, and saw some advantage in an agreement that would permit Alsace to remain within the German economic orbit. Certainly, a return to France would entail the risk of harming many interests, whose reaction was not necessarily the same as that of German nationalists who viewed such a move with extreme anxiety. Some of the Catholic clergy distrusted the secular French Republic and tried to explain this to their congregations. One priest had even argued that the conscientious duty of the Alsatian Catholic must be to oppose Alsace's entry into an atheistic country (*Elsässer*, 6 November 1918). Civil servants feared the loss of employment; the workers, that of the social laws that might pose problems for the Social Democrats, some of whom also saw some value in political independence (*Mulhäuser Volkszeitung*, 6 November 1918). Thus this group, which argued for a separate Alsace, incorporated those who hoped a plebiscite would help maintain a German Alsace, thereby safeguarding the economic and cultural links with Germany. But it also included some who wished for a return to France, but who hoped to utilize a pre-plebiscite period of independence to negotiate conditions that would be the most favourable to Alsace. Berlin, naturally, could only favour this separatist agitation, but its hopes were rapidly destroyed by events.

The Alsatian bourgeoisie, frightened by the introduction of the German revolutionary atmosphere into Alsace, abandoned their reticence to welcome the French troops with open arms. The return of revolutionary soldiers and sailors to Alsace had led to the formation of Soldiers' Councils in most towns, and also some Workers' Councils. In the eyes of the bourgeoisie, the situation could only breed social disorder. They viewed with disquiet the red flag hoisted on Strasbourg Cathedral (*Freie Presse*, 13 November 1918). But this standard, the symbol of socialism and internationalism, in fact masked within the Councils the same divisions over the national problem as in the whole population: proponents of reconciliation between peoples and states, German nationalists, those who preferred to see Alsace integrated into a revolutionary German republic rather than a capitalist French state, francophiles who made the flag a symbol of the liberty that had been regained through the victorious French armies, and those who hoped that a French Alsace would enable the revolution to be extended to France (*Strassburger Post*, 18 November 1918). The Council movement was revealing not only of the divisions of opinion on the national question, but also of the importance of the latter in the innumerable debates. As Richez (1979) stresses, it would be wrong to see it only as a German diversion, or as confined to German soldiers. Alsatian soldiers were major participants, and in the larger towns they had taken over working-class districts. The Soldiers' Councils were more clearly revolutionary and germanophile than the Workers' Councils established by local Social Democrats, who were anxious to retain control of the movement. But however complex and ambiguous the latter, it did contribute to the growth of a pro-French sentiment among the middle class: France emerged as a sure bulwark that could defend them against the revolutionary menace.

A major consequence of the formation of Soldiers' Councils was the transformation of the second Landtag chamber into a National Council. For the deputies it was a question of managing affairs in place of the now-defunct imperial government in order to hinder the Councils themselves from forming a government. The deputies too were divided on the future of Alsace: to let the peace conference take the decision, to secure a special status for Alsace-Lorraine within France, to opt for a straightforward fusion with

France, to endow the region with its own government — the possibilities were many. Several versions of recent events were in circulation: was the republic proclaimed in Strasbourg a socialist or a French republic? The National Council formed a provisional government, which was immediately recognized by Germany. The division of power with the Soldiers' Councils made for difficulties, rather than true opposition and defiance. Resistance to the Council movement was manifested in town councils, where francophile bourgeoisie and socialists tried to hasten the entry of French troops into Alsace with numerous delegations and appeals. German officers made similar requests, and Hindenburg telegrammed Foch to send troops into Alsace as soon as possible. These appeals surprised the French army chiefs, but they prepared to face up to possible problems and entered Alsace three days earlier than planned. Their headquarters had already announced that the military authorities did not recognize the Soviets. However, the withdrawal of the German troops had contributed to the progressive disintegration of the Soldiers' Councils. This was the end of the period of indecision where Alsace could question itself about its future.

The entry of French troops swamped the hesitators in a wave of enthusiasm. The 'French fever' seized the whole of Alsace, and the extent of the emotion was noted by all observers (*Strasburger Bürgerzeitung*, 23 November 1918). For France, 'the plebiscite had already occurred': Alsace had chosen to return to its lost fatherland. Alsatians displayed the symbols of their French nationalism: the national costume, the veterans of 1870, Napoleonic mementoes, military songs, flags. The well organized celebrations released a frenzy that surprised even the most devout partisans of a French Alsace. This enthusiasm, even though due also to relief at the ending of the war, the return of the conscripts and the reappearance of commodities, nevertheless corresponded to a nationalist aspiration in Alsace — albeit absent before 1914 — that stressed a sentimental attachment to France. Alsace, once again French, escaped both defeat and revolution: there was no longer any question of a provisional 'independence'. France solemnly promised to respect Alsatian rights, traditions and beliefs: Joffre, Poincaré and the generals all gave such assurances (Behe 1920). Any disquiet was swept away in the general mood of delight.

Discrimination and discontent

The apparent accord between France and Alsace was built upon a double illusion: each had a mythical and idealistic image of the other, which disappeared very quickly under the pressure of the problems of reintegration. France, despite its assurances, was not prepared to make any political or administrative concessions: the commitment to the unity and indivisibility of the republic was quickly asserted. Military reports indicated the astonishment of some officers confronted with a population that had elected to be French, yet could not speak the language: they thought Alsace to be germanophile, or at least indifferent.[1] The euphoria of rediscovery did not last, and the reciprocal misunderstandings explain the feelings of those who delicately referred to the Alsatian 'malaise', which almost inevitably generated a sturdy autonomist mood, especially between 1924 and 1932.

The accumulation of integration problems made the situation more difficult. Many Alsatians were offended and discontented, which annoyed Paris. Instead of searching for the causes of rancour and friction, France was content to accuse Alsatians of holding ill will against the centre, and frequently to denounce it as the result of German machinations. Some Alsatian deputies and senators lacked any commitment to defend their province and too frequently yielded to Parisian decisions. Alsatian opinion and its political representatives did not always speak the same language, and in fact neither Paris nor the Alsatian representatives took into account the dissatisfaction and the widespread atmosphere of strain and crisis.

The first measures very quickly revealed to Alsace the limits of French goodwill. Despite the timid reminder by the National Council of the promises of the 'mother country', Alsace was scarcely handled with tact. It was subjected not only to a centralized bureaucracy, but also to an exceptional degree of control. Some Alsatians protested against the hardships of a 'dictatorship'. Those suspect of sympathy towards Germany were treated harshly: Ricklin, the former Landtag president and head of the 1918 provisional government, who had wanted the Alsatian question to be handled at Versailles, was exiled over the Rhine, because his influence was feared to be too great to allow him to participate in the 1919 election. Commissions classified people according to their supposed degree of attachment to France. The repression and censure were contrary to the image of a generous and democratic France that Alsace had held in 1918.

Alsace had to give up its hopes of autonomy and decentralization. The elected Council that had been established by the 1911 Constitution was replaced by a bureaucratic board with only a consultative role. The powers of the High Commissioner attached to the presidency of the Council were constantly amputated: he could rarely take any initiative. It was evident that France wanted to reconstitute the three 'départements' as quickly as possible. Alsace was not able to administer its own affairs: the province was treated by the military and newly installed French civil servants more as occupied territory than a liberated region. The absence of any animation in the 1919 election campaign and an abstention rate higher than in 1911 and 1912 indicated the degree of disillusionment, and the pro-France Bloc National only captured 53 percent of the votes on the first ballot in Bas-Rhin and 63 percent in Haut-Rhin.

Alsace rapidly lost all hope of special treatment. The abolition of the High Commission was soon announced, and its functions transferred to Paris. Unable to secure anything more, Alsace wished at least to retain the advantages gained under the German regime in the fields of local government and social welfare, and its civil service statute. Among these 'local laws' were the Concordat of Napoleon I and the 'loi Falloux' which the German Empire had never abrogated. The state paid the clergy, and maintained theological faculties at Strasbourg and a system of confessional education. It was concerned at least as much with defending religious interests as with preserving the Alsatian 'personality', for the latter was better embodied in the local law. The local civil law was recognized by legislation in 1924. It applied to both Alsace and Lorraine: outside these regions, descendants of the citizens of the two provinces could also claim its benefits. The confusion was extreme because it concerned not only the conservation of fundamental elements anchored in Alsatian tradition, but also German legislation. In the eyes of Paris, the same ambiguity arose when Alsatians demanded the use of German in administration and law. In 1931 9.3 percent of the population spoke only dialect, and 43 percent only dialect and German: over one-half of the population could not understand French (Dollinger 1979). To the extent that these demands were wholly or partially unsatisfied, Alsatians were discontented, the more so as many found themselves in humiliating situations. Within the administration, civil servants from 'the interior' relegated Alsatians to subordinate positions, especially those who had an insufficient command of French, and, moreover, operated in a manner similar to that of their counterparts in the colonies.

The economic integration of Alsace also encroached upon

numerous interests. It required a restructuring of the economy because France was not able to absorb the same products as Germany. Furthermore, Alsace was poorly linked to the rest of the country, and, despite the rail links and passes through the Vosges, trade with France remained poor. Alsace had benefited greatly during the five years of its franchise from its exports to Germany, although inflation prevented the accumulation of real profit. In subsequent years German custom tariffs became prohibitive, and provoked numerous protests. On the morrow of the armistice, internal competition had not seemed excessive because of the destruction of factories in northern France and the need for reconstruction: but difficulties rapidly cropped up. Agriculture was a victim of over-production of cereals and a decline in dairy farming and viticulture. The farmers had lost the German market, and were unable to export to the French. Certain industries, such as engineering and brewing, were favoured, but neither the quality of its product nor the opening of the colonial market could save the textile industry. Economic slump bred unemployment. The plan to extend the Rhine axis with a Rhône-Rhine canal permitted some progress to be made in 1924 with the idea of a free port at Strasbourg. While the harbour was completed, the other projects were postponed because of strategic reasons and worsening international politics. The world depression further aggravated the situation. In particular, the crisis affected the banks which, deprived of German capital in 1918, had had to orientate themselves to France, and the traditional local banks had suffered from strong French competition. Only the Raffeisen and Sogenal savings banks, which could draw upon an extensive network in Germany and Switzerland, were able to survive. In short, discontent and disillusionment accumulated. Comparison with the prosperity of pre-1914 days could not but go against France, despite the improvement in living standards between 1924 and 1928.

After the enthusiasm and hope of 1918, the difficulties caused by changing nationality in unfavourable social and economic circumstances could only rouse uncertainty and confusion. Alsatians came to hold the French government responsible for everything. As early as 1919 they demanded the retention of the 'ancient rights' and 'local laws'. These claims led to the formation of a strong autonomist wave in subsequent years, precisely because French concessions appeared insufficient. The stages in the march towards autonomism were marked by the strikes of 1919 and 1920, demonstrations against the Cartel des Gauches in 1924 and the formation of the Heimatbund in 1925. After 1927, repression and the international climate halted the advance of the movement, which fell into the trap of collaboration with the Nazis in 1940.

The 1919-20 strikes were not just pro-Alsatian, but also pro-working class. They indicated the strength of Alsatian socialism and its roots in a powerful trade unionism, both of which had an influx of adherents in 1919.[2] The strike wave began in 1919 in the textile industry, and culminated in two general strikes a year later. These strikes were remarkable in their extent, at times affecting up to 90 percent of the workforce, and received widespread support in the local press and population, because of their regional character.[3] They did not coincide with national strikes, and had very specific objectives. In addition to demands for wage increases and anti-capitalist slogans, they were impregnated with Alsatian symbols (Cecchini 1971).

The defence of 'ancient rights' was central: these related to the treatment, promotion, social insurance and pensions obtained under the German regime. A major theme, defence of the Heimats-recht (the right bound to the native soil), encompassed not only the 'local' laws and rights, but also, and more vaguely, the 'rights of the people', their 'existence' and their 'future' (*Freie Presse*, 20 April 1920). Alsatians deeply resented the persecution and harass-ment by French bureaucrats who had replaced the German person-nel: they felt they should have priority in employment, especially in senior positions from which they had been excluded. The more general slogan of 'Alsace for Alsatians' appeared after 1919 among the protesters (cf. ADBR: D286 (355)). This slogan referred not only to employment, but also to an Alsace that would be free from French militarism and capitalism, especially after the French army brutally repressed the strikes. Alsatians were weary of an 'occupa-tion' regime, and no longer accepted the 'looting' of their resources by French capitalists. The Socialist newspaper, *Freie Presse* (28 April 1920), declared that the Alsatian people could not accept the substitution of militarism and capitalism from beyond the Rhine with that from beyond the Vosges. The ambiguity of the phraseology was such that it could equally mean affirmation of attachment to France, which had been accepted by the moderate socialist and trade union wings.[4] The moderates had not wanted the strikes: rather, they tried in vain to secure a rapid return to work in return for obtaining promises from Paris. The significance of the strikes emerges more clearly when we consider the position of the principal protagonists. The Alsatian socialist movement had been deeply affected by the German organization: many of its members had served in the German army, experiencing the revolutionary movements in central Europe as well as in Alsace. Not entirely insensitive to the separatist propaganda of 1918, they were the nucleus of an original kind of communism typified by one of the leading activists in the 1920 strikes, Charles Hueber. They were strong in Bas-Rhin, with 4,600 members by December 1920, though weaker in Haut-Rhin (1,500 members). These represented the working-class element, pro-dialect and committed to the Heimatsrecht, that was one component of the future autonomist wave.

It was easy to see in this strike wave the effects of separatist propaganda and German influence. Neither the authorities nor the employers' federation of Bas-Rhin could tame it (ADBR: D286 (366)). While the strike of April 1920 did coincide with the distribution of 'separatist' pamphlets issued from Baden-Baden, the latter did not contribute either to stimulating the strike or to the choice of slogans. It was merely a coincidence, as was the similarity of objectives of separatists and workers. Nevertheless, these strikes were the first mass symptom of the Alsatian 'malaise', reflecting the territorial preoccupations of workers and employees. The demands were repeated in subsequent years by the Alsatian Communist Party, which claimed that Alsace was treated and exploited like a colony: 'decisions are taken in Paris without consulting us, just like in the black colonies' (*Die Neue Welt*, 1 October 1921).

The electoral victory of the Cartel des Gauches gave Alsace a fresh opportunity to express its 'malaise'. After the 1924 election and the resignation of Millerand, a real crisis grew between France and Alsace. In June the Cabinet announced the end of special legislation for the three départements and the introduction of 'republican' laws. The promises of 1918 to respect customs, rights and beliefs were forgotten. The introduction of 'republican' and 'secular' laws in Alsace would mean the end of the 1802 Concordat and the loi Falloux of 1850. The clergy and the confessional schools, deprived of state financial support, risked losing their influence over the population. Since 1921 the Catholic clergy had provided one pro-Alsatian instrument, the Catholic League of Alsace. They could also rely upon the Union Populaire et Républicaine (UPR), the large Catholic party which also owned an important press network. In 1924 the church, UPR and Catholic League were able to mobilize the region against the government's proposals. A wave of meetings and a massive rally in Strasbourg testified to the emotions aroused and the influence of the church.

The Conseil d'État in Paris did not follow the Cabinet on the Concordat question, and the educational issue remained unresolved at the beginning of 1925. An Action Committee for Religious Defence launched a petition to maintain religious schools: it was signed by 63 percent of the electors in Haut-Rhin and by 45 percent of those in Bas-Rhin. The Catholic League and UPR also organized a schools' strike in March 1925. This received its strongest support in the countryside and smaller towns, though it affected less than one-half of the population in the larger centres. Herriot was replaced as prime minister in April, and his successor discreetly abandoned the inflammatory area of the secular laws.

Alsatians thus reacted much more massively against government policy to defend their religion than to defend the accumulated rights (Heimatsrecht). The display of reticence by the churches towards the government was not only due to a concern for safeguarding Alsatian traditions. Culturally and politically, the clergy remained close to the German tradition, Lutherans because of their creed and the use of German in their worship, Catholics because they had been trained either in Germany or at the theology faculty at Strasbourg, which had been charged with the task of germanizing the clergy. The UPR, descended from the Zentrum and committed to Christian Socialism, was at the right of the French political spectrum. The party formed not a nationalist opposition to the French state, but a political opposition to a secular France, especially to the French left.

Indeed, the lack of enthusiasm towards the entry of French troops in 1918 among some Catholic clergy and some Germanic Protestant villages had earlier been remarked upon (AMG: 6N 285, GQG, 25 May 1918), but this attitude had not been general. The churches had not only feared that the return to France would entail a reduction in their powers: they were also apprehensive over a fall in revenue, and even more, because of the intrusion of the French language, of their influence over the population. On the Protestant side the clergy had viewed with a jaundiced eye the arrival of French pastors to take over the running of the church (Dollinger 1979). In short, the churches had two essential preoccupations: the fight against a secular and left-wing regime, and the protection of the faith against the presumed pernicious influences from beyond the Vosges. The best guarantees against these influences were language and religion.

The political aspects of the religious claims should not be ignored. For the whole population they affected, above all, local traditions. Religion occupied a central place in Alsatian life. The churches were stable institutions that provided a socially privileged cadre which filled a gap at the political and administrative level. Alsatians, tossed from one state to another, had found their identity in religion rather than in their social and political trappings. It was not surprising that the threats hovering over religion in 1924-25 hardened the Alsatian stance. It was the middle class and rural milieus that were most concerned with these issues, the former through political as much as religious commitment, the peasants because of loyalty to their faith. The churches were less influential in the larger towns, but even so they succeeded in mobilizing half the families behind their schools' strike. Political circles also reflected the anti-secular struggle, because clergymen were prominent within them. The crisis of 1924-25 indicated the extreme sensitivity of Alsace to all suggestions that implied limitations upon their rights: the religious questions also demonstrated the deep differences that separated the province from France.

It was this self-consciousness and the coming together of the discontented that led, almost immediately, to the autonomist crisis. Since the end of the war there had already been several tentative attempts to establish autonomist parties and movements. Father Sigwalt and Joseph Hümmel had wanted a Federalist Party with the objective of making Alsace a French protectorate; the 'Executive Committee' of Baden-Baden desired a plebiscite and a Rhine federation;[5] and Zorn de Bulach had founded an Alsatian Party in 1922. These first manifestations of hostility towards French centralization had not been successful; either because their efforts were outlawed, or because they were not taken seriously. It was otherwise in 1925. The religious question was the springboard for the autonomists because the alarm of 1924 obliged all parties to take a position on French policy in Alsace. At the same time, Paris considered that the regained provinces should be départements like any other, and abolished the Commissariat Général. It was replaced by a Direction Générale for Alsace and Lorraine under the direct authority of the prime minister. Nothing remained of Millerand's regional inclinations. The years 1925 and 1926 were a political as well as an administrative turning-point: autonomism became a concrete possibility.

'Autonomism' had several strands: a right clerical wing, that of *Die Zukunft* and the *Heimatbund*, and two less prominent left-wing threads representing communism and an Alsatian radicalism.

These were the major components of what has been called Alsatian autonomism.

In 1925 there appeared *Die Zukunft*, a weekly dedicated to rallying Alsatians in the defence of their rights, culture, language and, especially, religion, and stressing the grievances of Alsace and Lorraine against the French administration. One year later the autonomists formed a movement, the Heimatbund. The success of this enterprise was apparent, despite the violent anti-autonomist campaign conducted by the very official *Journal d'Alsace et Lorraine*.[6] The partisans of *Die Zukunft* and the Heimatbund were the major component of the autonomist movement. A majority, coming from the Christian milieu and the UPR (including several priests and pastors), belonged to the middle classes: especially numerous were civil servants and teachers who followed Joseph Rossé, who had refounded a Christian trade union. But farmers and winegrowers were also among the signatories of the Heimatbund manifesto. The autonomists therefore were recruited from those circles whose interests had been most affected by the return to France: victims of the verdicts of the 'clearance' committees; functionaries relegated to subordinate posts; teachers and professors — especially the religious — who feared the loss of employment; farmers, merchants and industrialists who had not been compensated inside France for the markets they had lost in Germany.

These interests came together in a common self-consciousness, expounded by the Heimatbund manifesto, in which the Christian component was strong. The demand was for political autonomy within France, to be based upon a popularly elected legislative assembly in Strasbourg with budgetary powers and an executive 'embodied in the Alsace and Lorraine people'. The manifesto likewise asserted the need to safeguard the 'indefeasible and inalienable rights of the people of Alsace-Lorraine', and to maintain the confessional statute. It also wanted to preserve German, the province's mother-tongue, alongside French in the schools, the administration and the courts. Other interests were not forgotten by the manifesto: the signatories affirmed themselves as pacifists, opponents of 'chauvinism, imperialism and militarism in all their forms'. In short, they hoped to unite the whole population within the Heimatbund.

There was no question of a complete rupture with France, but it was easy to imagine, no matter how much care was taken, that the French nationalists would feel themselves threatened. In any event, the manifesto's tone appeared very hard to a France accused of plundering, reneging on its promises, disrespect for rights and freedoms, oppression, injustice and backwardness. The term 'national minority', based upon the criteria of the League of Nations, which Ricklin applied to Alsace, could only offend the Parisian power and French opinion. *Die Zukunft* also envisaged in 1926 the evolution of France from centralization towards federation. This could be one solution, since Alsace, governed by its own laws, could be 'an integral part of a federal French state'. The demand, therefore, was not for simple devolution, but for autonomy, involving a total restructuring of the French system.

Alongside this 'Christian' autonomist activity was the communist strand rooted in the working class. The Communist Party had complained about the special regime installed in Alsace and its harassment of party members. After 1921 it depicted the Alsatian situation as 'colonial', and criticized the excesses of French 'imperialism' in the same language as that adopted by the party's central committee. Between 1921 and 1924 the Alsatian Communists concentrated on social and economic problems, and did not participate in the religious battle, though this demonstrated to them its importance to the Alsace-Lorraine problem, and in their turn they engaged in the defence of the rights of a country that was the victim of French capitalism. Hueber, a parliamentary deputy since 1924, pleaded to keep German as a second official language. The congress of workers and peasants held in Strasbourg in 1925 also released a manifesto demanding a referendum preceded by the evacuation of French civil and military authorities from Alsace-Lorraine. Pending such a referendum, it insisted upon the maintenance of German, the withdrawal of French bureaucrats, the end of special taxes and administrative autonomy. These claims, while less radical than those of the Heimatbund, nevertheless strengthened the latter, especially as the Communists admitted that the right of Alsace and Lorraine to self-determination could lead to independence.

These two autonomist movements, divided by historical political traditions, ended by merging with each other. In 1926 the Alsatian Communist Party decided upon a policy of unity of action. This bore fruit in the electoral pacts of 1928 and 1929.

The third autonomist element was radicalism, a paradox because the Radical Party in Alsace, founded in 1919, was specifically anti-clerical and had aligned itself behind a rapid assimilation of Alsace by France. However, this party, an inheritor of the German liberal tradition, paid much attention to Alsatian problems. It wanted to protect the Alsatian economy, particularly by a customs exemption for exports to Germany, Alsatian civil servants and the use of dialect. It tried to be moderate and reformist. Its recruits came primarily from the urban middle classes, and Protestant cantons supplied the largest concentration of supporters.[7] Jews were also prominent among its activists. Although marginal in Alsace, it was the third autonomist strand. It was directed by Georges Wolf, a pastor and former Landtag deputy. 'Moderately' autonomist, he would have been content with a regional assembly with an essentially consultative role; secularly 'moderate', he accepted religious instruction in schools on condition that it was made optional. He was, however, more intransigent on the language issue, wishing to preserve German as the dominant language. These positions were adopted by the party by the end of 1925.

The other major Radical 'autonomist' was a Protestant, Camille Dahlet, editor of the newspaper, *La République de Strasbourg*. He argued in defence of those adversely affected by France (groups in fact that constituted the traditional Radical support), criticized an administration that had slowly surrendered to centralization, and argued that the imposition of French as the official language did not serve Alsatian interests. But in 1926 both Wolf and Dahlet left the Radicals to found the Progressive Party, which declared its allegiance to both radicalism and autonomy, and defined Alsace as a national minority. The consequence of the split was the collapse of the Radical Party. Dahlet, whose stance fitted better with the Alsatian mood, was elected as an Independent Radical deputy in 1928. Dahlet, described by Bernstein (1970) as a regionalist, wanted an autonomy similar to that of 1911, with an elected parliament responsible for the budget, law, culture, economy, education, the mail and the railways. But such an autonomy could be possible only through a profound transformation of the French system. The president of the League of Human Rights declared in 1928, 'The Alsatian is a European: he has no sentiment of national attachment to the French fatherland. He cannot have any, because he considers the French idea of fatherland as idolatry and mysticism' (Zind 1979: 527). Dahlet also dreamed of the end of the nation-state and the formation of a 'United States of Europe', and saw the League of Nations as defending the rights of national minorities. These attitudes conferred upon Dahlet a more open-minded attitude towards the problems of other minorities in France and Europe. Dahlet wanted the separation of church and state, but not secularist legislation. For him, these questions could not be incorporated into the Heimatsrecht, but neither could they be resolved without the consent of the Alsace-Lorraine population.

Of these three components of Alsatian autonomy, only the Christian had links with a regional party (UPR): the other two arose within national parties of the left. Alsace's political life was transformed by this convergence, so distinct from the traditional left-right cleavage, which approached the aim of the Heimatbund, to unite all Alsatians across political divisions in the defence of their rights: Georges Wolf too proposed to bridge the gap between clerical and secular, and the Communists compared the tactic of the 'united Alsatian front' with that of the class struggle. Such a political realignment could not occur without some party fracturing. Only the SFIO, consistently pro-French and 'assimilationist', stayed outside the debate. It refused to tolerate particularism, and autonomist endeavours even less. It is at the national political level that we must look for the real causes of disruption.

To the right, the leagues and Action Française at first looked favourably upon the Catholic opposition to the Cartel. But when the Bishop of Strasbourg, Mgr Ruch, begged for the support of French Catholics, they responded with the slogan, 'always Catholic and French', rejecting any support for autonomy. The Bishop, anxious not to disrupt Catholic unity and worried by the collaboration with Communists and Radicals, in turn condemned autonomy. The passion of French Catholics persuaded many in Alsace to abandon nationalism (Gras 1977).

The UPR accent was on religious defence: only later was it carried away by the autonomist wave (ADBR: D286 (353)). After the formation of the Heimatbund, presided over by Ricklin, the exile of 1918, the UPR hesitated and compelled members of its executive committee to withdraw their signature from the manifesto. Even so, the latter was published in three Catholic newspapers. The UPR had clearly opted for prudence, but the appeal of the Catholic press, with a daily circulation of 100,000, and a weekly of 30,000, was influential. The UPR, above all, did not want to be isolated. In 1926 its eight deputies in Paris proposed a regional administrative organization: a similar plan for administrative decentralization was introduced in the Senate. These gestures were determined by the circumstances, but conformed to the moderate regionalism of the UPR, which rejected the notion of 'national minority'. There was a deep cleavage between it and the Heimatbund.

The party was internally split between an autonomist wing headed by Father Haegy and a French nationalist faction. It was this that led almost all the party's senators to push for a break with the autonomists and to rebel against any 'rapprochement' with the Communists. The quarrels lay more at the leadership level than among the membership, among whom the nationalists found a limited audience. Haegy, a former Reichstag deputy, came to exercise a deep influence upon the clergy and faithful now that he leaned towards the autonomists.

Despite the condemnation of *Die Zukunft* by Mgr Ruch, the French nationalist wing could not engineer the exclusion of the autonomists. Furthermore, the 1928 election was unfavourable to the French nationalists: they had three deputies, compared to the moderates one and the autonomists' six. They decided to secede from the UPR in December to form the Action Populaire Nationale d'Alsace (APNA). Their deputies, four of six senators and one-half of the district councillors, left the UPR for this new party; however, the mass base was more limited. In the 1932 election the UPR gained two additional seats, and won 21 percent of the vote in Bas-Rhin and 39 percent in Haut-Rhin. The APNA gained only 11.8 and 10.2 percent respectively in the two départements. In 1932 and again in 1936 the UPR won nine seats. Catholic nationalism was not attractive: its leaders came more from groups within the upper middle classes linked to industrial circles and the French administration, while the autonomists were much more strongly rooted in the broader middle classes and among the peasants.

For the Alsatian Communists, the unity of action with autonomism, adopted by the 1926 congress, was especially profitable at the local level. Hueber was elected mayor of Strasbourg in 1929, thanks to the Heimatfront formed with the UPR. In Colmar, nine Communists were elected from a common list with the UPR. The French party (PCF) was furious: 'Autonomists? No! Communists? Yes!' (*L'Humanité*, 3 December 1927). But in 1929, despite an investigation that demanded the resignation of those elected in Colmar, the party in Alsace followed Hueber. So began the expulsions: the Colmar representatives, one deputy elected in Strasbourg in 1928, Hueber and others were forced out of the party. There were thereafter two communisms in Alsace. The orthodox position was not very clear. On the one hand, it gradually abandoned the defence of Alsatian rights to approach the Socialists and Radicals by 1934; on the other, it continued to support Hueber, the expelled autonomist, in the 1936 election as long as he opposed other autonomist candidates. The vacillation ended in

1936, when Thorez resolutely condemned all separatist attempts as playing fascism's game. Autonomist claims were then abandoned and the 'entente' with the Socialists cemented.

Those expelled from the party, followed by a majority of members in Strasbourg and Colmar, formed a Communist opposition that was influenced by Brandler and Thalheimer in Germany and by French Trotskyists. To abandon autonomy would be to betray the masses: they saw nothing 'immoral' in their clerical alliance, regarding autonomy as being opposed to the capitalist state (Fourier 1928). They believed that the working class would spearhead the struggle for minority rights, and that the direction that the PCF wished to impose upon Alsace was in defiance of democracy (Mourer 1929). Excluded from the International Union of Communist Opposition in 1934, they formed in 1935 the Alsatian Workers' and Peasants' Party (PAOP). It adhered to the Popular Front in 1936, gaining two representatives, Hueber and Mourer. Electorally, autonomism was profitable for the PAOP, as it was for the UPR.

Political life in Alsace was not only turned upside down by the splits within the UPR, Communists and Radicals. It was also affected by the autonomist Heimatbund and its paramilitia: though rarely used, the Schutztruppe of 1928 was sufficient to scare French opinion. At the outset the Heimatbund regarded itself as being above the parties, diffusing its ideas through the journals, *Die Zukunft* and *Die Volkstimme*. But it soon accepted the need for more solid structures. The Autonomist Party of Alsace-Lorraine (PAAL) was formed in 1927. Autonomy, envisaged as part of a United States of Europe, went much further than that of Dahlet: Alsace-Lorraine would be the mediator between France and Germany. Its slogan was simply 'l'Alsace-Lorraine aux alsaciens-lorrains'. PAAL immediately participated in the creation of a co-ordinating body, 'The Central Committee of the National Minorities in France', along with Corsican and Breton autonomists. This provided the Alsatian movement with a new breadth: a co-ordination of minority movements could bear more heavily upon the French state. Federalism was central for PAAL. The new orientation and party did not occur without causing some disruption. The Heimatbund's secretary, Jean Keppi, resigned. Carl Roos became the new secretary of the Bund and PAAL. Son of a teacher and a professor himself, Roos had taught in various institutions in both Alsace and the Ruhr before 1914. After the war he published a pocket Alsatian-French dictionary, and in 1924 had become an inspector of French schools in the Saar. He had also signed the Heimatbund manifesto.

In 1927 PAAL allied itself with the colourful leader of the Alsatian Opposition Bloc, Claus Zorn von Bulach, whose paper, *Die Wahrheit*, had a circulation of 70,000 (Zind 1979: 434). The movement and journal had a violent tone. From the fusion came a new body, the Independent Regional party (PRI), which demanded complete autonomy, though more moderately than von Bulach. Thus the several 'tendances' had found a consensus for working together to defend Alsatian rights.

In 1929 and 1930 closer links were formed between the parties that had worked in common for the 1928 legislative and district elections. The PRI and Progressives had a joint 'working committee' with a single newspaper, the *Elsas-Lothringer Zeitung*(ELZ), which in 1936 reassumed the name *Die Zukunft*. These efforts indicated simultaneously the autonomists' desire to fight in common their difficulties, divergencies of opinion and rivalries. The aim was to build local organizations to maintain closer contact with supporters and ensure a regular flow of subscriptions. But they could not prevent the wrangling and splits, which worsened towards the end of 1934. Other agreements characterized elections: the Action Committee in the 1928 legislative election, the United Alsatian Front in the 1928 district and 1929 local elections, the People's Front in the 1930 local election. The election results cannot be easily interpreted because of the attitudes of the UPR, which simultaneously backed the autonomist and French nationalist cards. This makes it difficult to evaluate the autonomist support after 1932.

But until then the results had appeared clear. In 1928 two autonomists among 16 Alsatian deputies represented a success at a time when they had lost heavily in Lorraine. The United Front, linking the PRI, Progressives, UPR and Communists, captured 23 of the 61 district council seats: even without a majority, it constituted a significant force. A similar success occurred in the 1929 local elections: 30 seats in Colmar and 53.3 percent of the vote in Strasbourg. It was a defeat for the 'nationalists': SFIO, Democrats and APNA. The autonomists even established an Association of Alsatian Mayors, which offended the government. In 1931 the People's Front increased its number of district councillors from 23 to 28, gaining a slight majority of the votes in Bas-Rhin. At that point there was no denying the electoral progress of Alsatian autonomy.

In 1932 the UPR attitude changed. Father Haegy, heavily

defeated in the 1928 senatorial election, had died. Electoral agreements were no longer wholly respected. The People's Front did have 11 seats, but these were UPR candidates: those of the PRI were heavily defeated. The same situation was repeated at the 1934 district elections: of 30 seats, 20 went to the People's Front, but not one was held by the PRI. The gap between PRI and UPR deepened in the 1935 local elections. In Strasbourg the People's Front won 51 percent of the vote and 16 seats: the Socialists and Democrats each won 8. The UPR then switched its allegiance, and its 5 councillors voted not with the autonomists, but with the Democrats. A new majority took shape and chose an anti-autonomist mayor.

The ambiguity persisted for some time afterwards because among the UPR deputies elected in the 1936 legislative elections were several prominent autonomist sympathisers. In fact, while the UPR had followed the autonomist thrust, to which many of its supporters were sensitive, its leadership had always been divided. To avoid disintegration, the party had shirked a clear position on the question. It was essentially opportunist, and this explains the apparent break of 1935. The UPR maintained its links with the People's Front as long as the latter was electorally successful: it loosened them when the period of decline began after 1932.

The decline was due not only to French policy, but also to the international situation, which made autonomy more suspect. Repression was not surprising in a country dedicated to the myth of the unity and indivisibility of the Republic, most especially as, with the exception of the PCF and the other minorities, this myth was endorsed in its entirety by public opinion. The French parties were disquieted by the scissions that appeared in their Alsatian ranks. Autonomy threatened the unity of French Catholicism, while Radicals and Socialists refused to admit the critique of national unity, regarding autonomy as a play of the clerical Right. For the Right, the entente with Communism was offensive, as were pro-autonomy arguments: on the one hand, because autonomy was an insult to all the soldiers who had died in the effort to regain Alsace and Lorraine; on the other, because it represented a serious external danger. It was feared that only Germany would profit. The *Echo de Paris* (3 January 1926) explicitly stated that under Locarno Germany was only pledged not to use force, that it could hope to regain the territories through negotiation. Federalism was alarming, as was the idea of a Rhenish state that might embrace Alsace. Above all, there was a dread that the autonomists would play Germany's game and be the instigators of war. From there it was only a short step to consider as traitors those who spoke German and did not wish to be French. The nationalists attacked the autonomists frontally, provoking several violent incidents, of which the most notorious was Colmar's 'Red Sunday' in August 1926.

The government too strived to bring the autonomist leaders into disrepute with this kind of argument, to which the climate of espionage in Alsace easily lent itself. The anxieties of the Laval government appeared in administrative sanctions against the signatories of the Heimatbund manifesto and the appeal of the Communist Action Committee. Rossé was dismissed from office, and district councillors were ordered to withdraw their signatures. In 1928 the local council at Haguenau, accused of autonomist intrigue, was dissolved. After these administrative sanctions came legal proceedings against newspapers. In November 1927 *Die Zukunft, Die Volkstimme* and *Die Wahrheit* were banned because they were printed in a foreign language.

In December a wave of arrests swept the province, first of Rossé, then of the other promoters of the Heimatbund. In early 1928 other newspapers, including that of Dahlet, were suppressed: later, while Ricklin and Rossé learned in prison of their election to the Chamber of Deputies, the Colmar trials began. Ricklin, the former Landtag president and one of the founders of *Die Zukunft*; Rossé, leader of the discontented civil servants; Schall, the organizer of the para-military guards; and Hauss, son of the former secretary of state of the Land of Alsace-Lorraine and publisher of *Die Zukunft* — all were accused of conspiring against the security of the state, as were several others: Father Fashauer was accused of having received German money. The prosecuter defined the conspiracy as an attempt to separate Alsace from the national territory. Eleven of the accused were acquitted, Ricklin, Rossé, Schall and Fashauer were found guilty and each sentenced to one year in prison and five years' local banishment. Poincaré had wanted to strike at the head of Alsatian autonomism, and especially at the two deputies, Ricklin and Rossé, whose election was invalidated. But two of the acquitted were elected in their stead. The attitude of the electorate and street demonstrations contrasted with the prudence of the parties. The UPR's protest was timid, arguing that the two deputies be freed for the duration of their mandate. The Communists were more vigorous, but in general political circles feared further sanctions.

Judgement on the absentee accused was given in 1928. Typical of the heavy penalties inflicted was the 15-year sentence imposed upon

Carl Roos, the PAAL secretary. Roos, who was in Switzerland, published a brochure (Roos 1928) and tried to draw international attention to the trials. After giving himself up, he was tried the following year and acquitted. The others who had been sentenced were pardoned in 1928. France condemned, but displayed generosity. Accommodation had been especially marked in education. The Pfister memorandum of 1927 had permitted three hours of German weekly in primary schools, an important concession for a country that relentlessly asserted the primacy of French. With the enthusiastic demonstrations in Strasbourg and Colmar that greeted the return of Roos, the amnesty was forcefully claimed. But it was only when Pierre Laval signed a general pardon in 1931 that the condemned regained all their civil rights.

France, through the courts, had repeated that it would not tolerate autonomy, though without undue severity. The trials had allowed a distinction to be made between autonomy and separation, and had demonstrated the importance of dialect and German to Alsace. But the 1928 and 1929 elections showed that the trials had not halted the autonomist impulse, and its pursuit was continued, both in and outside Parliament. However, none could move the government from its position, for whom the accusations of German money and German influence loomed large. Undoubtedly, the autonomists received German funds, and Poincaré could justly complain about German propaganda. The emigrés from Alsace-Lorraine had many organizations and their own newspaper. One of their leaders was Robert Ernst. With an Alsatian father and German mother, Ernst had for long lived in Alsace before settling in Germany. According to his memoirs (Ernst 1954), he helped the non-Communist autonomists and secured subscriptions in Germany for their publications. But the autonomists were not the only ones to receive help; so did Catholic and secular organizations in both Alsace and Lorraine (Zind 1979: 484). This German aid smacked of treason, and it was a weighty argument that could be used to discredit the autonomists, who often fell into traps set by spies infiltrated into their ranks by the French government.

More serious was the Nazi temptation. Though the connections were loose and perhaps doubtful, some Alsatian groups adopted a style close to that of the Nazis — but so did many others throughout France. The Jungmannschaft of Herman Bickler joined the Heimatbund, and Bickler, a Strasbourg lawyer, joined the central committee of the PRI. His movement wore an evocatively brown uniform. Its ideology, insisting on 'ethnic security', aimed at ridding Alsace of 'foreigners'. It was anti-Marxist, anti-capitalist, and opposed to party plurality. However, it broke away from Roos's party to form the Party of Alsace-Lorraine. Two other organizations, the National Labour Front (FNT) and the Alsace Peasant Union (UPA), showed ideological links with Nazism (Reimeringer 1980). The FNT, started by Joseph Bilger, had three elements: the Werkbund, the UPA and the Jung-Front. Its newspaper was the *Volk* and it obeyed a Führer. The similarities went further: green shirts, the Hitler salute, and the armbands left no doubt as to the source of its ritual inspiration. The ideology of Bilger's movement was at once anti-Semitic, anti-capitalist and in favour of a new state based upon Christian, familial and corporatist values. Its autonomist stand was clear: Rossé, moreover, helped it to seize control of the winegrowers' union. It was only in 1938 that Bilger broke with the UPR, which he felt had not pushed for autonomy with sufficient vigour, and which had successfully run a candidate against him in the 1936 legislative election.

Bilger enjoyed great success among the peasantry, with almost 200 unions in 1934 and perhaps 10,000 members by 1938. His links with Germany were apparent, and in foreign policy the UPA argued for reconciliation, at least economically, with Germany. Bilger was independent of the major autonomist groups, concerning himself mainly with peasant defence. But his movement can none the less be classified as autonomist and fascist. Was a similar orientation to be found in the PRI? Hueber, the UPR and Dahlet had sunk some of their differences, and the latter had quit the ELZ. At the same time, Carl Roos was photographed in Hannover making the Nazi salute, an embarrassing occurrence for his former allies. The consequence was a cleavage within the People's Front and defeat in the 1935 local elections. Autonomism paid the price for this identification with Hitler's Germany.

For a time the Popular Front seemed to restore Alsatian unanimity over the education question. Two decrees had prolonged school attendance in Alsace and Lorraine to compensate for the time spent on religious instruction and German. There were major demonstrations in the large towns, and new petitions were launched by the 'Committee of Religious Defence' against these provocative measures. Agitation led to an autonomist renewal, which subsided only with the resignation of Léon Blum and the reversing of the decrees.

It was perhaps in the wake of a rediscovered consensus between the autonomists and the broader Alsatian milieu that a new initiative by Rossé enjoyed some success, the 'Alsatian Vigilance and Economic Action Group' of 1938: the district councils, chambers of commerce, banks and unions collaborated with it to draw up a programme of economic reform and assistance (*La France de l'Est*, 30 August 1939). It was an Alsatian attempt to improve the regional economy and surmount the party barrier. But it failed in its attempts to establish a credit society to aid industry, to secure state orders for local textiles, and to push for commercial agreements with neighbouring countries, including Germany.

The Alsatians preferred to be politically neutral and to defend the autonomy of their schools, local economy and language, but hesitated to remonstrate against French policy in a more tense international situation. Events in Germany pushed many Alsatians closer to France. The Progressives splintered, with Dahlet leading a democratic and pro-French wing in opposition to a fragment that stayed close to nazism. A new Landespartei was built around this Progressive fraction which had opted for Alsatian demands over French democracy, Bickler's party and all those who, whether attracted by National Socialism or not, hoped that the European turmoil caused by Hitler would enable the party to extract autonomy from France: alongside them were the partisans of a German cultural renaissance, whose symbol was the Hünenberg youth centre, financed by Ernst and designed to revive Alsatian traditions, where there were held several rallies to demonstrate the unity of youth on both sides of the Rhine.

The French response was predictable. In April 1939 a decree banned all movements suspected of separation. Their newspapers ceased to appear. Carl Roos, accused of spying, was arrested. After the declaration of war, other leaders joined him in prison. In October Roos was tried by a military tribunal in Nancy. The proceedings were quickly over, and led him to the execution stake the following February. Was the death of Roos meant as an example? Was he guilty of espionage? He could be censured, at least, for envisaging an Alsatian uprising in the event of a German invasion. After his death, some of the other prisoners were held in secret while awaiting trial: the remainder were distributed among several prisons in the south of France.

The autonomist period drew to a dramatic end. For the government, the war had turned all autonomists into potential traitors. The time no longer lent itself to clemency, as in 1927-28. While Alsace had obtained some satisfaction in religious and linguistic education, the state had not conceded the essential point: it had not admitted even the least possibility of administrative and political autonomy.

Back to Germany: incorporation and resignation

After victory in 1940, Germany promised to respect French territory: but it annexed Alsace, ignoring the protests of the Vichy government. French civil servants departed, and were replaced by those of Baden. German legislation and law were progressively introduced, while many Alsatians received German nationality. There began 'the drama of Alsace' (Mey 1949), a period that Alsatians would want to forget and that was for long very largely forsaken by a historiography concerned not to reopen the rift that had separated Alsace from France after 1919. The desire for silence was partially lifted after 1970 over the question of those drafted into the German army, but the majority of studies continued to shun any allusion to Oradour or the Liberation trials. The silence was incomprehensible, in that the extent of collaboration was confined not to Alsatians, but also applied to other Frenchmen. But Alsace remained very sensitive about this period, and for a long time the memory of the war served to enclose Alsace in a 'psychological exile'.[8] France, too, misunderstood the Alsatian behaviour, taking for collaboration what was often only passivity.

At the beginning Alsace benefited from preferential German treatment. Prisoners of war were quickly returned, and Alsatians previously evacuated to central and southern France were greeted with a fanfare by the Germans, who used the occasion for propaganda purposes. Many accepted administrative posts and joined Nazi organizations which, according to Gauleiter Wagner, encompassed 63 percent of the population by 1944. The DAF, the labour front combining both workers and employers, counted over 230,000 members. The frequency of meetings in 1940 enabled the autonomist, Schall, to say, like the French before him in 1918, that the plebiscite had been held.

The autonomists played an important role in this apparent acceptance of the German regime. After the armistice Robert Ernst increased resources for the development of cultural life in Strasbourg and for preserving a facade of autonomy. The 'Nancy group', freed after pressure from the armistice commission, were persuaded to sign an appeal to Hitler requesting the reincorporation of Alsace into the Reich. This 'Manifesto of Trois Epis' (Kettenacker 1973) had 13 signatories. All had been traumatized by the death of Roos and the fear that they would suffer the same fate. Freed through the efforts of Ernst, they remained 'in quarantine' at Trois Epis until the manifesto was signed, succeeding only in securing a slight revision of the text, but nothing more. They were then given relatively important positions in the local administration. Ernst and Wagner made them intermediaries between German authority and the population, and treated them with respect: in November 1940 the Nancy group were hailed in the Reichstag. Ernst, who became Oberstadtkommissar in Strasbourg, founded the Elsassische Hilfsdienst (EHD), a mutual aid organization concerned with the return of evacuees and prisoners, and with the participation of Alsatians in political life. Wagner was suspicious of Ernst and his organization, seeing the EHD as a simple springboard for the Nazi Party. Did Ernst really seek to preserve some freedom of manoeuvre for Alsace? He affirmed after the war that he had wanted to create an Alsatian Party. In this case, he lost. One could interpret in this sense his opposition to the drafting of Alsatians in 1942: he accompanied them to the Eastern Front, and Wagner quickly dissolved the EHD sections. At any rate, the Nazis regarded the transition as being over as early as autumn 1940.

Hopes of autonomy quickly evaporated: there remained only germanization and collaboration. Carl Roos became a martyr, and a mass ceremony at Hünenberg celebrated the return of his ashes in 1941. One had the impression that Alsatians in their entirety had followed the path shown by the leading autonomists. To protect themselves, as is often said? Or under coercion? It is undeniable that the organizations and administration functioned successfully. In the same way, factories produced for Germany, which was very lucrative for their owners. One can explain this attitude by fear, respect for order, habits of obedience: on the other hand, those Alsatian leaders most hostile to Germanization had not returned to the province. Germanization per se was perhaps not unpopular, but the Nazi party did not inevitably fire enthusiasm. The earlier strongly autonomist areas of Altkirsch (Ricklin's home area), Haguenau, Wissembourg and Saverne would be more refractory to German organizations than many others (Dollinger 1979).

Relationships with the Nazis deteriorated with the introduction of an obligatory six-month labour service in May 1941, and even further in August 1942 when conscription drafted 100,000 men into the German army. The latter generated some resistance, but it was only later that the Alsatian Youth Front was formed, an organization that strove to send the most recalcitrant across the Vosges. Escapes and intelligence activities essentially constituted the sum of resistance in Alsace. Sabotage was rare, and the 'maquis' appeared in the Vosges only in 1944. The resisters came from the traditional nationalist milieus, Action Française, and the numerous reserve officers in Haut-Rhin. From these circles hostile to autonomism there came in 1944 the National League of Alsace, whose aim was to maintain attachment to France. The Communist resistance had many recruits among the railway workers of Strasbourg: it distributed pamphlets and newspapers, and helped many to escape, as did Catholic youth organizations. These movements, necessarily small because there was no organized maquis, were not without influence. Until the spring of 1942, German surveillance in Alsace was not too strict, permitting the success of several spectacular escapes, such as that of General Giraud in April. But the latter led to a more systematic repression. Several death sentences were pronounced, though not always executed: the Communists, however, were victims of a systematic harshness. As elsewhere, there were collaborators, incontestably numerous, in Alsace, even if they acted carefully, as historians of the period have often affirmed (cf. Maugué 1970: 122). Others helped the resistance and entered its ranks. Alsatians did not behave differently from the other victims of German occupation.

With the Liberation came reprisals. There was talk of 40,000 internments, but the number is uncertain.[9] There was no popular revenge against collaborators: Alsace did not experience summary executions. The balance-sheet of sentences is not easy to establish: many of those convicted were pardoned individually or under a more general amnesty. The cleansing operation hit the autonomists, even when they had not obviously collaborated. The state also used the occasion, as it did elsewhere, to rid itself of its prewar adversaries. The trials resulted in about 8,000 sentences: those of the leading autonomists were in 1947. Their defence was that they had wanted to protect Alsatian interests. Mourer was condemned to death and executed by firing squad. The death sentence was also passed on the absentees: Hauss, Bickler and Schall. Rossé was 'spared', perhaps because of his connection with the UPR, but was given 15 years' hard labour. All were regarded as traitors.

The autonomist cause was discredited for a long time. Even regionalism was suspect, despite the existence of autonomists who, like Dahlet, had not collaborated. Once again, Alsace was wounded by the trials, misunderstood by the French and often also by the former evacuees. This period left poisoned memories, which were hardened in 1953 during the Bordeaux trial of those responsible for the Oradour massacre in 1944. Twelve from Alsace-Lorraine had taken part as members of the SS Das Reich division. The PCF demanded their heads, which caused a backlash among its Alsatian members. A Communist councillor in Strasbourg was expelled from the party; one of the defending lawyers of the Bordeaux accused resigned from the party. Both appeared in the next election on a common list that included the president of the association of draft-evaders and those conscripted by force. The whole of Alsace rose up to support the accused, for whom the Alsatian deputies eventually secured a pardon. Alsace, despite the atrocities that had been committed, refused to regard these men as guilty. Through them it felt itself also accused; hence its indignation. It bore a grudge against France, which had abandoned it in 1940, for reproaching it with its faults. It wanted to be considered not as guilty, but as a victim. After 1953 silence on this unhappy period became the rule. But despite the trials, the return to France after the war no longer generated resistance. Alsace renounced its autonomist perspective and resigned itself to its fate, to gallicization.

Assimilation and accommodation

Today, Alsace has apparently changed a great deal, both politically and economically, and the old desire for autonomy scarcely seems to be present. France has avoided the prewar mistakes. It has been more aware of regionalism, setting up a network of regions in 1955, and regional councils in 1972, with a certain decision-making power. Possessing some limited regional authority could be satisfactory, and Alsace, without apparent difficulty, has been integrated into national political and economic life. The Concordat and religious statute have been retained, and German reintroduced in primary schools. Election results confirm the impression of satisfaction: the national parties dominate. This change in attitude was eased by Franco-German reconciliation and Alsace's place in Europe. The border location, so uncomfortable before 1939, was henceforth a trump which allowed the region to give up its introspection and play a positive European role. The choice of Strasbourg as the home of the Council of Europe and European

Parliament revived the vocation that the city had had under the German Empire, that of a capital. Attachment to the region remains, but is based less today on the past and its traditions than on pride in seeing the region surmount its ordeals and prosper. Before the war the misfortunes of Alsace had reinforced local patriotism. Today the latter comes from its high living standard and confidence in a European future for Alsace, which believes less in the assimilative will of France.

Two economic facts dominate: industrialization as a factor of radical change, and a slowing of expansion after 1974. The border region could offer a shining facade to its neighbours, because of the co-ordinated efforts undertaken under the auspices of economic planning: 35,000 new jobs were created between 1955 and 1968, with even greater acceleration afterwards. The demographic growth rate reflected industrialization. Between 1900 and 1936 the population had increased by only 50,000: since 1946 Alsace has gained almost 400,000 inhabitants (Kleinschmager 1974). With a population of 1,520,000, Alsace has a higher medium density than France, 1.5 percent of the national territory and 2.9 percent of the French population. Geographic mobility contributed to the uncloistering of the region. Industry, benefiting from old traditions, the labour force, capital and the facilities offered by the region's crossroads position, today has 46 percent of the active population (the French figure is 39 percent).

There has also been structural transformation. The old textile sector, already overtaken by the world crisis before 1939, did not rise again, despite postwar rationalization. It employed 53,000 workers in both 1936 and 1954: by 1979 there were only 38,000 — 14 percent of the labour force as against 31 percent in 1954. Engineering came to dominate the industrial sector; textile decline was accompanied by a geographical redistribution of industry. The Vosges was 'de-industrialized'. It is the Rhine plain that has attracted the new establishments: the plain has been transformed, forests have disappeared, the rivers polluted. Despite agreements with the countries bordering the Rhine, France has continued to dump wastes into the river. Industry has reversed the natural balance, and plans for atomic power stations disturb the inhabitants.

The concentration of industry and the association of Alsatian enterprises with French companies has effectively transferred economic decision-making to Paris or other French regions. Foreign capital, especially German and Swiss, also flowed to

Alsace, creating 38 percent of the new jobs between 1955 and 1968, and helping to retool declining industries in difficult areas that could not attract French capital. The pattern has entailed some risks: in times of crisis these enterprises are disfavoured, while under full employment the profits leave the region.

One inevitable consequence of industrialization and the rapid advance of the tertiary sector was the rural exodus that emptied the countryside. The three major urban centres — Strasbourg, Mulhouse and Colmar — hold over one million inhabitants, or 72 percent of the population. Agriculture had not been able to over-come the difficulties that affected it between the wars. Small units have dominated: in 1970 44 percent were less than 5 hectares, against 29 percent in France. Too often, they practise a less pro-fitable mixed farming. Moreover, despite the progress of co-operatives, the young often leave the land, leaving behind an ageing work force. Alsatian agriculture has a lower income than that of its French counterpart: it has also invested too little, has made too little use of new techniques, and its products have not really con-quered internal or export markets. Employment in agriculture has declined from 31 percent of the active population in 1949 to only 5 percent, lower than the national average. This could only transform the regional mentality, for traditions and the use of dialect were more deeply anchored in the countryside: cultivators were quite numerous in the autonomist ranks before the war.

Since 1974 Alsatian development has suffered from the Western world's economic downturn. The latest economic plan has not been wholly realized, and credit is less easy: the planned Rhone-Rhine canal link has been placed in question; the autoroute link to the south is unfinished; industries are not at full production; and export outlets are less certain. This is serious for a region that ex-ports finished products, and imports more than it exports. The most visible consequence was an increase in unemployment. Its activity rate declined slightly in the 1970s, and with 39.7 percent of its population active in 1979, Alsace was below the French average. In 1978 the region ranked eighth on unemployment. This unemploy-ment is partially masked by the presence of 'frontier workers', around 25,000, attracted by employment and higher wages in Ger-many and Switzerland: it is, moreover, often the skilled workers and the 'brains' who work over the border.

Alsace suffers from inferiority with its neighbours, which is pre-judicial to the ambitions of its towns, which aspire to become the capitals of multinational regions. Anxiety has a real basis. The average household income is only slightly superior to the French average, and over the past decade middle and high incomes, and those of farmers, have progressed less rapidly than those in the whole of France. Economic transformation has produced social change. Before the war, the middle classes, discontented either because of poor integration with France or because of economic difficulties, were among the most autonomist. This is no longer true, at least of the middle class as a whole. But to the extent that expansion has slowed down and optimism is in retreat, the attempt to lay the blame on inadequate national subsidies and excessive taxation is frequent (*Rot un wiss*, October-December 1975). In fact, the French bureaucracy has not consolidated the economic links between Alsace and France (Egen 1969). France remains attached to the idea of a national economy, and has only an imperfect conception of multinational economic regions. The policy pursued towards Alsace has faltered between the two types of development, and the risk of a lack of confidence in Parisian promises exists especially as the future now seems less promising. However, Alsace has displayed scarcely any regional sensitivity to being partly deprived of economic decision-making. Since 1967 there have been conflicts with the reactivation of workers' movements, but these have not contained particularist themes.

The linguistic domain appears to be one of the last refuges of Alsatian claims. But it is a relatively neutral area which has not generated great political debates or rallied Alsatians of different parties and social backgrounds. Philipps (1978) points out that, though the Alsatian identity is based upon dialect, it is vital to be able to speak French. The linguistic problem consists of regrets and a bad conscience over a language in retreat. Demands are more timid than before the war: they relate primarily to bilingualism in

TABLE 1
Newspaper circulation in Alsace

	Total bilingual and German press	Dernières Nouvelles d'Alsace		L'Alsace	
		German	French	German	French
1946	317,000				
1964	219,000	100,000	30,000	50,000	40,000
1977	144,000	79,000	143,000	36,000	95,000

schools, not to administrative bilingualism. A 1979 survey indicated that 75 percent of those aged over 15 spoke dialect, a decline since 1962, when the figure was 87 percent. The levels are impressive, but the current use of dialect is much lower among the young. A 1975 survey among 1,024 pupils in Bas-Rhin showed that in the countryside 28 percent most often spoke French: in Strasbourg this figure was 92 percent (*Dernières Nouvelles d'Alsace*, 14 February 1980). Other data show that 61 percent watch television in French, while 71 percent read a daily newspaper in this language (*Sondages*, 1972, no. 2). The decline of the German language press is significant (Table 1), which helps to explain the progress of French, since dialect and German are mutually supportive.

Despite its decline, the German dialect can be considered as the native tongue of most Alsatians. Its origins lay in the Alemanni who crossed the Rhine in the fifth century. From their language came the popular speech that progressively implanted itself in Alsace, while being merged with Latin. On the whole, the dialect is low German. However, the extreme north-western corner speaks Francique Rhenan like Moselle; the far south is high German, as in Switzerland; and the central valleys of the Vosges are French-speaking. Above this diversity of dialect, the written, literary and business language is high German, and distinct from the spoken language.

French had advanced during the Second Empire, especially as a language of culture. But its progress in Alsace was restrained by the churches. The pastors spoke only German, while the Catholic clergy saw in French the vehicle par excellence of revolutionary ideas. After 1870 German, the official language, and the only language of education, grew. In the 1920s some limited provisions were introduced for religious instruction in German or dialect and for the teaching of German. This was revoked in 1945. Simultaneously, German publications were banned. Only bilingual editions, with their title and at least 25 percent of their text in French, were permitted: all sports items and those aimed at youth had to be written in French.

These measures did not provoke any reaction: in 1945-46 who could defend German without risking the accusation of being pro-Nazi? The SFIO supported gallicization, since one of its deputies from Strasbourg was Minister of Education between 1944 and 1948. French and French culture prevailed. Alsatian, with German as its written form, was not officially considered to be a regional language, but allogeneous — the same as Flemish. Neither Alsatian nor German was affected by the laws of 1951 and 1975, which introduced instruction in regional languages and cultures at all educational levels on condition that the confrontation of cultures did not damage national unity. The state's linguistic policy in Alsace has scarcely changed since 1945.

But Germany was no longer a threat, and the question of bilingualism re-emerged. Christian Democrat (MRP) and PCF deputies from Alsace requested the renewal of instruction in German. The arguments in favour were several: it was used in the family, at work and in relations with neighbouring countries. Bilingualism also appeared necessary for the cultural and psychological development of Alsace, since it was through German instruction that dialect could survive. In 1952 and 1953 the government conceded limited German instruction in the final two school years, though this entailed employing part-time teachers. The decrees were well received by parents, who approved of them by some 75 to 80 percent, but they did not ameliorate the linguistic situation. They were, in fact, opposed by the teachers' unions, with the consequence that such teaching was marginal. This opposition was governed by the idea that young dialect-speakers would not be able to master both languages, and so would enter practical and professional life at a disadvantage, an opinion that has survived, though less distinctly. The other reason was the churches' fear that a bias towards dialect and German might increase religious discrimination. The new bilingualism therefore developed in a climate of reluctance.

In 1968 the newly founded Cercle René Schickelé participated in the Council for the Defence of Languages in France, which had fought since 1960 for instruction in regional languages. The Cercle argued that bilingualism was essential in a united Europe. It contributed to the introduction in 1971 of the experimental Holderith method, whereby German would be taught with the help of dialect: by 1975 some 3,500 classes were taught German by voluntary teachers under this plan. A 'Comité pour le droit au dialecte à l'école maternelle' was formed in 1977, arguing that children should be plunged into a 'linguistic bath' upon entering school; and in 1978 the district council of Bas-Rhin commissioned a study of the conditions that surrounded bilingualism. The latter had been rehabilitated, at least in theory. Since 1968 young teachers have been more tolerant of dialect. But while the Holderith method has been defended by SFEN, the teachers' union affiliated to the

CFDT and by a leftist tendance within the National Educational Federation, the most important union of primary teachers (SNI) has rejected the idea of a special linguistic statute for Alsace (*Bulletin du SNI du Bas-Rhin*, May-June 1978). The parents often appeared hesitant, preferring that their children become proficient in French (*Rot un wiss*, January 1979).

The political background of this linguistic debate was not at all clear: many statements, electorally motivated, were soon forgotten by their authors. At times the arguments in favour of dialect appeared so embarrassing that one feared that the campaign would again be linked to regionalism or, even worse, autonomy. Refuge was taken behind the 'Rhenish' or 'European' vocation of Alsace (the 1972 manifesto of Alsatian senators); in the necessity to rediscover old traditions; the charm of the homeland (*Dernières Nouvelles d'Alsace*, 3 June 1974); or in practical arguments that the rehabilitation of dialect in education would increase proficiency in French (*Rot un wiss*, April 1979). However, the tendency not to mention the real argument, the defence of Alsatian identity and culture, has gradually diminished over the past decade. This evolution of attitudes went beyond the question of bilingualism in the schools. The 'appeal of writers' of 1980 demanded 'official status' for Alsatian, which must be present in cultural life so that Alsace could control its own destiny.

The state appeared to have officially recognized a cultural identity in the 'cultural charter' of 1976. Bilingualism was recognized, and the aim was the promotion and diffusion of Alsatian culture, with an agency financing local initiatives and providing necessary resources. The Charter, a contract between the state and local bodies, could nevertheless be influenced by the state's representatives, who could use the subsidies to favour French culture at the expense of the Alsatian. Negotiations over its renewal in 1979 came to a dead end, and the project was reduced to 'efforts in favour of safeguarding dialect ... a few fair words' (*Dernières Nouvelles d'Alsace*, 6 July 1980).

The population seemed to be little interested in this debate, which was confined to intellectuals and political militants. There was no overwhelming interest in the often deplored decline of dialect, a decline that was less important quantitatively than qualitatively. In effect, 'to speak' did not mean fluency. In addition, the young used dialect less, especially in the towns. Parents have spoken dialect more at work than at home, believing that French would be a better preparation for their children in professional life. This was particularly a bourgeois trait, introducing a linguistic cleavage between the classes: the workers remained more loyal to dialect. The strongest users of dialect remain the peasants, a much diminished group, and pensioners. The young have been more aware of the utilitarian arguments: dialect-speakers had a poorer scholastic record and a less satisfying mastery of French. In Alsace only 33 percent of children continued their education beyond the age of 16, ranking nineteenth regionally: the French average was 44 percent. The *Dernières Nouvelles d'Alsace*, the most widely read newspaper in Alsace, stressed the difficulties of the dialect-speakers: but it still refused to publish a text prepared by the Cercle René Schickelé intended as propaganda for the use of German as the premier language at college.[10] This made parents more anxious about the projects for bilingual schools. It was this reticence, rather than the statistics, that alarmed the partisans of bilingualism, and explained their prudence. When Zeller, an Alsatian deputy, proposed such a law in 1980, he stressed that bilingual education would not be obligatory, and that the primacy of French must be preserved. The impression of a sentimental attachment by Alsatians to their language lingers on, but they appear to be influenced more by the inevitability of its decline and of gallicization.

There was, simultaneously, an Alsatian literary renaissance, supported after 1968 by leftist militants and students. While it had a certain snobbish air, much of the preference for folklore, which affected all French regions, was not entirely artificial. Theatre in dialect, popular since the end of the nineteenth century, attracted even greater audiences in the early 1970s (*Dernières Nouvelles d'Alsace*, 9 October 1974). This wave of dialect expression in popular entertainment was accompanied by an increase in more literary activities. In 1971 the Council of Alsatian Writers was formed to struggle against Parisian centralization and to promote Alsatian literature in Alsace and beyond. Twice before, in 1918 and 1945, writers had known the difficulties of reintegration. Those who used German, faced with losing their public, often chose to live in Germany. Grouped in 1924 around the review, *L'Homme de Fer*, the Alsatian writers demonstrated a will to write in the language of their choice, to defend the culture of their choice: the review wanted especially to promote Alsatian literature in German, and exalted the Alsace-Lorraine 'patrie'. An editorial in 1926 demanded a break with Paris to mark the awakening of Alsace. Police reports considered *L'Homme de Fer* to be autonomist and germanophile (Gras 1975). An analogous situation appeared in

1945. The German writers felt themselves isolated, and political as well as linguistic problems delayed their reacceptance for a long time. Only the francophones were integrated into French literature.

While those who write in dialect or German have regained popularity in Alsace, they feel that Parisian dominance has limited their appeal: many have fallen back upon their Alsatian roots. The Cultural Front of 1974 brought together not only writers, but also singers, ecologists and social militants who wanted 'cultural self-management' (*Dernières Nouvelles d'Alsace*, 20 October 1974). This meant the recognition of bilingualism, the promotion of Alsatian culture, the study of local history and the effective regionalization of television and radio.[11] The Council of Writers accused them of obsession with the past and archaic traditions, though the Front responded by asserting the total irrelevance of autonomism and separatism. In general, Alsatian writers and intellectuals have felt that, because of the disquieting regression of the language, Alsace has reached a turning point in its history. In 1980 they again launched an appeal for dialect and literary German to be given official status: though the signatories envisaged 'the Alsatian question' to be only linguistic, their tone was much harder than in 1974.

Could this be the prelude to more intense Alsatian 'agitation?' The province has seemed quite apathetic, but appearances can be deceptive. The linguistic debate did not gain the same prominence as the ecological fight. The question of local law appears completely forgotten today (*Le Monde*, 29 March 1974): the left in particular regards it as 'depassé'. It is, in fact, 'congealed', lacking any evolution since 1918, which explains why it has few defenders (*Rot un wiss*, January 1980). The true situation — an Alsace deaf to the many appeals, apparently indifferent to the dilemma — is surprising for a region so attached to its traditions. The middle classes are largely satisfied with the regionalization of 1972, the subsidies and the attention of the centre, and are proud of a cultural renaissance that gives it self-respect. The majority of the population — lacking perhaps a political conscience, but above all because they do not feel oppressed — hardly express themselves: 'nationalist' fever appears more after the victories of their football team. Accusations of the 'treason' of their elected representatives, too docile towards Paris (Fiscus et al. 1975), of 'alienation' or of 'Elsass: Kolonie in Europa', hardly roused them.

The cultural 'awakening' contrasts strongly with the discretion that was the rule in the political domain. The national parties, apprehensive of the possible electoral success of autonomism, supported the linguistic demands with a fair degree of unanimity, but continued to condemn all ideas of political autonomy. The autonomist groups and parties have been marginal, ignored by the local press and hindered by a constant recall of a past that made them vulnerable to all kinds of accusations.

The four major political families have confirmed national unity and the necessary primacy of French. Their schemes, affected by the nuances that follow from their ideological stance, are limited to a greater or lesser regionalism. The four also share the contradictions that emerge between local conceptions and Parisian plans. The Gaullists, who overshadowed the others between 1962 and 1973, won all the Alsatian seats in 1967 and 1968: in 1978 they gained just over 33 percent of the votes. The Democratic centre, built upon a Christian Democracy that disintegrated after the coming to power of de Gaulle in 1958, stabilized at around 23 percent of the vote in 1973 and 1978. The left is no longer represented in the National Assembly, but since 1973 the new Socialist Party (PS) has outstripped the PCF. While all the parties are solidly attached to the nation and the state, they have also claimed to be partisans of a stronger regionalism.

The strongest Gaullist support has been in the most Lutheran rural regions (Dollinger 1979), which had most strongly supported autonomism before the war. The party has also won some of the Communist electorate among groups that were equally favourable to autonomy before 1939. The shift of these groups from autonomism to French nationalism merits a more intensive study. De Gaulle perhaps seduced Alsace by pandering to its self-respect: in 1945 he had accorded to Alsace, and especially Strasbourg, a central role in the French 'revival', and pointed to the Rhine as the 'bond of western Europe'. The General's Catholicism was a supplementary guarantee to Alsace, sensitive to all efforts at a Franco-German reconciliation.

Gaullism in power did not intend to move beyond decentralization. Pompidou even claimed that the 1972 law creating regional assemblies would go too far (Peyrefitte 1976). He saw no place for regional languages and cultures in a France that would be central in Europe (*Dernières Nouvelles d'Alsace*, 16 April 1972). It was the time when triumphant Gaullism preferred a 'grand' eastern region, essentially economic, to a more limited Alsace. The Gaullists did not oppose bilingualism, but they did not see it as necessary to an Alsatian identity. Their motivations, essentially practical, were built upon the potential exchanges with Germany: a minister estimated in 1969 that the development of French in Baden and the Palatinate would counterbalance the diffusion of German in Alsace. Since 1973, with an electoral decline of almost 20 percent in Alsace and the defeat of their presidential candidate in 1974 the Gaullists have been more attentive to regional preoccupations. In 1978 a deputy from Lorraine proposed a law in favour of bilingualism and the regional 'personality'. In 1980 the seven Gaullist deputies from Alsace presented a plan for a regional council with enhanced power which would contain nationally and locally elected deputies and councillors, as well as 'socio-professional' representatives — a less democratic formula than that of the 1972 law, which would have given power to the Gaullists no matter who governed in Paris, while still safeguarding the powers of the state through the prefects. Their preoccupation with national politics risked trapping them in an Alsatian option.

The PCF, very suspicious of autonomism, and nationalist in the face of the USA and Europe, was still attentive to the problems of minorities. The PCF has always claimed bilingualism for Alsace, in administration and law as well as education (*Le Monde*, 29 June 1971). It has equally favoured bilingual regional television. Georges Marchais affirmed in Alsace in 1977 that regional cultures could express themselves without damaging the pre-eminence of French (*Dernières Nouvelles d'Alsace*, 16 June 1977). Educational bilingualism would not only rescue the regional languages, but would also permit a better integration of the young to their milieu. Understandably, the PCF paid attention to socioeconomic questions, accusing the centre of destructive politics. It also wished to grant extensive powers to the region, so that it could direct its own territory and resources (*L'Humanité*, 21 May 1980). Politically, however, the PCF has opposed any idea of separatism; only France could determine the democratic resolution of the regional problem, of 'making France a coherent construction of autonomous and jointly responsible communities'.

Socialism has had the same faith in national unity and the same concern for helping the regions and respecting the 'right to be different'. The PS has also rejected separatism. But while wanting to preserve the essential powers of the state, it has fought against excessive centralization and the insufficiencies of the 1972 reform, not an easy task. The dominant tone within the party is 'Jacobin', though there are those on the left who favour a real decentralization. The party also proposed in 1980 an elected regional council with enlarged powers. The PS is thus very far from the centralizing position and the desire for systematic gallicization of the prewar SFIO. For Alsace, the 'right to be different' implies bilingualism in primary schools,[12] Alsatian culture, the maintenance of certain local laws and a secularity 'adapted to the Alsatian body' (*Frontière Alsace*, November 1975), a very enigmatic formula. The questions of the Concordat and schooling have not been raised by the PS, though its opponents accuse it of wanting to suppress the former (*Le Nouvel Alsacien*, 25 January 1976). The party has gained many Christian recruits, both Catholic and Protestant, and is no longer as secular as the SFIO, which explains its electoral progress since 1968. The several Independent Socialists (PSU) who joined it after 1974 have injected their concern for the respect of minority rights and self-management in regional matters. Alsatian socialist deputies have actively participated in the linguistic and cultural 'battle'. The Christian and leftist influences contributed towards limiting the centralizing tendencies of the PS, but in Alsace as in other regions, socialist positions are not always clear because of the contradictions between the state, even democratic and socialist, and the regions' aspirations to autonomy. The socialism 'respectful of difference' (*Le Quotidien*, 24 June 1980) has not clarified sufficiently its will to develop regional autonomy. The left, therefore, supports Alsatian demands, albeit prudently. Its solutions perhaps lack boldness. The tiny PSU is the only one firmly to support the minority movements of the French hexagon, but it is barely active in Alsace.

The fourth party group, centrist and clustered around Giscard d'Estaing, has adopted various positions according to its various components. The President himself contrasted subsidies for 'good Alsatians' with the criminal courts for 'bad' Corsicans and Bretons. He accepted instruction in dialect, 'the regional blooming', and in German, 'the language of our illustrious neighbours', to further European cohesion: however, he did not believe in regionalisation, relying upon the départements and communes to lighten the tasks of the state and to contribute towards its strength and efficiency.[13] During a meeting of Giscardian deputies at Colmar in 1980, it was stressed that the movement accepted decentralization, but nothing more. But many of its Alsatian representatives are the spiritual heirs of the prewar UPR, regionalist with some autonomist leanings. From an essentially Catholic tradition, they have defended Alsatian distinctiveness. The manifesto issued by its Alsatian senators was typical: 'In the face of all challenges, Alsace must awake, Alsace must assert itself'

(*Dernières Nouvelles d'Alsace*, 5 December 1972). In 1976 a more reformist-minded deputy, Zeller, founded a group with a symbolic title, 'Alsatian Initiatives': several local councillors were attracted by his campaign, which at least in its stress upon environmental problems found a strong echo among the population. Politically shrewd, Zeller avoided the word 'autonomy', contenting himself with opposing the 'Parisian bureaucracy'. Overall, then, the centre representatives also display Alsatian and regionalist sympathies. In this they are similar to those from the other parties: all are conscious of the greater popularity of the idea of regionalism that emerged in the 1970s.

Before the war, the churches were vigilant in defence of the language and particularism. Today, they are more discreet and slightly less influential. Neither the Concordat nor the status of confessional schools has been threatened. However, the Bishop of Strasbourg sounded an alarm in 1980 — 'they want to expel God from the schools in Alsace' — directed against the partisans of secular education. The Bishop stressed that 'the religious fact is part of our Alsatian identity'. But 1924 is far away, and this anxiety did not stir public opinion. For the churches, however, as for the parties, the affirmation of commitment to 'alsacianité' reflects the concern for following the preoccupations of the faithful and the electorate.

The autonomist fringe

Immediately after the war, autonomy and even regionalism would have been ill-received in Alsace. The first attempt was fostered by the 1953 Bordeaux trial. An Alsatian Popular Movement in Haut-Rhin preached decentralization and the defence of local rights and traditions. In 1958 Camille Dahlet began the federalist *La Voix d'Alsace-Lorraine*, which also carried information on linguistic minorities throughout Europe. In 1966 it was accused by *Elan*, the review of Christian Social intellectuals, of having ties with neo-Nazis: it was still difficult in 1966 openly to defend Alsatian traditions. But after 1968 there was a political thaw. There emerged a new regionalism, provocative in tone and action, represented by the journals *Uss'm Follik* in Bas-Rhin and *Klapperstei 68* in Haut-Rhin.

Shortly afterwards came the ecological wave, strong in Alsace. The defence associations were numerous, and they were Franco-

German — a sign of the times. They were concerned not only with industrial pollution, but also with the decline and disappearance of rural communities (*Le Nouvel Alsacien*, 1 July 1974). There is now an ecological area of demands, not to be confused with regionalism or autonomism, though it links up with them in the defence of the region against accelerated industrial development. The successful campaign of the Marckolsheim affair, a protest against the building of a lead stearate factory, won many Alsatians to the ecology cause. They had gained a new courage through being able to impose their will upon the prefect and Paris. Some momentum was maintained later by the campaigns against nuclear installations. By 1980, however, the prominent leftist input had gone, and ecology had taken on the most classic forms of electoral protest.

Other bodies persist with the struggle for bilingualism and a discrtee regionalism. But intellectually, the history of Alsace suffers from certain silences. *Rot un wiss* (July 1979) stressed that the Institute of Arts and Popular Traditions of Alsace had rewarded all intellectual and artistic disciplines in 1979 except history, and that 'it is only colonial peoples who are refused the right to know their own history'. The same point was made by other journals.

A 'regionalist and federalist' journal, *Elsa*, appeared in 1969, and the regionalist Movement of Alsace-Lorraine decided to promote a 'free Alsace-Lorraine within a European federation of free peoples'. Neither the journal nor the movement escaped accusations and divisions. Its founder, Dr Iffrig, a former MRP member, wished a 'decolonized' economy, a regional parliament and an autonomous radio and television network. His movement immediately attracted Gaullist hostility and several accusations of being too accessible to former collaborators. Iffrig himself did not try to conceal his prejudices. He affirmed 'We are German' (*Elsa*, 1975, No. 34), and did not hesitate to display his anti-Semitism. His apologies for Carl Roos and Joseph Rossé caused much offence. His activities provoked an important split in the movement, and led him to be convicted in 1978 of inciting racial hatred.

The secessionists from Iffrig's movement formed the 'EL Autonomist Front of Liberation', which publishes *Rot un wiss*. Autonomist and federalist, it seeks to defend a crumbling Alsatian identity. Its objective is clear: autonomy for Alsace-Lorraine within a federal Europe. Its activity has taken several forms, exposing it too to countercharges and physical assaults. The monthly *Rot un wiss*, open to several tendances, has castigated the Alsatian deputies for being too subservient to Parisian power, and tries to demonstrate the economic 'colonial condition' of Alsace. A central place is reserved for movements from other regions, especially Brittany and Corsica. EL has participated in the ecological and anti-nuclear campaigns because for its leaders ecology is necessary not only to preserve the environment, but to change a dehumanized society into one where the individual can protect the right to his language and culture (*Rot un wiss*, March 1978). EL tried to rally other organizations to it in 1977, and the Federalist Party of Alsace-Lorraine and the Movement for the Liberation of Alsace-Lorraine responded favourably to an appeal in *Rot un wiss*. A similar call for the formation of an Alsatian Party came in 1978. Two further 'umbrella' organizations appeared in the same year: the federalist Movement of New Autonomists, and a more moderately autonomist Popular Alsatian Movement.

The autonomist movement has been extremely fragmented. The old prewar militants, their spiritual heirs and the young leftists do not always live happily together. The groups, furthermore, also suffer from the various cleavages of French political life. It is difficult to estimate their true strength: electorally, however, they are no longer convincing. With slender resources, their small organizations have not been able to challenge the monopoly of the major national parties. The autonomists' electoral performances in Alsace have been comparable to those in other regions. At the 1976 district elections, a candidate won 11 percent of the vote in a Strasbourg ward with the battle-cry, 'Alsace must be liberated'. Another autonomist candidate could win only 4 percent of the vote in the 1978 national election. The movements persist with candidates, with almost no success. Their militants come primarily from the lower and middle bourgeoisie. The upper middle classes remain hostile to the whole autonomist movement, while the workers and farmers are content with their attachment to dialect and Alsatian traditions. But all have remained faithful to the national parties. The federalist and autonomist movements have to deploy huge efforts to carry conviction. They are heard with a certain sympathy and with less indulgence, but they have not been powerful enough to gain support or influence. Rejecting violence, it remains to be seen whether they can find original modes of expression that can give them weight in Alsatian political life.

Conclusion

The contrast between particularism, the importance of dialect and the moderation of regional claims in Alsace is striking, especially when it is compared with the situation in Corsica or Brittany. To explain the calm in Alsace, one could point to the fact that the region has preserved some 'local rights' and some freedom in local law. Because of its relative wealth, Alsace has been able to accommodate its citizens, who have not, except in 1871, had any need to emigrate. A large population and the absence of mobility have allowed the retention of traditions and a living culture: Alsatians do not feel themselves dispossessed of this culture.

The war left enduring memories, and resentment against the leading autonomists who were led astray into collaboration, even though a good part of the population had expressed confidence in them to ameliorate the situation in Alsace. Misunderstood before the war, this population felt itself disliked in 1945 and has since preferred to run the risk of gallicization rather than feel itself completely isolated. Autonomy has lost the popular foundation it had before the war because the churches no longer support it and because it no longer can rely upon those workers who brought the Communist opposition to it. The middle classes are gallicized, with few causes for resentment. Moreover, Alsace has been favoured by regional prosperity for most of the post-war period, eliminating the earlier economic discontent: France, moreover, has understood the lesson of the past and has treated Alsace and its regional susceptibility more competently.

There has been in Alsace, therefore, a retreat of the old self-consciousness. Although it is easy for cultural demands to lead to political demands, Alsace, after having desired autonomy, has fallen back on the cultural claims. The autonomists saw Alsace before 1939 as an oppressed 'national minority', which in fact shared their views. Today, except for a small minority of activists, Alsace is a region like any other. No doubt there could be a forgotten or dormant sentiment that merely requires a favourable conjunction of circumstances for its expression. Nevertheless, it seems more likely that the cultural and linguistic demands will retain a prudent form because of the fear of a repression, which is always possible, of political aspirations that therefore dare not manifest themselves too openly.

NOTES

This paper was written in 1980.

1. AMG (19N673, report by Colonel Calla, 18 January 1919).
2. In 1912 the German Socialists had won almost 32 percent of the votes in Alsace-Lorraine. In 1919 the reconstituted SFIO gained 36 percent of the first ballot votes in Bas-Rhin, and 37 percent in Haut-Rhin. The three départements of the east provided French socialism with 6.5 percent of its total vote.
3. In 1920 there were 42 strikes and 46,500 strikers in Bas-Rhin, 23 strikes and 65,420 strikers in Haut-Rhin (*Annuaire statistique du Bas-Rhin, du Haut-Rhin et de la Moselle*, new series, 1919-31).
4. This Socialist wing had 33 percent of the votes from Bas-Rhin in the federal congress, and a slight majority (52 percent) in Haut-Rhin.
5. In 1919 it had launched an appeal for an independent state of Alsace.
6. By January 1926 *Die Zukunft* had grown from 8,000 to 24,000 copies. By 1928 its circulation was 35,000.
7. In 1924 the party won 10.8 percent of the vote in Bas-Rhin, and 6.3 percent in Haut-Rhin.
8. An expression employed by J.C. Guillebaud about the conscripts in *Le Monde* (21 April 1974).
9. According to Wagner, there had been in Alsace some 650,000 members of Nazi organizations. Most of these, however, were of an obligatory character, especially in the professional organizations. A narrower estimate would take into account members of the Nazi Party, about 30,000 including Germans, and of parallel organizations reserved for those who were not citizens of the Reich — perhaps 145,000.
10. *Rot un wiss* (April 1980) was indignant over this, characterizing it as typical of the 'repression' in Alsace.
11. A collection of writings, signed Jean, and published in 1977 by German journalists.
12. Letter from Mitterand to the Cercle René Schickelé, 13 May 1974.
13. Letter to the Cercle René Schickelé, 19 April 1974; speech at Thann, June 1979.

REFERENCES

ADBR Archives départementales du Bas-Rhin. Strasbourg.
AMG. Archives du Ministère de la Guerre. Paris Vincennes.
Annuaire statistique du Bas-Rhin, du Haut-Rhin et de la Moselle. New series, 1919-31. Strasbourg: INSEE.
Behe, M. (1920). *Heures inoubliables. Novembre-Décembre 1918*. Strasbourg: Le Roux.
Bernstein, S. (1970). 'Une greffe politique manquée: le radicalisme alsacien 1919-1939,' *Revue d'Histoire moderne et contemporaine*, January/March, 78-103.
Catalogue (1978). Catalogue de l'exposition 'Paris-Berlin', Paris: Centre Georges Pompidou.
Cecchini, M. (1971). 'Les Grèves de 1920 dans le Bas-Rhin et leur signification'. Memoire de maîtrise, Strasbourg.

Dollinger, P. (ed.) (1979). *L'Alsace de 1900 à nos jours*. Toulouse: Privat.

Egan, J. (1969). 'L'Alsace en quête d'une identité'. *Le Monde diplomatique*, November.

Ernst, R. (1954). *Rechenschaftsbericht eines Elsässer*. Berlin: Bernard & Graefe.

Fiscus, G., Sperber, O., and Colme, J.J. (1975). 'l'Alsace trahie', *L'Expression de l'Est*, no. 5.

Fleurant, J. (1907). 'L'Idée de patrie en Alsace', *Revue politique et parlementaire*, 324-45.

Fourier, M. (1928). 'Particularité de l'Alsace-Lorraine', in *La lutte des classes*. Paris.

Gras, C. (1975). 'Sur le régionalisme alsacien en 1926: un éditorial de *l'Homme de Fer*'. *Pluriel*, 00: pp. 51-6.

Gras, S. (1977). 'La Presse française et l'autonomisme alsacien en 1926', in C. Gras and G. Livet (eds), *Régions et régionalisme en France de la fin du XVIIIième siècle à nos jours*. Paris: Presses Universitaires.

Hoffet, F. (1951). *Psychoanalyse de l'Alsace*. Colmar: Alsatia.

Kettenacker, L. (1973). *Nationalsozialistische Volkstrum politik im Elsass*. Stuttgart: Deutsche Verlag-Anstalt.

Kleinschmager, R. (1974). *L'Economie alsacienne en question*. Colmar: Alsatia.

Maugué, P. (1970). *Le particularisme alsacien, 1918-1967*. Paris: Presses d'Europe.

Mey, E. (1949). *Le Drame de l'Alsace*. Paris: Berger-Leviault.

Mourer, J.P. (1929). 'Discipline et démocratie dans le parti', *Contre le Courant*, 28 July.

Peyrefitte, A. (1976). *Décentraliser les responsabilités. Pourquoi? Comment?* Paris: Documentation Française.

Philipps, E. (1978). *L'Alsace face à son destin, la crise d'identité*. Strasbourg: E.S. de la Basse-Alsace.

Reimeringer, B. (1980). 'Un Mouvement paysan extrémiste des années trente: les chemises vertes', *Revue d'Alsace*, no. 106: 113-33.

Richez, J.C. (1979). 'Novembre 1918 en Alsace: revendication de classes et revendications nationales', Memoire de maîtrise, Strasbourg.

Roos, C. (1928). *Politique et politique de violence en Alsace-Lorraine*. Zürich.

Zind, P. (1979). *Elsass-Lothringen, Alsace-Lorraine une nation interdite 1870-1940*. Paris: Copernic.

NEWSPAPERS AND REVIEWS

Rot un wiss	*Le Monde*
Elsässer	*Dernières Nouvelles d'Alsace*
Mulhäuser Volkszeitung	*Dernières Nouvelles de Strasbourg*
Freie Presse	*Sondages*
Strassburger Post	*Bulletin du SNI du Bas-Rhin*
Strasburger Bürgerzeitung	*Frontière Alsace*
Die Neue Welt	*Le Nouvel Alsacien*
L'Humanité	*Le Quotidien*
La France de l'Est	*Elsa*

9

Urban Politics and Rural Culture: Basque Nationalism

Marianne Heiberg
London School of Economics

Spain, or rather, 'the Spains in their pluralistic unity' (Vicens Vives 1970: 32), was born out of the Reconquista (718-1611), and an unstable alliance of independent Christian kingdoms stretching in a line from Galicia to Catalonia and pushing southwards against the Islamic invaders. The initial unifying idea behind Spain was simply that of Christian opposition to the Muslim threat. Unlike Navarra, Leon, Aragon and Catalonia, Castilla, the core of the future Spanish state, was a product of the Reconquista, without any prior separate political existence. Castilla was part of the repopulated areas, and the dominant classes consisted of 'colonizers' — Basques, Cantabrians, Aragonese and so forth. In short, Castilla was an amalgam of the peripheral ethnic and regional groups over which it later attempted to assert authority.

Neither the Catholic kings (1469-1516) nor their descendants, the Hapsburgs (1516-1700) tried to force a unified royal administration upon the Iberian peninsula. The Basque provinces — Navarra, Catalonia and Aragon — maintained distinct legal codes ('fueros') and autonomous political institutions. This disparate political arrangement was institutionalized in the Pacto Monarquico, which granted formal recognition to and respect for the autonomy of the different regions under the Castillian crown. Moreover, the Spanish regions were regarded as having a contractual rather than subordinate position to the central royal authority. However, while Spain's political institutions gave substantial guarantees for regional autonomy, the degree of autonomy that the regions actually exercised has frequently been exaggerated.

Imperial expansion generated considerable social and political

Author's Note: This paper is based on research originally funded by the Social Science Research Council and supervised by Julian Pitt-Rivers. I should also like to thank Ernest Gellner for his valuable comments on previous drafts.

integration in Spain. It imposed a type of organic solidarity. A critical element in the imperial endeavour of benefit to all Spain's regions was military control over the seas and colonies. This requirement demanded the mobilization of human and material resources well beyond the capacity of any single region. The Spanish state in general and Madrid in particular owed in large part their prestige and coherence as a political framework and centre to their ability to fulfil this requirement. But each region in Spain had different, specialized imperial ambitions. The Basque country was oriented towards the Americas. Castilla looked towards North Africa, America and the centre of Europe. Aragon and Catalonia turned toward the Mediterranean. However, the realization of these specialized overseas ambitions and interests required co-operation and social integration between the regions. The requisites and benefits of empire forged an extensive urban and administrative network throughout Spain that linked the country socially, economically and politically.

When Spain began to lose control over the mechanisms of imperial power, and thus over the composite empire, the bases of Spanish unity and prosperity were also eroded. The need for a more explicit administrative unification became acutely apparent. However, a more formal, stable unity was possible only if Spain's diverse regions were able to perceive their specific economic interests in terms of metropolitan Spain itself. A successful political centralization was dependent on the pre-condition of some form of Spanish internal economic unity. It was to this goal that the Bourbons in the eighteenth century and, more energetically, the liberals in the nineteenth addressed themselves. The fact that a satisfactory economic and political alternative had not emerged to substitute for the empire when it finally collapsed in 1898 lay close to the heart of what was to become Spain's intractable regional problem.

In this context the majestic failure of Spanish liberalism is of particular importance. The central vision of liberalism was to construct a political and economic framework for a modern bourgeois society. Politically, it sought constitutional unity as a means of transferring power away from the realms of sectoral and regional privilege into the hands of Spain's middle classes. Economically, liberalism insisted upon the establishment of absolute individual property rights as a means of stimulating production and creating a viable internal market. Influenced by the 'ideas and armies of the French Revolution' (Carr 1966: 79), the main ideological assumption underlying these considerations was the idea of national sovereignty, of Spanish nationalism.

Liberalism did achieve a certain partial success. A uniform educational system was established and fiscal and administrative uniformity was imposed. Limited social welfare legislation was enacted, and Spain received the rudiments of a national communication system of roads and railways. But on the whole the fiasco of Spanish liberalism was undeniable. Among the more basic reasons for its failure was that the centrepiece of a bourgeois revolution — an increasingly influential national middle class based in industry and modern commerce — did not exist. This type of bourgeoisie developed only in Catalonia and the Basque country. The lack of capital, entrepreneurial skills and technology made the emergence of a solid Spanish — as opposed to local or regional — middle class impossible. These same factors severely restricted the creation of an internal market. However, an additional major obstacle lay in the agrarian problem. Without agrarian reform the majority of Spaniards simply could not be brought into a market economy. Peasants had to be transformed into consumers in a modern sense. And agrarian reform required a political power and unity of will that the liberals lacked.

Although liberalism strengthened and broadened the functions of the Spanish state, it could never overcome the fragmented nature of Spanish society. The idea of a cohesive Spanish nationalism crumbled upon a new form of intrusive state aimed at the protection of shifting conglomerations of local vested interest groups (see Romero-Maura 1977: 53-62). With the empire lost, this state, centred in economically stagnant Castilla, was no longer viewed as a source of economic benefit or as an essential political requirement for the country's more dynamic peripheries.

The origins of the Basque 'national' problem

The two regions that developed the most aggressive and confident nationalist movements in Spain — the Basque country and Catalonia — shared many features which set them apart from the rest of Spain. Both enjoyed a relatively prosperous agricultural base characterized by medium-sized landholdings, security of tenure, polyculture, dispersed residential patterns and an inheritance system that transmitted the rural farmstead intact to only one heir. Both were in direct geographical contact with Europe and developed powerful mercantile classes which were enthusiastic recipients of European intellectual and technological innovations. Finally, during the nineteenth century both experienced an industrial take-off, where industrialization was managed by a native industrial bourgeoisie and generated a politically militant proletariat of mixed regional origins. Yet despite these similarities, Basque and Catalan nationalism are very different political movements.

Basque nationalism first emerged in Bilbao — the Basque industrial heartland — during the last quarter of the nineteenth century. For the Spanish Basque country, especially Vizcaya, this period was one of severe social crisis during which the traditional pattern of government by local Basque political institutions and an economic system dedicated to commercial and agricultural enterprise were superseded by a new Basque society founded on a unified Spanish constitution and heavy industry. Basque nationalism was a political reaction to two compelling integrative forces: state centralization imposed from Madrid, and industrialization imposed from within. The feature that sets Basque nationalism apart from other European ethnic protest movements is that its principal attack was not centred on Madrid: the main concern was with the new social and economic relationship inside the modernizing Basque country.

The fate of Euskera

Although the Basque language, Euskera, was a constant element in nationalist arguments, Euskera was more a symbol of the nationalist cause than a fundamental issue. When Basque nationalism first made its appearance in Bilbao, Euskera was hardly spoken in the city and was rapidly receding throughout the rest of the Spanish Basque country. Already by the seventeenth century the political and economic élites, mostly urban dwellers, were thoroughly Hispanized, and Basque urban politics and business were conducted in Spanish. In general, the Spanish language had unremittingly radiated out from the pathways and nodal points of long-distance communication. The retreat of Euskera was most notable in those areas that had direct geographical links with major roads, administrative and commercial centres and, later, the industrial complexes with their large, Spanish-speaking immigrant populations. The rough estimates that are available put the total number of Basque-speakers in the Spanish Basque country during the mid-1800s at between 400,000 and 600,000 (or between 55 and 84 percent of the population). Currently, about 600,000 individuals

out of the present population of some 2.5 million are assumed to
have some working knowledge of the Basque language (Table 1).[1]

TABLE 1
The linguistic situation of the Spanish Basque country in 1975

Province	Population	Basque-speakers	%
Alava	238,262	18,863	7.9
Vizcaya	1,152,394	174,366	15.1
Guipúzcoa	682,517	307,279	45.0
Navarra	483,867	53,340	11.0
Total	2,557,040	553,848	21.7

Source: Siadeko (1979: 68).

The vulnerability of Euskera was largely due to its lack of a
literary heritage and production. Although the first book in
Euskera was published in 1545, by the start of the nineteenth century only 111 publications — mostly obscure pamphlets and
religious tracts — had been issued in the language (Payne 1974:
268). The proportion of Basque-speakers declined markedly after
1876, when primary schools were established throughout the
region. These schools provided instruction in Spanish only, and the
rise of literacy and the regression of Euskera became parallel processes. Euskera was regarded as the language of unsophisticated
rustics in contrast to Spanish, the language of refinement, culture,
education and urban success. However, the fate of Euskera and
Basque traditional culture must be placed in the wider context of
pre-industrial Basque society in general, and in the evolution of the
Basque bourgeoisie in particular.

The foral regime

Unlike Catalonia, the Basque country had never formed a single
political entity.[2] Each Basque territory was enclosed within distinctive political arrangements. With the exception of Navarra, which
remained a viceroyality until 1841, the various Basque territories
were associated with the Castillian monarchy by the fourteenth cen-

tury. Guipúzcoa in 1200 and Alava in 1331 were integrated into Castilla through voluntary treaties. In 1379 title to Vizcaya fell through inheritance to the wife of the Castillian king. Although under Castillian sovereignty, each territory retained its separate political framework based on 'fueros'.

Although their reinstatement has always constituted a major nationalist demand, the fueros were not a unique attribute of the Basque country, or of Spain in general. This system of local laws was a common feature throughout medieval Europe, although the Spanish Basque country retained its local constitutions longer than other parts of Europe. The fueros consisted of local laws and economic and political privileges underwritten by the kings of Castilla (and later Spain) in return for political allegiance to the monarchy. Generally speaking, there were two types of fueros. One type comprised common law, which arguably evolved directly out of Basque social practice and formed part of a constantly evolving oral legislation prior to codification in the thirteenth and four-teenth centuries. However, the second was the result of direct interventions by the kings of Castilla in response to strategic needs arising out of the Reconquista.[3] Fueros were granted to various frontier districts in order to stimulate population growth, usually through the extension of privileges such as exemption from taxa-tion, military conscription and arbitrary arrest.

The Basque foral regimes were founded on three main institu-tions: the Junta General, the 'corregidor' and the 'pase foral'. The Juntas Generales (General Assemblies) were the governing provin-cial assemblies that formed the apex of a political structure, the base of which lay in local municipal councils. The flow of delegated authority in this structure ran from the local level up to the provin-cial. Although the Basque traditional political system has sometimes been described as 'a peasant democracy' (Douglas 1971), from the late sixteenth century onwards rules concerning eligibility to elected office became increasingly strict. In both Viz-caya and Guipúzcoa stringent property qualifications and literacy in Spanish — thus effectively barring peasant participation — were requirements for elections to the Juntas Generales.

The corregidor and pase foral were institutions through which the different Basque territories were politically articulated into the Castillian monarchy. The first functioned to affirm and impose royal authority; the second, to limit it. The corregidor was the king's principal agent in the territories and exercised far-reaching judicial, administrative and fiscal powers. He also presided over the Juntas Generales and had the power of veto over all proposed legislation. In contrast, the pase foral was the mechanism by which it was impermissible to implement any royal law or decree without previous approval of the foral authorities. Because of the contradictory effects of these two institutions, it is difficult to establish what degree of political autonomy the Basque country actually enjoyed. In any case, until the advent of Basque nationalism, relations between the Basque political authorities and Madrid had tended towards collaboration rather than confrontation.[4]

However, during the eighteenth and nineteenth centuries, when the foral regime came under attack both from within and outside the Basque country, the main concern was not with these political institutions: the conflicts centred on the economic aspects of the regime. The foral regime provided the Basque country with two economic advantages that were the lynchpins of its entire economic evolution. These were exemption from both state imposed taxation and Spanish customs duties. The fueros established the Basque region as a duty-free zone. Merchandise could be freely imported into and exported from the area. Spanish customs lines were in the interior, along the Ebro.

The evolution of the Basque bourgeoisie

The Basque country has enjoyed three major economic resources that have patterned its whole economic and social history: abundant reserves of lumber and high-quality iron ore; natural protected harbours in the Bay of Biscay; and a strategic geographical location. The Basque country was a cross-roads between two important economic areas: England and Flanders, and the Castillian meseta, rich in wool and grain. Against these advantages, the mountainous Basque countryside was burdened with an arduous terrain and chronic overpopulation. Helped by the economic privileges granted by the foral regime, Vizcaya and Guipúzcoa oriented themselves towards exporting Basque goods and people abroad and acting as commercial brokers between Castilla and Europe (and later the Americas) in return for importing foodstuffs, manufactured items and capital.

A Basque mercantile bourgeoisie had already begun to emerge by the fourteenth century. By the late sixteenth century this bourgeoisie formed a powerful class whose impact was especially visible in Bilbao. Its economic power was supplemented by landholdings purchased with wealth accumulated in mercantile enterprise. From the late seventeenth century onwards the Basque bourgeoisie was — arguably along with its Catalan counterpart — the most culturally advanced sector in Spain, especially when viewed alongside the languid and stagnant Andalusian and Castillian aristocracy.[5]

In terms of identity, this class often described itself as the most 'Spanish' in Spain. Its members regarded themselves as the direct descendants of those — 'uncontaminated by either Jewish or Moorish blood' — who had reconquered Spain from the infidels and restored civilization and Christianity to the peninsula.[6] Euskera, together with other aspects of Basque traditional society — all parts of a rural mode of life — were deemed largely irrelevant to the contingencies of modern Basque society.

Perhaps one of the most salient features of Basque history has been the antagonism between the urban centres and the rural hinterland. Town and country were perceived as two culturally, politically and economically opposed orders (Caro Baroja 1974: 31-51). Culturally, the hispanized, literate and internationally oriented towns were in direct contrast to the 'euskaldun' (Basque-speaking), geographically isolated rural areas, which lacked roads, rivers and literacy. Economically, linkages between the towns dedicated to commerce and administration and a countryside only marginally tied into a monetary economy were minimal. Politically, the two orders were enclosed within different legal frameworks. From the seventeenth century onwards, and culminating in the nineteenth, the Basque region was the scene of numerous peasant rebellions against urban political and economic encroachment.

Rural defeat, urban ascendancy:
the collapse of the foral regime, 1808-1876

The beginning of the nineteenth century marked a major turning-point for the Basque country as well as for the rest of Spain. Spain's monarchy and 'ancien régime' had disintegrated under the thrust of the French invasion of the peninsula. The loss of the American colonies — Cuba and Puerto Rico excepted — had meant the loss of vital markets. By 1829 Spain's foreign trade had plummeted to only one-third of that in 1785 (Carr 1966: 35), and the impact of collapse was especially severe in the Basque country.

The tensions that these events generated were to erupt twice into civil war, and result in the abolition of the foral regime.

Shrinking drastically as a consequence of the ruin of the Basque mercantile adventure, the Basque urban bourgeoisie split. One sector became increasingly tied to the land and merged with the rural notables in conflict with the urban centres. The other sector turned its eyes to the rich potential of a modernized Basque metallurgical industry in a future Spanish market. The realization of the ambitions of this aspiring Basque industrial bourgeoisie, however, required radical changes inside the Basque country.

Three aspects of the foral regime exercised a stranglehold over the possibilities of industrial growth. First, the foral provincial governments were based on municipal representation. Each municipality had an equal vote. Thus Bilbao and San Sebastian were consistently outvoted by the rural municipalities on those issues that involved a conflict of rural and urban interests. Second, the main resources of the Basque country — iron ore and lumber — were municipal rather than private property. Moreover, the Vizcayan provincial junta had prohibited the export of ore. Third, because the Basque region was a duty-free zone, incipient Basque industry was cut off from the rest of Spain while simultaneously subjected to an unrestricted inflow of competing European products. The issue of internal customs lines was a point of particularly fierce contention. The rural notables and peasantry were adamantly opposed to integration into a national market, which would mean the free importation of Castillian cereals and livestock and, hence, a decrease in the price of Basque agricultural produce as well as a dramatic increase in taxation on imported consumer goods.

In the ever-increasing bitterness between urban and rural interests, liberalism was an obvious creed for the urbanites. Thus the centralizing tendencies of the various Madrid governments combined with the aspirations of the Basque urban bourgeoisie to attack frontally the fueros. In contrast, the political attitude adopted by the rural notables was belligerently anti-liberal. Moreover, alarmed at liberal intentions to subordinate it as a political and educational institution, the church consolidated its support for the rural notables in their foral cause. In the Basque countryside an alliance was formed that united notables, peasantry and clergy under the slogan, 'God and Fueros!' which would last throughout the nineteenth century.

The growing confrontation exploded into war in 1833 and again

in 1872. Inside the Basque country the Carlist wars were Basque
civil wars.[7] The liberal, anti-fuero, urban centres were pitted
against the Carlist, pro-fuero, rural areas. With the final Carlist
defeat in 1876 the foral regime was abolished by Madrid decree.
The modern Basque country was a product of liberalism and the
abolition of the fueros. And it emerged deeply divided. Foral abolition meant the political defeat and alienation of the rural population. It also meant the political, economic and cultural ascendency
of the urban centres. One immediate result was the spectacular
growth of Basque industry after 1876.

Basque industrialization, 1876-1900

Basque industrialization was based on the massive export of ore,
extracted from the supremely rich mines near Bilbao, and the
investment of the accumulated profits in iron and steel, shipbuilding, banking and a host of subsidiary industries. Foreign
observers in Bilbao at the time reported that the pace of
industrialization in the city was probably unequalled in European
history. Within a span of 25 years, Vizcaya came to produce nearly
20 percent of the world's total output of iron ore. By 1900 Basques
owned 45 percent of Spain's merchant fleet, and Basque — or,
more accurately, Bilbao — industrialists and bankers formed the
most important economic interest group in Spain, controlling
some 30 percent of all Spanish investment capital.

This industrial take-off was considerably aided by government
policies. Generous subsidies were given to steel production,
railroad construction and shipbuilding. Moreover, while the Basque urban elite had fought for foral abolition, they were dedicated
to preserving within the new system the one aspect of the fueros
that had been of considerable importance to their interests: fiscal
autonomy. In 1878 the Spanish Parliament approved a special
fiscal and administrative regime — the 'conciertos económicos' —
for the Basque provinces. Essentially, this enabled the Basque provincial governments to negotiate their taxes with Madrid and pay a
fixed sum, raised in whatever manner the provincial government
deemed suitable, into the Madrid treasury.

Basque heavy industry developed with extreme velocity, and so
did the two new social classes that were the prime movers of this
industrialization: the Basque industrial oligarchy and the non-
Basque urban proletariat. By the end of the nineteenth century a

new, tightly knit industrial elite, descendants of the former commercial and landed elites, had gained control over most of Basque heavy industry and banking. If 30 percent of Spanish capital was in Basque hands, around 50 percent of Basque capital was in the hands of five families.[8] The economic power of this oligarchy was translated into near monopoly control over the political institutions of Vizcaya. At the other end of the ladder, a vast influx of immigrants, mainly from the impoverished rural areas of southern and central Spain, flooded into Bilbao in order to fill industry's ever-increasing need for cheap, unskilled labour. From 1876 to 1900 the population of Bilbao almost trebled, and in several industrial neighbourhoods it increased some thirtyfold. By the turn of the century less than 20 percent of Bilbao's inhabitants had actually been born in the city. The demographic crush and the intense rhythm of industrialization placed an intolerable strain on the entire social fabric of Bilbao. For the middle sectors of Bilbao society, social life deteriorated rapidly. Inflation, urban congestion, crime, lack of social services and pollution became serious problems affecting daily life. In the 1880s and 1890s Bilbao had the highest mortality rate in Europe and the highest crime rate in Spain.[9]

Politically, the 1890s had two dominant features. One was widespread electoral corruption as the Basque oligarchy consolidated its political position. Votes were purchased on a grandiose scale and economic force became the major determinant of election results for the next 25 years. The political machine operated by the industrialists was popularly known as 'La Piña' (the pineapple). Its official title was Unión Liberal.

Among the immigrant workers there was a growing resentment and rancour in response to appalling living and working conditions. This resentment erupted into a general strike in 1890 — the first in Spain — which required the imposition of martial law to defeat — and, subsequently, a series of violent, partial strikes. Organization of these strikes was quickly taken over by the socialists and the Unión General de Trabajadores (UGT). Bilbao provided Spanish socialism with its first real mass support, and Ramiro de Maeztu, a member of the famous literary generation of 1898, called the city 'the Mecca of socialism'.

Politically, economically and socially entrapped between, and threatened by, these two major forces of Basque industrialization were those middle sectors of Bilbao society that had been marginal to the main thrust of economic change: small-scale artisan producers (many of whom had gone bankrupt), merchants, shopkeepers, skilled workers, the clergy, and professionals. These had witnessed the shattering of their economic and social world. Bewildered and anxious, they lacked any real political power. These entrapped middle sectors of Basque industrial society were to become the main social base of Basque nationalism. In 1893 *La Lucha de Clases* (The Class Struggle), the socialist newspaper, and *Bizkaitarra* (The Vizcayan), the magazine of an embryonic group of Vizcayan nationalists, appeared for sale in the streets of Bilbao almost simultaneously.

The first nationalism:
the politics of culture, 1893-1937

The ideology of Sabino de Arana-Goiri: Basques and anti-Basques

The ideology of Sabino de Arana is central to an understanding of Basque nationalism.[10] Arana was the architect of the Basque nationalist ideology and the founder of the Partido Nacionalista Vasco (PNV). Basque nationalism evolved out of Vizcayan nationalism, and spread only later to the other Basque provinces. The early nationalists, like Arana, lived in Bilbao and generally referred to themselves as 'bizkainos' (Vizcayans) and not Basques. Despite its rural imagery, the ideology of Basque nationalism was constructed by urbanites in order to deal with the problems of industrializing Bilbao. The spread of nationalist sentiment has closely followed the spread of industrialization. Despite its archaic tone, the ideology of Arana currently, though in an updated version, lies at the heart of the nationalists' version of their cause.

Arana held a visceral dislike of modern Basque society. He perceived Vizcaya — or rather Bizkaya, according to his orthography — to be in an advanced state of moral, political and ethnic decay, and he laid the blame for this decadence on hispanization. To Arana, Vizcaya was 'mangled by the foreign fury and dying, but not dead which would have been preferable; rather humiliated, trampled and mocked by that weak and miserable nation, Spain' ('El discurso de Larazabal', 1893): true Basque society, as opposed to modern Basque society, was essentially democratic and egalitarian. Social injustice and class conflict were 'foreign' imports. The Basque historian, Corcuera (1977: 108), has written 'The first cry of Sabino is the cry of Basque traditional society destroyed by industrialization and centralism. It is the mythification of an archaic, democratic and happy past destroyed by Spain and factories.'

Central to Arana's nationalism was his notion of the Basque race, the foundation of the nation. Although the most visible sign of a member of the race was in Basque surnames, a proof of Basque descent, Arana described the race more in moral and spiritual terms — which he defined in opposition to things Spanish — than in linguistic or cultural terms. Euskera, traditional culture and territory were viewed as serving to differentiate and protect the Basque nation rather than as being essential attributes of it.[11] Basque nationalism was depicted as the historic obligation of all Basques in order to reassert the absolute right of the Basque race to govern itself independently of all other races (Larronde, 1977: 123). With independence the Basque race, governed by the harmony of the fueros, would recuperate the original Basque democracy, liberty and egalitarianism. Arana coined the term, 'Euzkadi', to describe an independent confederation of Basque states joined in a unity of race and religion.

Not only had the race been corrupted by 'españolista' tendencies — for example, liberalism, 'that son of Satan' — but more importantly, the race was under a monumental attack launched inside the Basque country on two fronts: one Spanish in origin and the other Basque. One front was composed of the invasion of 'maketos', Arana's derogatory label for the Spanish immigrants, who were condemned as foreign, anti-Catholic, disrupters of social order, socialist subverters of Basque values and, by definition, anti-Basque. Arana vehemently opposed intermarriage and urged the expulsion of all maketos from Basque soil: 'If it were possible to have an 'euskaldun' [Basque-speaking] Vizcaya, governed by the fueros, but inhabited by the maketo race, its realization would be the most odious thing in the world...' (Arana-Goiri, n.d.: 197). The second front was formed by the Basque liberal industrialists who had encouraged the influx of immigrants and whose economic and political activities were responsible for all the 'immorality, blasphemy, crime, free-thought, socialism, anarchism...that is corrupting the Basque soul.' (Arana-Goiri, n.d.: 441). The core of the problem was the process of industrialization itself. Condemnation of capitalist industrialization led to an exaltation of traditional rural society: 'If Bizkaya were poor and only had fields and livestock, then we would be patriots and happy' (Arana-Goiri, n.d.: 441).

Arana's ideology, based on the notion that races had a natural right to self-government, conceived of the industrializing Basque country as morally cleavaged. The moral divide separated 'us' from 'them', 'abertzales' (patriots) from 'españolistas', Basques from anti-Basques. Not only were both the socialist immigrant working class and the liberal Basque industrialists denied 'national' status; they were excommunicated from the moral universe. The core of Arana's preoccupation with race — and of contemporary nationalists' emphasis on Basque culture — is not a matter of descent or language. Douglas (1971: 180) has pointed out that many of the early nationalists had 'shaky genealogical claims to Basque descent'. The overriding concern was one of political loyalties. Knowledge of Euskera, ardent Catholicism and Basque patronyms were not defining elements of the true Basque. Ethnic identity and awareness were subordinate to political action defined in terms of the moral duties of the race. The Basque nation was a moral and political community, and the pivotal qualification for membership was 'abertzalismo' (patriotism). Arana concluded — and this conclusion is fundamental in the social logic of Basque nationalism — that the only true Basque, the only Basque with 'national' status and rights, was the Basque nationalist.

Arana transformed elements derived from Basque history and preindustrial society into symbols of national legitimacy. On the one hand, the elements of Euskera, religion and folklore symbolized the Basque mode of being as opposed to the Spanish mode, and were designed ethnically and ethically to differentiate one section of Basque urban society from the others.[12] These symbols were destined for use principally in a competitive struggle over economic and political resources inside the Basque country —or, rather, in an internal nationalism, directed at other Basque social and political forces. On the other hand, the fueros and the notion of original Basque sovereignty were utilized to separate the Basque country as a whole from the rest of Spain. These symbols were the basis of an external nationalism, directed at Madrid, which concerned the argument over the distribution of political and economic resources between the central state and the Basque country (see Heiberg and Escudero 1977 for a discussion of the nature of symbols in Basque nationalist ideology).

El Partido Nacionalista Vasco

A prolific writer of intensely emotional appeal, Arana's ideas had a terrific impact. In 1894 Arana, together with a group of followers, mostly recruited from the Bilbao petty bourgeoisie, formed the Euskeldun Batzokija (Basque centre). Faithful to Arana's primitive nationalism, the Batzokija was strictly confessional, open only to those of pure Basque descent, and closed off from any contact with parties whose platforms were either españolista or insufficiently religious (Larronde 1977: 184-6). One year later the Euskeldun Batzokija, under the overarching slogan of 'Jaun-Goikua eta Lagi-Zarra' ('God and the Old Law'), became the Partido Nacionalista Vasco. If the PNV had rigidly adhered to the political line and policies of its main theorist and initial membership, Basque nationalism may well have remained confined to a group of Basque urban romantics with only limited political leverage.

The year 1898 was the turning-point. War with the United States and the loss of the remnants of Spain's colonial empire was a calamity of such magnitude that the viability of Spain as a modern state was placed in question (Payne 1974: 128). Although the Basque financial aristocracy never questioned the political unity of Spain, significant sectors of the middle bourgeoisie had for some time seen the need to disentangle the Basque provinces from the Spanish political disaster. However, these, which adhered to a political creed known as foral liberalism, had been unable to establish an influential, independent political presence. In the popular mind nationalism was coupled to the personality of Arana. Moreover, Bilbao lacked a politically sophisticated, commanding and 'autonomous' middle class similar to that in Catalonia. Arguably, a more moderate formulation of the nationalist cause was not feasible. There is little indication that the foral liberals found Arana's extreme nationalism any more attractive than Arana found the liberals, who, according to Arana, had the souls of traders and the corrupting fever of industry. None the less, in the elections of 1898 the foral liberals, headed by Ramon de la Sota, the only Basque oligarch to give support to the nationalists, gave ardent support to Arana's candidacy. Arana was duly elected as the PNV's first deputy to the Vizcayan provincial government. Shortly thereafter the foral liberals entered the ranks of the Basque Nationalist Party. The event marked the birth of the PNV as a modern political party.

This fusion of the political skills and the modern economic thinking of the middle bourgeoisie with the nationalist fervour of the petty bourgeoisie gave the PNV a hybrid, and in many respects contradictory, ideology. It also gave strength and the capacity of capturing mass support. But the coexistence of the two groups was never to be harmonious. The conflicts and tensions, often violent, generated by this fusion became a major characteristic of the Basque movement from then on. On one side the ultra-Catholic, fiercely anti-Spanish intransigent nationalists defined a political line whose main aim was the establishment of an independent, religious Basque state. They absolutely excluded any alliances with non-nationalist political parties: the modern equivalent of this line can be found in parties and groups that circulate within the orbit of ETA. On the other, the moderate, reformist-minded nationalists, economically dependent on Spanish markets and protectionism, aimed at wresting an advantageous autonomy statute for the Basque country. To achieve this autonomist aspiration, the moderates were willing, at propitious moments, to ally with Spanish parties — in practice, usually the conservative and liberal parties.

The increasing electoral importance of the PNV, together with the power vacuum that existed after Arana's premature death in 1903, gave a stimulus to fractional rivalries, separations and reunifications between the intransigents and the moderates inside the PNV as both attempted to achieve control over the party.[13] In terms of political platform and initiatives, the PNV was effectively divided. The effects of these bitter disputes could have reduced it to impotence. However, quite the reverse was the case. The intransigents — who have always comprised a majority inside the PNV — captured an increasing hold over the ideological apparatus of the party. *La Patria*, their newspaper and defender of doctrinal purity, became the official voice of nationalism. In public meetings and demonstrations radical nationalism took pride of place. However, the pragmatic political skills of the moderates insured them increasing control over the party's political structure.[14] As a result, the ideological statements of the PNV rarely corresponded to the party's actual political activities. This duality — the separation of political power in the hands of the middle bourgeoisie from ideological power under control of the petty bourgeoisie — has been a salient feature of Basque nationalism. It enabled the PNV to weld together divergent — at times antagonistic — sectors of Basque society. Basque nationalism has been used in quite different ways by quite different people. This contradiction, although a source of constant political instability, has provided the bulwark for a wide-ranging popular support. However, the PNV never viewed itself as just a political party. The nationalists' vision of a new 'Basque' society was translated into a proliferation of organizations which would provide the infrastructure for such a society.

In 1901 Euzko Gaztedia (EG; Basque Youth) was formed. In 1908 the Mendigoitzale Bazkuna (Mountaineering Association) was created. These Basque mountaineers became the shock troops of intransigent nationalism during the Second Republic. During the civil war and the repression that followed, the Mendigoitzale Bazkuna provided the core of the Basque militias and later organized the resistance. In 1911 the nationalists founded their own trade union in order to combat the growing strength of the socialists. The leadership of the Solidaridad de Obreras Vascos (SOV; Solidarity of Basque Workers) was strongly influenced by the church, and membership was recruited mainly from the ranks of white-collar employees and workers in small industrial workshops. The SOV gave Basque nationalism a type of working-class base that was to be of increasing utility. In addition to these organizations, the PNV established its own press, schools (the 'ikastolas'), cultural bodies, theatre groups, farmers' associations and women's organizations. The latter, the Emakume Abertzale Batza (EAB; Association of Patriotic Women), was one of the nationalists' more significant achievements. The EAB was explicitly founded so that women could function as transmission belts for the nationalist ideology inside the domestic sphere. They would form the first offensive for the formation and mobilization of a new generation of nationalist youth — 'the preparation of future patriots' (EAB document). By 1936 the EAB could count approximately 28,000 well-disciplined militants.

Rural nationalism

Nationalist penetration into the rural areas was slow. The first obstacle lay in the nationalist ideology itself. The idealized view of rural life in which the 'baserritar', the Basque peasant, was depicted as the noble and free repository of Basque traditions was designed to mobilize and consolidate an urban political following and had little in common with rural conditions. The vast majority of peasants were impoverished tenant farmers dominated by the local 'caciques', who were usually Carlist in political affiliation.

The Carlists, with staunch spiritual support from the clergy, used their economic hold over the peasantry to bar any competing ideologies or parties from gaining a popular foothold. Moreover, nationalism, together with other forms of republicanism, was branded as an urban phenomenon and, hence, treated with suspicion by the peasantry whose traditional resentment of the cities had increased since 1876.

This situation gradually changed. Spanish neutrality during the First World War brought considerable prosperity to the Basque country, particularly to rural areas on the margins of the industrializing zones, where many peasants, enriched by increased agricultural prices and lucrative supplementary jobs in industry, found themselves able to become small landholders. Although the baserritar stereotype continued to be lauded as the antithesis of the socialist agitator (Elorza 1976: 484), the PNV slowly became aware of the extent of the agrarian problem, and began to advocate the purchase of land by the peasantry as the means to their economic and political emancipation. Several political parties had expressed the need for agrarian reform; but it was the PNV that took decisive action. It established a whole series of schemes — insurance policies, legal and technical aid, credit facilities, etc. — that were of direct benefit to the peasantry. From 1917 to 1936 the number of peasant landholders increased dramatically, a trend for which the 'social programme' of the PNV was given ample credit.

Whereas the economic improvement of the peasantry had cleared the way, in a sense it was the Basque cultural revival that furnished the urban nationalists with a common language by which to communicate to a potential rural following. Accompanied by the publication of numerous Basque journals and magazines and constant, at times massive, folkloric performances, this revival helped remove the stigma attached to rural culture. Euskera was granted élite status as the language — symbolically, at least — of an urban political vanguard (few of whom actually spoke the language). The cultural revival, which always carried heavy political messages, helped gain many converts to nationalism especially among the rural youth, for whom it often provided the only source of entertainment and diversion. Especially during the Primo de Rivera dictatorship (1923-30), when political nationalism was prohibited (as were all political parties) but cultural nationalism tolerated, the activities surrounding these cultural events in the rural areas offered a convenient platform for enthusiastic youth to press home the nationalist doctrine. By the time of the elections of 1931, which marked the inauguration of the Second Republic, the nationalists were the principal force in rural Guipúzcoa and Vizcaya. The Basque peasantry who lived on the margins of the rapidly expanding industrial areas became — along with the Basque middle and petty bourgeoisie — the third unwavering pillar of nationalist support.

However, the meaning of Basque nationalism was distinct in the euskaldun rural areas. The urban nationalists tended to be by language, education and life-style more akin to other urban Spanish citizens than to their rural Basque counterparts. Nationalism was basically, although by no means entirely, a particular type of political strategy. Urban nationalism implied the creation of a differentiated ethnic identity. For the peasantry nationalism meant a reaffirmation of underlying cultural norms. Rural nationalism was at its core neither politicized nor, in a true sense, nationalist. Although the peasantry voted consistently for the PNV, they did not see their primary political loyalty in terms of a Basque fatherland or Euzkadi. In the rural areas nationalism was more a sentiment and organizational vehicle than a concrete political ideology, and it functioned mainly in terms of local loyalties and interests. Nationalism became a major vehicle by which sectors of the peasantry organized to dismantle resented local power structures dominated by the Carlists — and later reimposed by the Franco regime — and to replace them in many cases by genuinely popular organizations. In many areas nationalist ideas served as the basis for self-help organizations by which the rural population sought to solve many of the daily economic problems that confronted them.

The Second Republic and civil war

After the semi-clandestine years of the Primo de Rivera dictatorship, the PNV resurfaced in the Second Republic under the leadership of a young Bilbao lawyer, José Antonio Aguirre. The party was firmly dedicated to negotiating a statute of autonomy and augmenting its influence at home. However, the constant failures of the Basques to negotiate a statute were in marked contrast to the Catalans, who were granted home rule in 1932. The PNV displayed an open hostility toward the Republic, and in return its intentions were deeply mistrusted by Madrid. In the elections of 1931 and 1933 the PNV achieved major electoral triumphs by capitalizing on

TABLE 2
Parliamentary elections during the Second Republic
in the Basque provinces and in Bilbao

Election results 1931

Province	Turnout	Right*	PNV*	Left	ANV**	Communists
	%	%	%	%	%	%
Alava	81.3	30.6	17.7	32.7	—	—
Guipúzcoa	85.5	49.0	—	35.4	—	—
Navarra	83.5	53.1	—	30.7	—	—
Bilbao	76.8	—	31.9	44.8	3.8	6.8
Vizcaya	80.0	—	61.3	18.6	—	—

* Except in Alava the PNV and the right were joined in an electoral alliance
** ANV = Acción Nacionalista Vasca, a party of liberal nationalists who split from the PNV in 1930.

Election results 1933

Alava	%		Navarra	%
Carlists	52		Carlists	71
PNV	29		PNV	9
Radicals	6		Radicals	4
Left coalition	12		Socialists	14
Radical-Socialists	0.2		Republican Left	2
Communists	0.2		Communists	1
Turnout	71.6		Turnout	80

Guipúzcoa	%		Bilbao	%
Union of Right	19		Right	15
PNV	36		PNV	41
Radicals	2		Socialist Republicans	37
Socialists	36		Communists	7
Republican Left	6		Radical-Socialists	1
Turnout	78		Turnout	78

Vizcaya	%
Right	28
PNV	57
Left Coalition	14
Communists	1
Turnout	78

Election results 1936

Province	Alliance of Right	PNV	Popular Front
	%	%	%
Alava	57	20	22.5
Guipúzcoa	33	36.9	30
Navarra	69.5	9.5	21.3
Bilbao	22.7	32.6	52.2
Vizcaya	34.7	50.6	14.6

the widespread Basque fear of many of the policies enacted or proposed by the Republican government, especially those concerning
church and land reform (see Table 2).

Difficulties concerning the statute were made more severe by the
intricate Basque political situation. Unlike the rest of Spain, which
was broadly polarized between left- and right-wing political forces,
in the Basque region the situation was triangular. The PNV was a
type of centre wedged between its two traditional 'anti-Basque'
competitors. On the one hand, the right-wing parties were
lukewarm to any statute and totally opposed to one endorsed by the
republican government. On the other, the socialists and
republicans, while strongly in favour of Basque autonomy, refused
to endorse the statutes prepared by the PNV which, they argued,
was attempting to turn the Basque region into a 'Gibraltar vati-
canista'. The strength of these internal divisions was underlined in
the critical elections of 1936 which brought the Popular Unity
government to power. The PNV captured one-third of the Basque
vote, and the remaining two-thirds was equally divided between
Popular Unity candidates and the right-wing alliance.

Basque autonomy was granted by the new government after it
had become embroiled in civil war. On 7 October 1936, José
Antonio Aguirre was sworn in as president of the Provisional
Government of Euzkadi, and nationalists were given most of the
key portfolios as a means of insuring the PNV's support in the armed conflict.[15] The Spanish Civil War presented the PNV with a
substantial dilemma. Some nationalists argued that it was a war
between Spaniards and as such did not concern the Basques. The
PNV's first reaction to the uprising of the insurgent military was to
declare neutrality. This position was changed a few days later to
one of reluctant support for the republic. However, the pro-
government position of the PNV in Guipúzcoa and Vizcaya was
not shared by the PNV in the other two Basque provinces. In

Navarra the Carlists had risen in favour of the insurgent military and the PNV voted against support for the government. In Alava, which fell almost immediately to the rebel forces, the PNV maintained a position of neutrality. In June 1937, nine months after its belated inception, Euzkadi was overthrown, and the Basque region was the first industrialized area to fall under Franco's military crusade.

The second nationalism: from opposition to power, 1959-1980

The postwar years

With the collapse of Euzkadi the Basque government went into exile, first in Paris and later in Latin America, where its activities were financed by the wealthy resident Basque community. Inside the Basque country thousands of nationalists were imprisoned and many were executed. However, Franco needed to maintain the goodwill of the industrialists, and therefore the physical punishment of the Basques was restricted, especially when compared with the atrocities committed in Andalusia. But Franco viewed Basque culture to a large extent as an excuse for and sign of separatism, and the regime therefore unleashed a thorough campaign of cultural repression. The public use of Basque greetings, traditional garments, folklore, Basque names, publications and the teaching of Euskera were strictly forbidden. These measures were enforced by a vast array of new pro-Franco officials who exercised a stringent control over all aspects of public life.

In 1946 the United Nations formally censured the regime and a total economic and diplomatic boycott, broken only by Argentina and Portugal, was imposed on Spain. The Basques were euphoric. The fall of Franco seemed imminent. However, partly through Churchill's fear of a socialist country flanking the rear of England and the growing Cold War fervour in the United States, Franco did not fall. In 1953, with the signing of a military and economic treaty with the United States, Spain was readmitted as a fully fledged member of the international community. Although the early 1950s also marked a cautious liberalization of the Spanish regime, the PNV had already entered a period of organizational decline. The Basques, like most Spanish citizens, had become largely apathetic to political affairs.

In part this apathy stemmed from economic factors. The economic boycott had helped stimulate the second great wave of Basque industrialization. While they — together with the majority of Spaniards — were politically oppressed under Franco, economically the Basques as a whole and the Basque middle classes in particular benefited considerably. By the mid-1960s Guipúzcoa and Vizcaya enjoyed a per capita income close to double the Spanish average and the highest in Spain.[16]

In many families nationalism had become a sentiment imbibed 'with the mother's milk'; but during the 1950s nationalist activity was mainly confined to three groups. First, certain sectors of the socially committed lower Basque clergy continued to expound nationalist ideology and aspirations, especially to young people. Second, the PNV's vastly reduced youth branch, EG, carried on a modest campaign of 'pintadas' (political graffiti) and distributed propaganda. Finally, in 1952 a small group of young middle-class Bilbao intellectuals, few of whom spoke Euskera, founded a magazine called *Ekin* ('To Do'), which published historical essays together with abstracts of earlier PNV writings. During 1955 and 1956 many of the EG youth, disenchanted and impatient with the stagnation of their parent organization, fused with the *Ekin* group. In 1959 all ties with the PNV were broken and 'Euskadi ta Askatasuna' (ETA; Euskadi and Liberty) was founded.

The evolution of ETA

The initial ideology of ETA reflected the multifarious and frequently vague views of its leaders, who were inspired by a fusion of Sabino de Arana's primitive nationalism and social aspirations. ETA defined itself as 'abertzale' (patriotic), democratic and, unlike the PNV and despite the seminary education of many of its supporters, strictly aconfessional. Probably no factor was more important in influencing ETA's ideological shift toward the extreme left than the models provided by Cuba, Algeria and Vietnam. These revolutionary wars of liberation were viewed as both nationalist and progressive and as combining three essential elements that ETA adopted in a package deal: armed struggle, independence and socialism. Frederico Krutwig, ETA's chief ideologue and author of the influential book, *Vasconia*, argued that Euskadi was a conquered colony of Spain and that the national and social repression of the Basques arose from this reality. Therefore, ETA concluded,

Basque nationalism belonged to the revolutionary nationalisms of the Third World. The 'colonialist model' gave ETA its ideological framework and military justification.

In the latter part of the 1960s, ETA, financed by bank raids, began its attack on the symbols of Spanish domination. Radio transmitters, war memorials and railroads were bombed. In 1968 ETA extended its offensive to the hated Spanish police, who were widely regarded as an army of occupation.[17] Although the regime was at first unsure as to how publicly to deal with ETA — officially, nationalism had ceased to exist — the police responded with a strategem of massive and random repression. Constant road controls, arrests, house searches and the widespread use of torture became features of Basque life, and Basque political prisoners filled the Spanish gaols.

This campaign was counterproductive. Stirred by the militancy of ETA, the ever-accelerating counter-reaction of the police, the arrests of Basque clergy and, just as importantly, another enormous inflow of Spanish immigrants, popular support for nationalism was rekindled. ETA members were praised as the vanguard of the anti-Franco struggle even by the non-nationalist Basque opposition and those antipathetic to ETA's Marxism.

In spite of its political and military success, however, serious dissension was growing inside ETA. In 1967 a process of fragmentation began that by the end of the 1970s had left the left-wing nationalists scattered over innumerable different political groups with little, if any, effective leadership. ETA's internal problems stemmed from two basic causes. First, ETA was constructed upon an ideological contradiction: Basque economic and social reality and the 'colonialist model' clashed head on. During the second period of Basque industrialization, which stretched roughly from 1950 to 1970, capital was again largely in Basque hands and cheap, unskilled labour was supplied by a vast new influx of Spanish immigrants which brought the total non-Basque population of the region close to 50 percent. The intrusion of these immigrants into the newly industrialized zones was crucial to the groundswell of nationalist sentiment in these areas. Relations between the immigrants and the Basques were based on mutual resentment and hostility, and the two groups led separate social lives. As Minogue (1969: 138) has remarked: 'Marxism vertically divided the world into exploiters and exploited; nationalism horizontally cut it into many distinct nationalities.' In the Basque case, where the majority of the urban proletariat was non-Basque, the two creeds intersected disastrously at a particularly sensitive point in the social matrix. The new breed of Basque socialist nationalists, therefore, had either to limit their socialism to Basque workers only, which grossly diluted any claims to socialist purity; or to extend their nationalism to everyone who lived in the provinces, which seriously undermined all nationalist arguments based on ethnic distinctiveness and logically led to a doctrine of regionalism rather than nationalism. In the efforts to bridge this contradiction, many new political groups were formed, each of which possessed its own special mixture of socialism and nationalism.

Second, ETA was in any case vulnerable to fragmentation. ETA's leadership was unstable and weak owing to constant arrests, restricted means of communication and endless ideological and political disputes. Although formally highly organized, in practice ETA had almost no military or party structure. Links and loyalties between members were determined more by personal considerations of friendship and kinship than by the organizational forms adopted by the various ETA congresses. To a large degree ETA functioned as a loose matrix of groups of friends. Moreover, its unity was further marred by a deep-seated cultural and political antagonism between the leadership and the social base upon which it depended. The leadership tended to be urban, Spanish-speaking, professional and intellectual. In contrast, the social base tended to be semi-urban, petty bourgeoisie and peasant youth and was Basque-speaking and staunchly anti-intellectual. This base provided ETA with a logistic infrastructure, formed its military wing, had de facto operational control over the organization and could rarely be relied upon to follow the leadership's political directives.

The successive divisions of ETA — frequently ignited by personal rivalries, but always conducted in the terms of the conflict between socialism and nationalism — filled the Basque political scene with a potpourri of political initials — ETA 6, MC, ESB, LAIA, LAB, HASI, EIA, HB — almost ad infinitum. As the more politically adept elements of ETA were progressively hived off into these new political groups and labour unions — many of them short-lived — the anti-intellectual, poorly educated base of ETA increasingly assumed the political direction of the organization. ETA 'militar' — formerly the armed wing of ETA and now a separate organization — is currently the inheritor of ETA's prestige and popular legitimacy. The only nationalist group in which the leadership is recruited mainly from the ranks of the peasantry, ETA 'militar' remains faithful to the principles of armed struggle, Basque statehood and millenarian socialism.[18]

The various groups that still give explicit or implicit support to ETA form one of the two major blocs inside the nationalist movement. These 'abertzales de izquierda' (left-wing patriots) rest solidly upon the same social base that formerly supported the PNV intransigents. Ideologically, they are the direct descendants of Sabino de Arana. Like Arana, they demand complete independence from Spain and all that is Spanish, refuse all political contact with non-nationalist parties, and dismiss the Spanish parliamentary process as a sham. Whereas Arana insisted upon a Basque state based on the unity of race and religion, the abertzales de izquierda demanded a state founded on the unity of culture and socialism. As measured by the electoral results of the post-Franco period, the abertzales de izquierda seem to command the loyalties of some 15 to 20 percent of the population and currently form the second largest political force in the Basque country in terms of parliamentary representation.

The other axis of the nationalist movement is formed by the PNV, which re-emerged with vigour in early 1975 when the Franco regime was suffering its final crisis and the re-establishment of political parties as the main co-ordinates of political life was imminent. Ideologically Christian democratic with heavy populist overtones, the well organized and financed PNV probably has the most active and militant grass-roots support of any conservative party in Europe. Currently it is by far the most important party in the Basque area (see Table 3). Despite its occasional separatist rhetoric (a concession to the ideological power of intransigent sentiment) the PNV aims at reform and at obtaining the most ample autonomy possible for the Basque country inside the unity of the Spanish state. Reinforced by generational differences — the radical nationalists tend to be considerably younger than the moderates — this split in the nationalist ranks reflects the continued divergence of political lines between the Basque middle bourgeoisie and petty bourgeoisie.

The politics of nationalism:
abertzales and españolistas

Although the influence of nationalism is overwhelming, Basque nationalist parties represent the expressed political option of only about one-half of the resident population. Facing the nationalists

TABLE 3

Electoral evolution of Basque political parties in the post-Franco period*

	Percentages of votes cast		Municipal elections (3-4-79)	Elections Basque Parliament (9-3-80)
	1st general election (15-6-77)	2nd general election (1-3-79)		
	%	%	%	%
PNV	29	27	36	37
Herri Batasuna (HB)	4	15	16	16
Socialists (PSE-PSOE)	28	18	15	14
Euskadiko-Ezkerra (EE)	6	8	6	9.5
Unión Centro Democratico (UCD)	16	16	10	9
Alianza Popular (AP)	7	3	—	5
Communists (PCE)	4	4.5	4	4
Extreme left (ORT, MC, etc.)	2	3	4	3
Abstention (% of electoral census)	23	34	37	40

* Results from Navarra have not been included.
Notes:

HB — electoral alliance of left-wing, radical nationalists, supported by ETA 'militar'.

EE — electoral alliance of more moderate, left-wing, radical nationalists, supported by ETA 'politico-military'. Together, HB and EE form the 'abertzales de izquierda'.

UCD — government ruling party.

AP — right-wing Spanish party headed by Manuel Fraga.

are a large array of non-nationalist parties which, like the nationalists themselves, range from the extreme right to the revolutionary
left. The most important is the Partido Socialista de Euskadi
(PSE), the Basque branch of the Partido Socialista Obrero Espanol
(PSOE). In the first post-Franco elections of 1977 the PSE, much
to its surprise, gained more votes than the PNV. But in the three
subsequent elections the socialists — like the other Basque non-
nationalist parties — witnessed their electoral support evaporate
and the level of abstention increase accordingly. The collapse of the
non-nationalist vote, in spite of the ample sectors of the population
that are antagonistic to Basque nationalism, must be traced to two
fundamental factors.

The first is Basque nationalist ideology itself. Because of the
excessive, deeply resented centralism of the Franco period in
general, and the repression inflicted upon the Basque country in
particular, the idea that the Basques form a distinct nationality to
be governed by Basques only is now profoundly acknowledged by
nationalists and non-nationalists alike. However, this ideology was
designed to operate on two levels: an external level, which supported arguments for a special relation with Madrid, and an internal one, which backed arguments for a differential relationship between certain sectors of the population and others inside the Basque
country. While the non-nationalists have now incorporated the nationalist premise (and like the nationalists demand ample Basque
autonomy), they have not understood its internal implications. For
the nationalists the defining credential of 'Basque', a full citizen of
Euskadi, continues to be 'abertzalismo' (patriotism). And an
'abertzale' is someone who supports exclusively the Basque nationalist parties. The non-nationalists, regardless of linguistic abilities
or origins by descent, are dismissed as españolista and, therefore,
for political purposes they are ousted from 'national' status. This
division between abertzales and españolistas, which nationalist
ideology enforces and the non-nationalists reject, is the pivot
around which all Basque political life operates. The Basque country
was the only significant region of Spain in which the political opposition was unable to form an unitary organization during the last
years of the Franco regime. The nationalists refused to ally with the
socialists and communists in such an organization, despite broad
agreement on the platform, on the grounds that the latter advocated Basque autonomy mainly for opportunistic reasons and in any
case were not true representatives of the Basque people. Refusal to
enter into an alliance was also a refusal to extend 'national' accep-

tability to the Basque socialists and communists. In early 1978 a Basque General Council was formed whose principal task was to negotiate an autonomy statute with Madrid. When Ramon Rubial, a Socialist and native of Bilbao, was elected president, the PNV effectively boycotted the Council. Again, the chief complaint was that the ill-fated Rubial could not be considered an adequate spokesman for Basque demands.[19] Implicitly, Madrid accepted the nationalist argument. The Basque General Council was bypassed. Basque autonomy was negotiated exclusively by the government and the PNV, even though the PSE, which equalled the PNV in parliamentary deputies, had serious differences with the PNV as to the nature of the relation between the Basque country and the rest of Spain and the Basque internal regime. In public opinion Basque autonomy was viewed as a nationalist achievement to which the non-nationalists contributed little or nothing. Simultaneously, the non-nationalists' obsession with the Basque 'national' problem has alienated their potential voters, who are much more concerned with inflation, unemployment and dismal social services, than, for example, with whether or not Navarra forms part of Euskadi.

The second factor is the nature of the nationalist support. Basque nationalism is not only a political movement. Far more importantly, it is a social movement. Despite bitter political rivalries, Basque nationalists conceive of themselves as a moral community bound by personal interaction and cemented by a common political loyalty and shared moral and social codes. Moreover, since the late 1960s the infrastructure for this community has been resurrected and extended. Within the Basque country the Basque nationalist community forms a parallel society with its own schools, recreational clubs, cultural activities, trade unions and increasingly its own economic and financial institutions such as the industrial cooperatives and the Caja Laboral Popular (Popular Workers Savings Bank). Nationalism is not just an electoral option: it is the basis of a life-long social identity. This social cohesiveness provides opportunities for a far-reaching, disciplined and personalized organization and unity which enable the nationalists to mobilize quickly hundreds of thousands of individuals in ardent, vocal support for nationalist policies. In contrast, the non-nationalist parties must appeal to a floating population of unconvinced and often apathetic voters who have no shared interests, ideology, identity or political ambitions.

The Basque nationalist community
in power

When the new post-Franco regime of Adolfo Suarez assumed power in late 1976, a fundamental plank in its consensual base was, along with social and economic reform, political devolution. After the waves of strikes and street violence in support of regional autonomies that had characterized the previous years, it was believed that a stable Spain could only be a decentralized one. While the degree of devolution regarded as necessary and/or desirable was a matter of grave doubt — especially in areas like Andalusia — the new Madrid government recognized that it could never gain even the reluctant collaboration of the 'historic regions' of Catalonia and the Basque country, where parties demanding autonomy had an overwhelming electoral mandate, without the extension of home rule. In the Basque case home rule was made more palatable for the centre because of the nature of the dominant Basque party. The PNV was, despite its nationalist populism, a Christian democratic party and socially moderate.

Moreover, the ruling party in Madrid, the UCD, and the PNV had a cogent political competitor in common: the Socialists. Suarez viewed a Basque devolved government under PNV tutelage as a means of insuring at best the implicit co-operation and at least the neutrality of the PNV while simultaneously seriously undercutting Socialist strength in the Basque country and, consequently, also in the centre. On this aspect of Basque autonomy the Madrid political centre and the mainstream of Basque nationalists were in total accord.

Since the referendum of February 1980, which gave massive approval to the Basque statute of autonomy, Basque nationalism has to a large extent achieved its strategic goals. On the external level the Basque country now enjoys a very generous political and fiscal autonomy. Moreover, the PNV has two potent instruments at hand which could enable it to extract further concessions from Madrid. It has recourse to extra-parliamentary political activities on the streets of the Basque country, which the nature of its social base and populist ideology encourages. But as importantly, the moderate PNV and the radical ETA, dedicated to armed violence, are both representatives of the Basque nationalist community. The PNV has consistently impressed upon Madrid that if nationalist demands as peacefully pursued by the PNV are not met, then the nationalist community will give its support to ETA's violent methods of obtaining these demands. On the internal level, all important public political bodies and all institutions with public functions are under direct nationalist control. Moreover, the Basque nationalist monopoly of political power has been used to establish differential access to economic resources. In the worsening economic crisis, access to industrial subsidies, academic grants, cultural subsidies and, very importantly, jobs in both the public and private sectors are increasingly reserved for nationalists only. These scarce resources are 'Basque' by national right. Many Basque non-nationalists are beginning to fear that this structural inequality of access to public resources may eventually generate tensions in the Spanish Basque country similar to those experienced in Northern Ireland — with equally tragic consequences.

NOTES

1. Of those who are of immigrant origin, very few speak Euskera. Of those who are Basque by descent, some 50 percent have some knowledge of the language. All these figures, however, must be treated with considerable caution. Because of the current popularity of the 'ikastolas' (Basque language schools) and the impact of nationalism, it is possible that the number of Basque speakers will increase.

2. For a brief period during the eleventh century, all the Basque-speaking areas on both sides of the Pyrenees were brought together under the personal rule of the king of Navarra. Many nationalists will argue that the Kingdom of Navarra during this period was in effect an independent Basque state.

3. In the Basque country various urban centres were established and granted special privileges by the kings of Castilla in order to open direct and protected routes to the coast and to fence off Navarra from southward expansion.

4. Especially from the sixteenth to the eighteenth century, Basques exercised an influential role within the state administration. Basque state officials served to defend Basque interests at the royal court while simultaneously ensuring that royal interests were well protected inside the Basque country.

5. The main agency for the dissemination of the ideas of the Enlightenment was the Societies of the Friends of the Country, first established in the Basque country in 1765.

6. Against an intricate background of civil war in the Basque country and the wider military and bureaucratic ambitions of the Basque élites, the Castillian crown in the sixteenth century used this mistaken argument concerning the purity of Basque blood to confer the legal status of 'hidalguía' (nobility) to all free born Basques. See Otazu (1973) and Greenwood (1977) for two different views of Basque collective nobility.

7. The Carlist wars are a complicated affair. Nominally, they concerned a conflict over succession to the Spanish throne with the champions of Don Carlos, brother of the deceased king, Ferdinand VII, arrayed against the defenders of the king's infant daughter. However, Carlism actually implied a revolt against the economic, political and religious policies of the liberal regimes in Madrid.

8. These families were the Ybarra, Martinéz de las Rivas, Chávarri, Gandarias and Sota y Llano. The capital accumulated by these five families provided the foundation of the powerful Basque banking sector.

9. The comparative mortality rates for 1894 were: London, 19.4 per thousand; Nancy, 35.3; Madrid, 41.4; Bilbao, 45.1 (Echevarría 1894: 171).

10. The background of Arana is typical of many of the early nationalists. He was born in Bilbao, had a strict Jesuit education and was a lawyer by profession. His father, an ardent Carlist, had come close to bankruptcy because of his financial support for the Carlist cause. Arana was ignorant of Euskera until later in his career.

11. The role of Euskera, which is similar to the role of Basque culture and territory, in Arana's ideology is illustrated by the following quote:

> Bizkainos are as obliged to speak their national language as they are not to teach it to the 'maketos' or Spaniards. Speaking one language or another is not important. Rather, the difference between languages is the means of preserving us from the contagion of Spaniards and avoiding the mixing of the two races. If our invaders learnt Euskera, we would have to abandon it . . . and dedicate ourselves to speaking Russian, Norwegian or some other language unknown to them. [Arana-Goiri, n.d.: 404]

12. The opposition between the moral Basque and immoral Spaniard — an immorality that is contagious by contact — is a constant theme in Arana's writings. In an article entitled, 'Who are we?' Arana states that the Basque is intelligent, noble and masculine; the Spaniard inexpressive, sullen and effeminate. The Basque is capable of handling all types of work; the Spaniard inept even at the most simple tasks. The Basque is a learner, born to be his own master. The Spaniard never learns and is born to be servile. And so on.

13. The PNV steadily increased its percentage of the votes from 1898 onwards although their principal strength remained in Vizcaya. By 1918 the PNV controlled the Vizcayan provincial government, dominated the Bilbao town council and had seven deputies in the Madrid Cortes.

14. Part of the intransigents' political weakness lay in their refusal to participate fully in elections, all of which they regarded as fraudulent. Almost all of the PNV's elected representatives came from the moderate bloc, and they in turn exercised considerable influence within the party itself.

15. There is no evidence to support the argument that the Republican government conceded Basque autonomy only with great reluctance. Although representation inside the Basque government was proportional to the 1936 election results, the PNV was granted the important portfolios in the hope of its continued support for the Republican government. However, after the fall of Bilbao the PNV made several attempts to negotiate a separate peace treaty with the Italians.

16. The nationalists have always maintained that during the Franco period the Basque country underwent a severe 'decapitalization'. From the available statistics, however, it would appear that the Basque country was a net importer of capital from other areas of Spain. Although the Basque provinces — like most of Spain — experienced a net capital outflow in the public sector, in the private sector the Basque country saw a considerable capital inflow.

17. In 1968 ETA assassinated the chief of the political police in Guipúzcoa. The police responded by a massive campaign of arrests. In 1969 alone 1,953 persons were arrested, of whom some 890, according to Basque sources, were subjected to some

form of torture. Fourteen of those arrested were brought up for military trial at Burgos in 1970: all 14 were found guilty and 9 were sentenced to death.

18. That rural nationalism should in general adopt a more radical and belligerent expression than its urban counterpart can be explained by a confluence of factors. Not only did the peasantry bear the brunt of the cultural and political repression, but Franco's agrarian policies — or lack of them — were disastrous for the agricultural sector. Few Basque peasants feel any wider unity of interests, either economically or culturally, with the rest of Spain.

19. One of the more effective arguments used by the PNV against Rubial was that he was not a Basque-speaker. Although Rubial frequently apologized for this linguistic ignorance, it was never mentioned that very few of the PNV representatives on the Council spoke Euskera either.

REFERENCES

Arana-Goiri, S. (n.d.). *Obras Completas*. Bayona: Sabindiár-Batza.

Caro Baroja, J. (1964). *Introducción a la Historia Social y Economica del Pueblo Vasco*. San Sebastian: Txertoa.

Carr, R. (1966). *Spain, 1808-1939*. Oxford: Clarendon Press.

Corcuera, J. (1977). 'Tradicionalismo y burguesía en la formacion del nacionalismo vasco', *Materiales*, no. 5 (Barcelona).

De Ihartza, S. (pseud.) (1962). *Vasconia*. Buenos Aires: Norbait.

Douglass, W. (1971). 'Basque nationalism', in O. Pi-Sunyer (ed.), *The Limits of Integration: Ethnicity and Nationalism in Modern Europe*. Amherst: University of Massachusetts, Dept. of Anthropology, Research Reports no. 9.

Echevarría, M. de (1894). *Higienización de Bilbao*. Bilbao: no publisher given.

Eloraza, A. (1976). 'El tema rural en la evolución del nacionalismo vasco', in J.L. Delgado (ed.), *La Cuestión Agraria en la España Contemporarea*. Madrid: Edicusa.

Greenwood, D. (1977). 'Continuity in Change: Spanish Basque Ethnicity as an Historical Process', in M.J. Esman (ed.), *Ethnic Conflict in the Western World*. Ithaca, NY: Cornell University Press.

Heiberg, M. and Escudero, M. (1977). 'Sabino de Arana, la lógica del nacionalismo vasco', *Materiales*, no. 5 (Barcelona).

Larronde, J.C. (1977). *El Nacionalismo Vasco: su origin y su ideología en la obra de Sabino Arana-Goiri*. San Sebastian: Txertoa.

Minogue, K. (1969). *Nationalism*. London: Methuen.

Otazu Y Llano, A. (1973). *El 'Igualitarismo' Vasco: Mito y Realidad*. San Sebastian: Txertoa.

Payne, S. (1974). *El Nacionalismo Vasco*. Barcelona: Dopesa.

Romero-Maura, J. (1977). 'Caciquismo as a political system', in E. Gellner and J. Waterbury (eds), *Patrons and Clients*. London: Duckworth.

Siadeco (1979). *Conflicto Linguistico en Euskadi*. Bilbao: Euskaltzaindia.

Vicens Vives, J. (1970). *Approaches to the History of Spain*. Berkeley and Los Angeles: University of California Press.

10

The Politicization of
Galician Cleavages

César E. Díaz López
University of Madrid

> I have never understood why the Catalonian and Basque Country's affirmative nationalisms cause so much worry, whereas the nihilistic nationalism of Galicia causes no terror.
>
> *J. Ortega y Gasset*

Stein Rokkan, in Chapter 1 above, has defined a periphery as being dependent, at the outskirts of a system, normally a conquered territory, with a poorly developed economy and a marginal culture: 'in all these domains, politics, economy, culture, the periphery depends on one or more centres'. Very few Western European regions can identify so perfectly with this definition as Galicia, the north-western area of the Iberian Peninsula. Although geographically, linguistically, culturally, economically and politically a periphery in its own right, Galicia is an intermediary case, half-way between the 'victorious' peripheries and the 'losers', succeeding, nevertheless, in maintaining its own cultural identity through the centuries.

Language and culture

> Our people, the Galician people, were able to create their own language. If we are distinguished as a different unit from among the European families, it is precisely thanks to the language. It is not because of our political sovereignty or for having performed collectively great historic deeds, or even for the possession and utilization of great economic resources that influence the international market. What makes us truly unique, what gives us our particular personality vis-à-vis the others, is the language, that is, the spirit. That genuine spirit which is ours and is genuinely reflected in the Galician language.
>
> *Ramón Piñeiro*

The Galician language, even though conserving a lingering Celtic linguistic substratum, evolved from Latin in the same way as the other Romance languages. The location of Galicia on the periphery explains its late and slow romanization in relation to other areas of the Iberian Peninsula. In 212 BC 'Galaecia' became one more Roman province, and by the sixth century its population spoke Latin. However, over the next 200 years the fundamental characteristics of the 'Galician-Portuguese linguistic domain' (Rodríguez 1976) developed. Galician-Portuguese is the correct denomination, considering that 'Galaecia' extended from the Cantabric Sea to the Duero River or, in other words, comprised what is today Galicia as well as the northern provinces of contemporary Portugal. Between 1200 and 1350 there emerged from this common Galician-Portuguese language an important lyrical poetry that was known beyond its linguistic borders.

From around 1400, however, we must speak of two different languages, Galician and Portuguese, as two branches of the same common genre. From then on the two followed different paths, in keeping with the historical and political fates of their respective communities. Portuguese became an official state — and even imperial — language, but Galician was to remain fundamentally an oral rural language. At the end of the fifteenth century Galicia was politically conquered by the Catholic kings of Castille. This double historical fact — Portuguese political independence and Galician dependence on Castille — brought as a linguistic corollary a continual phonetic and orthographic distancing between Galician and Portuguese on the one hand, and an approximation between Galician and Castillian on the other.

Galician disappeared as a literary language from approximately 1400 (around the same time as it emerged as a colloquial language) until well into the nineteenth century. The historical importance of this fact for the later development of the language is incalculable. Perhaps two of the most relevant cultural occurrences in Galicia in past centuries were the introduction of the printing press (around 1490) and the founding of the University of Santiago de Compostela (1506) at a time when Castillian, and only Castillian, was used in transmitting the written culture. The first well-known book entirely in Galician (*Cantares Galegos* by Rosalia de Castro) was not printed until 1863. This significant four-century absence of written Galician has been explained by some as a consequence of the lack of cultural production centres in Galicia, itself a result of religious and political centralization (Montero 1973: 56).

For years any reference to the coexistence of two languages in Galicia was automatically labelled as 'bilingualism'. However, bilingualism refers to purely individual conduct, rather than to social behaviour. In this sense bilingual individuals could exist, but not bilingual communities. On the other hand, bilingualism presupposes the coexistence of two languages, equal in status and prestige, which is definitely not the Galician case. Galician socio-linguistic reality is better defined by the concept of 'diglossia', in which there is a scission or linguistic superimposition between a high language, used in formal communication — literature, religion, teaching (official and public affairs), etc. — and a low language, used in conversations of a non-formal or familiar character (Gutiérrez 1978: 11-14). This is 'grosso modo', the contemporary situation of Galician and Castillian in Galicia: 'the possibility of their being used, within certain limits, as situational variants' (cited in Montero 1974: 139) has been favoured by the process of progressive affinity between both languages since the Middle Ages, a process that has facilitated mutual comprehension between their respective speakers.

During the Middle Ages Galician was the 'normal' language, although it was not standardized. The entire monolingual period, with its dialectical variants, was characterized by the absence of linguistic conflict. The toppling of the relative linguistic stability of this period was provoked by the penetration of Castillian elements, which coincided with the decadence of the old Galician kingdom and the consolidation of Castille as the hegemonic kingdom in northern Iberia. The existence of continual conflicts between town councils and peasants on the one hand and feudal nobility on the other caused an increase in the number and intensity of the Castillian monarchy's interventions in Galicia. The Trastamara dynasty introduced the first great contingent of non-Galicians: speaking only Castillian, these constituted the first external diglossic nucleus. Simultaneously, an accentuated process of assimilation of Castillian ways and linguistic habits by Galician nobility formed the second diglossic nucleus, in this case autochthonous and therefore, in opposition to the former, bilingual, since the Galician nobles would not immediately forget their language. The second external diglossic nucleus occurred in the second half of the fifteenth century once Galicia had been conquered. Royal decrees created the Royal Audience in charge of the administration of justice as well as the centralization of public administration. Castillian was raised to official status, thereby

FIGURE 1
Changes in the linguistic structure of Galicia

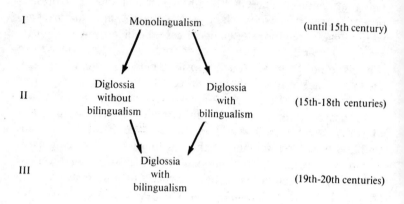

Source: Chacón 1978.

excluding the use of Galician from the writing of not only official documents, but also unofficial public ones. These trends proved to be irreversible.

From the end of the fifteenth century, then, Castillian, although spoken by a minority, had official status in Galicia and Galician was relegated to the oral language of the rural community. This was the prevailing situation at the fall of the Ancien Regime in the nineteenth century. The arrival of a constitutional monarchy only intensified and generalized the linguistic situation. The state's attempts at rationalization, tending towards the disappearance of personal privileges and 'señorios' (feudal estates), resulted in a greater presence of the Castillian-speaking bureaucracy in Galicia, especially after 1833, with the fragmentation of the region into four provinces without any organic connection between them and all directly dependent on the centre. This fact, combined with the actions of the Catholic Church, a strong agent of Castillianization in Galicia, and the creation of a unitary school system where Castillian was obligatory, contributed powerfully to the stagnation of the autochthonous language.[1]

In recent times the diffusion of Castillian through schools[2] and 'new' means of mass communication such as radio and, above all, television has only made the diglossic situation more acute, the population more dissatisfied. Educating Galician-speaking children in Castillian produces a devaluation of their own language, since they come to believe, because of constant correction at school, that in order to speak well one must speak Castillian, not Galician. Being incapable of distinguishing them as two different languages, they tend to consider Castillian as the educated variant. This schooling situation in a language different from the mother tongue —advised against by UNESCO—is to blame for the very low level of Galician children's educational achievement, a fact that is recognized even by the Ministry of Education.[3] It also generates a tendency to reject one's own language as uncultured and useless for incorporation in society or for use outside the family circle, ultimately generating rejection of one's self. This personal underestimation seems to be the base of a certain inferiority complex experienced by many Galicians, which has become a characteristic of the collective personality of the rural population.[4]

Whichever language is used, therefore, obtains a marked social character and is given the value of a 'status' indicator. Castillian (high language) is identified with power, wealth, education, culture, mass media, profession, social dignity, cities and luxury, while Galician (low language) is associated by its speakers with submission, poverty, lack of education, lack of culture, trades, low social esteem, villages. Two contrasting worlds are thus evoked: one of progress and the possibility of professional and social advancement, the other of backwardness and misery (Montero 1973: 30).

It has been said that, 'of all the external differentiating signs in a community, language is the most notorious and the one which confers on that community its cohesion and coherence' (cited in Montero 1974: 156). If this is true, and if the well-known Cartesian dictum that 'to be different is to exist' is also true, then the future of Galicia as a differentiated community depends on the future of its language.

The language, however, is threatened. In a polemical book published in 1973, one Galician intellectual warned of its immediate disappearance if a series of claims were not quickly conceded, for example, general education in Galician, and its greater daily use in the mass media. Seven years later these conditions have not, in practice, been met. Galician is losing speakers in its traditional realm: the countryside. For the first time, Galician peasants are raising their children to speak Castillian. But on the other hand, its acceptance in the cities is growing slightly. Linguists have claimed that the unstable and polarized character of all diglossic situations leaves only two solutions: substitution of the low language by the high one, or normalization (process of recuperation) of the low language for superior and formal functions previously reserved for the high language (teaching, administration, mass media, etc.)

The struggle for the survival of Galician is carried out on two complementary fronts: the cultural, which comprises linguistic-normative aspects and which works for the standardization of the language,[5] and the more specifically political, which aims at normalization. This distinction is more analytical than realistic, since the latter cannot exist without the former. Both fronts only form the two sides of the Galician coin, a naturally political phenomenon.

Society and economy

The economic situation of Galicia compared with the rest of Spain — with the exceptions of the also depressed Andalucia and Extremadura — is one of isolation and underdevelopment. This can be documented by referring to some comparative economic figures. Table 1 presents the percentages of employees in the different sectors in Galicia and Spain. The extraordinarily high percentage of Galicians employed in the primary sector, double the Spanish average, immediately draws our attention. But if we stop to examine the productivity of this voluminous sector compared with that of other Spanish regions, we find that the lowest index of productivity, measured in thousands of pesetas in 1960, pertained to Galicia, while the highest belonged to the Basque Country, with 40.9, the Spanish average being 27.2. As an example, we can compare the yield of the three most important crops in Galicia to those of Spain: corn (22 percent of the total area cultivated), potatoes (14.1 percent) and dried beans (12.6 percent), which together cover almost one half of the area cultivated, measured in kilograms by hectare. Only rye, the fourth crop, at 8.3 percent of the land cultivated, gave superior yields (115.4 percent) in Galicia than in Spain (CRPO 1978: 328-98).

TABLE 1
Employment by sector in Galicia and Spain, 1970

	Galicia	Spain
	%	%
Primary	49.0	24.8
Secondary	23.4	37.4
Tertiary	26.1	36.5

Source: INE (1976: 326).

TABLE 2
Agricultural productivity in Galicia and Spain, 1976
(kilograms per hectare)

	Galicia	Spain	% Galicia/Spain
			%
Corn, 1976	1,831	3,578	54.8
Potatoes, 1977	13,147	14,873	88.4
Beans, 1976	184	611	30.1

Source: CRPO (1978).

Scarce industrial development and the decisive weight of a primary sector not geared for competition explains the widespread Galician economic backwardness. But what can explain the low agricultural yield? To find out would mean delving into the reasons for Galician underdevelopment with respect to the Spanish average, and ultimately to pointing out the social-economic and centre-periphery cleavages: (1) the type of rural economy practised; (2) property structure, which, in turn, conditions the size and type of enterprise; (3) demographic structure. These three factors, as well as the ecological potential, will condition the possibilities of modernizing the rural economy and, therefore, its productive capacity.

Although it is changing, the rural Galician economy is still generally of a traditional nature as it fulfils the conditions of what

Thorner calls a 'peasant economy': predominance of the agricultural sector; separation of city and country by the economic isolation of the latter; the family as a unit of production and consumption; and an orientation towards subsistence rather than marketing and profits.[6] These characteristics of a semi-autarkic economy lead to a concentration on generalized subsistence crops, which obliges the land to produce 'a little of everything' rather than specializing in relation to its productivity and yield; and to a low level of monetarization and purchasing power as a result of the small surpluses generated by this system. Both circumstances have as a corollary a low level of income and low living standards, and are perfectly applicable to Galicia, with the following aggravating nuances:

1. *The reduced size of enterprises.* The hilly relief of Galicia resulted in fragmented plots or 'minifundia', in which 91.9 percent of the total of cultivated land is in units of less than 20 hectares, with 73.5 percent being not more than 10 hectares — all in a densely populated region having a limited cultivated surface area (27.8 percent of the total) and with a low productivity index. This situation contributed to creating an awareness of 'limited goods' among the peasants and an overvaluation of the land that achieves — its real price aside — almost mythical connotations (García Fernández 1975: 84; also MPWU 1978).

2. *The high degree of population dispersion.* 'In 1970, out of the total number of population settlements in Spain (63,393), 50.3 percent were in Galicia (31,882), and over the years this great number of settlements in Galicia has changed very little as indicated by the fact that in 1880 the number was 33,131' (MPWU 1978: 37). Comparing the average size of these settlements in Galicia and Spain can be rather clarifying, for while in Spain as a whole it grew by 31.6 percent between 1950 and 1970, in Galicia it remained almost stable (1.8 percent increase) (cf. Table 3).

TABLE 3
Population settlements in Spain and Galicia, 1950 – 1970

	Average population of settlements		
	1950	**1960**	**1970**
Galicia	79.6	80.3	81.0
Spain	406.8	446.4	535.6

3. *General demographic structure.* Compared with the rest of Spain, Galicia is densely populated. In 1970 there were 87.8 people per square kilometer in Galicia compared with 67.0 for Spain as a whole. In 1900, by contrast, the figures were 68.3 for Galicia and 38.8 for Spain. Furthermore, whereas at the beginning of the century the Galician population represented 10.6 percent of the total Spanish population, in 1975 it was only 7.4 percent. The uneven growth rates are explicit. Since the beginning of the century the Spanish population has grown by 93.5 percent while the Galician community increased by only 35.5 percent until 1950, and decreased thereafter until 1970. From 1970 to 1975 there was a slight increase, which coincided with the return of emigrants to their birthplaces (MPWU 1978: 80ff.; INE 1976: 55).

4. *Emigration.* This has altered the growth rates in Galicia. The system of intensive subsistence crops, insufficient for absorbing the active population — who can find no work in the cities either — has generated a labour surplus that has been eventually reduced through emigration. Between 1950 and 1973 Galicia lost half a million young people of family-raising age, mostly from rural areas, which released an 'authentic process of dissolution of the rural society', understandable, given that the contemporary population does not quite reach three million (cited in Montero 1973: 204; also Beiras 1972: 68-81; MPWU 1978: 41-80). A direct consequence of this exodus is the aging of the agrarian population: 35.5 percent of the peasants are aged 65 or over, while 62.8 percent are over 55 years. By contrast, those younger than 35 represent only 2.9 percent of the total (MPWU 1978: 40). This fact becomes most relevant when we contemplate the high degree of ruralism, the highest of all Spanish regions, with around 70 percent of the population living in rural areas. The corollary is that the percentage of inhabitants living in municipalities of more than 100,000 (15.0 percent) is the lowest in Spain.

5. *The peculiar mercantile system.* Alongside the low level of monetarization in Galicia, there exists a network of more than a hundred periodic rural markets having a perfect hierarchy and structure, remains of which can hardly be found in the rest of rural Spain (MPWU 1978: 29-30). Indirectly this fact explains, in Stein Rokkan's words, the existence of a 'subsistence economy outside the territorial [Spanish] network of exchanges'.

In conclusion, the socioeconomic cleavage between Galicia and Spain — or between the 'centre' and the 'periphery' — is real and omnipresent. It does not really matter whether we index equipment, services, consumption of energy or of luxury articles per inhabitant: the difference between the average Galician and the average Spaniard always manifests the inferior degree of Galician development. Table 4 illustrates as an example the number of automobiles per 1,000 inhabitants.

TABLE 4
Automobiles per 1,000 inhabitants in Galicia and Spain

	1965	1975
Galicia	11	79.2
Spain	25	120

Assuming Galicia and Spain to be units of comparative analysis, we might be led to believe in a certain internal homogeneity within both territorial areas. Nothing is farther from reality. In the interior of the 'rest' of Spain, as well as Galicia itself, there exist all kinds of inequalities. Galicia is distinguished by its dual economy: a relatively dynamic industrial economy, integrated into the Spanish one, superimposed on a traditional rural-based economy. The relationship between the two could not be more negative. The industrial economy is not tied to the rural, and the latter, because of its weight, makes integration of the Galician economy into the Spanish market very difficult as well as limiting the dynamic effects induced by the industrial one (MPWU 1978: 13).

This industrial economy, generally promoted from outside Galicia, contributed to reproducing within it some of the internal inequalities that exist between Galicia and Spain. Thus, it contributed to a widening of territorial inequalities between coastal and interior Galicia, owing to an increase in the population density in metropolitan and urban areas that, with only two exceptions, are all located on the coast:

The 56 municipalities that form these areas, in an extension of 3,196.8km², representing 10.8 percent of the total surface area of Galicia, is where, in 1975, almost 52 percent of the total population was settled, as well as almost all of the companies having more than 50 employees. Here, industrial employment created, between 1965 and 1977, 50 percent of the region's housing, and more than 80 percent of the total of hospital services in Galicia. However, even more outstanding is the fact that, during the period 1960-70, while Galicia lost almost 20,000 [sic][7] inhabitants, these areas grew by 143,000, representing an increase of 12.7 percent and from 1970 to 1975, while Galicia grew by 3.9 percent, these areas increased by 9.7 percent, or by a total of 123,556 inhabitants. [MPWU 1978: 64]

FIGURE 2
Population by sex and age in the four Galician provinces, 1970

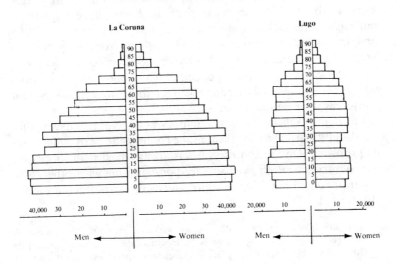

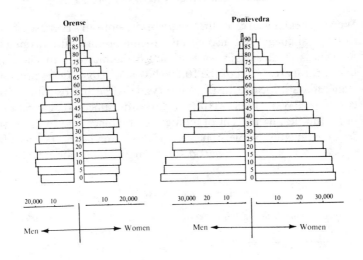

Source: Situación actual y perspectivas de desarrollo de Galicia, Vol. 3. Madrid, 1975.

The cleavage between the two Galicias — the urbanized and industrialized littoral provinces (Coruña and Pontevedra) and the agricultural and underdeveloped interior ones (Lugo and Orense) — can be seen in the population pyramids shown in Figure 2. It is worsened by the subsistence nature of rural agriculture, which cannot ensure the Galician cities an adequate food supply, thereby obliging them to acquire many food products from other parts of Spain. The fact that, on the other hand, the industrialized economy is greatly concerned with exporting out of Galicia obviously does not favour commercial exchange between these two economies, and reinforces their mutual isolation.

But if the Galician countryside does not serve the city, neither has the city learned to be useful to the countryside or to understand its problems. On the contrary, it has always lived parasitically on the countryside. Both exist side by side, ignoring each other. The contrast rooted in this city-country dualism, writes Ramón Piñeiro (1974a: 151),

> is not strictly economical but rather broader and more profound. In Galicia class differences are based above all on being from an 'aldea' [rural settlement] or from a city. This differentiating factor is much more spontaneous, profound and energetic than the purely economic one...peasants and city dwellers [are] in reality the two characterizing classes of social structure in Galicia.

We are speaking of two different worlds, not only because of two distinct languages but also because of the opportunities available: as regards medical assistance, although the majority of Galicians live in rural areas, out of the 10,796 hospital beds in the region only some 300 are located in the countryside while the rest are in the cities (MPWU 1978: 110); as regards unequal educational opportunities, 64 percent of all high school places are found in the seven main cities, while 70 percent of the young people between the ages of 15 and 19 reside outside these large municipalities (MPWU 1978: 108).

Language as a politicized cleavage

> ...the Galician language, so unknown by all and so despised still by the peasants themselves.
>
> *P. Sarmiento (1754)*

> Bringing about social justice in Galicia requires two things: spiritual redemption of the people's inferiority complex — so that they can actually be a people — and material redemption of their economic backwardness, so that they can live in dignity...The redeeming weapon is the language...
>
> *Ramón Piñeiro*

The Rexurdimento, or Galician Renaissance, of the nineteenth century was more poetic than literary because of the lack of importance given to 'Galicianizing' other genres. The first novel in Galician was not published until 1880, while the language was hardly used in theatre at all. All in all, the Rexurdimento was more of a literary and cultural phenomenon within the European current of regionalism engendered during the Romantic period; it meant an awakening of popular life, and of interest in the Galician language and culture, an occurrence more noticeable in Galicia because of the region's traditional degree of estrangement from its own language.

The first to desert the Galician language were the nobility, with the penetration of Castillian functionaries. Then in the eighteenth century there was an immigration of Catalonian fish canners, and the 'de-galicianization' of the urban society was accelerated. Here lies the fundamental difference between Galicia and Catalonia. In the latter there was an identification between the bourgeoisie and the Renaixença. In Galicia, however, this social class did not give its backing to any renaissance: 'the precariousness and dramatic situation of Galician culture — before and after 1900 — resides in the fact that the bourgeoisie of Galicia did not adhere and remained, in the past as well as in the present, reticent or ironic to the cultural claims of our men of letters' (Montero 1973: 85). Even the intellectuals and writers of the previous century found themselves, more than they do today, prisoners of the ambiguity of a diglossic situation of which they themselves were generally not even aware; the *spoken* language of Galician intellectuals at the end of the nineteenth century was not Galician, but Castillian. In 1888 the novelist Emilia Pardo Bazán alluded to the paradox that Galician 'is not spoken by those who write it'; and of course those who spoke it customarily did not read it. The urban middle classes were not interested in Galician. 'The degree of linguistic-cultural alienation is such in Galicia', wrote Robert Lafont in 1968, 'that in a polemic discussion in 1876-1877, four newspapers declared themselves against the use of the country's language' (Lafont 1967: 237-8).

This lack of interest was not the exclusive patrimony of the bourgeoisie or intelligentsia. The political leaders of the period, even those who were worried about Galicia and its people, were unaware of the language problem. Ricardo Mella (1861-1925), a great theoretician and anarchist activist, the renowned agrarian leader Basilio Alvarez and his 'Liga de Acción Gallega',

'Solidaridad Gallega' and the socialists not only used Castillian exclusively or almost exclusively in political interventions: they did not even register the linguistic issue as problematic in their writings and speeches.

Of the three historical-political periods into which Galicianism is usually divided — provincialism (1843-68), regionalism (1868-1916) and nationalism (1916-36), only in the last does the language appear for the first time as a politicized cleavage, capable finally of crystallizing sizeable movements around itself. The Irmandades da Fala of 1916 were the first Galicians to claim, not literarily but politically, the need to restore the Galician language. Their primary goal was to diffuse its use. They were fully conscious that their linguistic claim (and in this way they differed from many Rexurdimento intellectuals) was potentially a political weapon. Their second goal consisted of 'conquering a broad autonomy for Galicia'. The priorities were clear, and the first step was the language.[8] They dedicated their efforts to its cultivation: they began to organize lectures, art exhibits, plays and music in Galician, and to distribute leaflets and the magazine *A Nosa Terra*, the ideological mouthpiece of the 'Irmandades', which soon extended to other areas of Galicia.

The following years would be of decisive importance to Galician nationalism, which sought to re-introduce the language through two important institutions: the Seminario de Estudos Galegos, a scientific-humanistic research centre, and the *Revista Nós*, founded in 1920, which would give its name to an entire generation of highly regarded cosmopolitan Galician intellectuals who from their writings would transform 'an already literary language into a cultural language too. . . by using Galician in discussing all types of subjects (Galician or not) from socialism to cubism to Oriental mythology' (cited in Baliñas 1977: 11-12). These experiences, like so many others, were interrupted by the military uprising against the Republic in 1936.

The idea of the Galician language as a maximum identifying element goes back to the nineteenth century; for Murguía, 'it is the true national flag', since 'aside from being what identifies all the members of a group as a genuine cultural unit, it is what for the same reason distinguishes it from all the rest' (Piñeiro 1974: 97). At this point it would be useful to consider the alliance policy of the nation-builders and Galician periphery resistance to central power pressure, applying Stein Rokkan's (1970) scheme, adapted with certain variations to the case of Galicia. The efforts of the constitutionalist liberals in power, particularly noticeable after 1833, to ra-

FIGURE 3
A model of central and peripheral alliances in Galicia

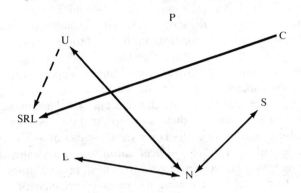

C = Catholic Church
S = Secular
N = Nation-builders
L = Latifundistas
U = Urban element
P = Galician nationalists
SRL = Small rural landowners

tionalize the state bureaucratic apparatus 'à la française' through the identification of Spain = nation = state met with difficulties. These derived primarily from their own weakness, but also from the more or less explicit resistance of the peripheries which had their own cultural identity. In the case of Galicia, active resistance was marginal compared with the passive resistance, the effects of which are still felt today. We can study the reasons for this weakness by inspecting the alliance patterns (Figure 3):

1. the central power alliance is the same for all the peripheries, N – L – S;

2. the influence of the Catholic Church on rural Galicians (mostly SRL) was through the parish priests;

3. there was a direct tie — in both senses — between U (urban element, generally commercial or functionary) and N (Madrid), the source of monthly income and therefore the centre of loyalty;

4. the temporary (only, or fundamentally, at election time) and indirect (through the local boss) influence of U on SRL;

5. the most outstanding phenomenon: the inability of P to create a counter-alliance capable of successfully resisting the centralist one, manifest in the slight or non-existent Galician nationalist

influence on U (in spite of their generally urban extraction) and on SRL, that is on the two largest sectors of Galician society. Because of that, these intellectuals found themselves isolated from the sector of society — and this is the most painful — which they desired to redeem, the small rural landowners. This explains the scant development of Galician nationalism until recent times, above all if we compare it with the Basque and Catalan cases: the only backers of Galicianism — impotent before the state and more powerful sections of Galician society — were the intellectuals.

For the first generation of Galician nationalists the language, aside from being the most radical and representative element of Galician culture, is what conferred on it the character of a differentiated people with a *right to political autonomy*. Language and autonomy were thus converted to synonyms until present times.[9]

'Why is it necessary to combine the peasant's claims for civil liberty and economic [justice] with that of language?' asked the skeptical Miguel de Unamuno. 'Because', replied Villar Ponte, 'in order to speak in the country, to inspire confidence, we have to speak, feel and think in Galician since the great majority of Galicians do not express themselves in nor speak any other language than their own' (cited in Villar Ponte 1971: 306). In spite of this fact, and the consideration that 'the language is the greatest problem posed by reality', the first generation of nationalists were tolerant of Castillian. Their compromise attitude is reflected in the Manifesto de la Asamblea Nacionalista de Lugo of 1918 (Risco 1976: 147) as well as the Partido Galeguista (1931-36) programme in favour of official parity of Galician and Castillian. It was also stated in the Autonomy Statute voted upon in a plebiscite in 1936, only a few days before the beginning of the Civil War, which aborted its ratification and implementation (Bozzo 1976: 170ff.; also Zubillaga 1974; Vilas Nogueira 1977; Cores Trasmonte 1976).

Under the authoritarian and obsessively centralizing and standardizing pressure of Francoism, Galician acquired a certain bulwark character, to whose defence the surviving Galicianists would rally.[10] The focus around which they gathered was the publishing house, Editorial Galaxia, founded in 1950 after the dissolution of the Partido Galeguista and the abandonment of any hope of an Allied intervention to support the democratic losers in 1939. During the dictatorship, speaking and writing Galician took on a marked political character.[11] But in an authoritarian system anything political must 'necessarily' remain circumscribed to the sphere of the dictator's personal power or it is proscribed and accused of subversiveness. Writing in Galician was just that, and in this way Galicianism became another form of democratic resistance and opposition to the dictatorship.

During the early 1960s there was a general attitudinal change towards the Galician language and culture. The Galician church, influenced by the renovating winds of the Vatican Council, opened its liturgy to the Galician vernacular, thereby timidly introducing it into the overwhelmingly Castillianized church.[12] There was also an increase in the number of books in Galician, the appearance of children's literature and even singing groups: 'This fact coincided with the growth in ideologies and progressive attitudes', explains Montero (1973). The University of Santiago, in the same way as other Western universities during that period, registered the presence of contestatory and anti-bourgeois movements that found their purpose in anti-Francoism and Galicianism. The association between these two sentiments in the minds of many young democrats helped to increase the use of the native language in urban areas. The association of either language with the respective social class of its speakers led to a shift of Galicianism towards Marxism. From the core of this dialectic, and because of a split in the Consello da Mocedade, there emerged the Galician leftist parties: the Union do Pobo Galego (UPG) in 1964 and the Partido Socialista Galego (PSG) in 1965. We have now reached the latest generation of nationalists, whose more radical general attitude is reflected in their attitude towards the language. While both parties are favourable to the use of and giving official status to only Galician, they can foresee the necessity of a transition period which would permit the incorporation of Castillian-speaking Galicians into the Galician culture. The distance travelled from the compromising attitude of their remote predecessors of the Irmandades, who were in favour of a co-official status, is great.

Galicianization of politics in the 1970s can be described as almost complete. The language is used as the basic mode of communication in nearly all political gatherings, not only in those organized by Galician parties, but also in those Spanish parties operating in Galicia. The same is true of leaflets, books and electoral programmes published by the various parties. This phenomenon, however, is more noticeable in leftist parties than in those to the right. The governmental party (UCD) and the smaller formations to its right use Castillian mostly or, in the case of the extreme right, almost exclusively. This relatively recent process of galicianization in local politics has even led to name changes in the majority of statewide parties acting in Galicia, especially on the left, where they have tended to add a 'G' for Galicia. Since 1969 the Communists (PCE) have been the Communist Party of Galicia (PCG), and other parties have followed suit: the Workers' Party (PTE) is the PTG, the Communist Movement the MCG, etc. The fact that this nominal galicianization is not always accompanied by true political autonomy for the Galician section of the party with a freedom to form policies complementary to Galician reality has always been a motive for left-wing Galician parties to attack them, accusing them of being 'mail-box' branches and Spanish-oriented parties that followed directives conceived, within a general strategy, in Madrid.

However, all of these formations today defend, to a greater or lesser degree, the necessity of an autonomy statute for Galicia and, with greater or less emphasis, the necessity for the normalization of Galician as well as its co-official status with Castillian. The presence of exclusively Galician parties, and particularly the most radical Marxist-Leninist UPG, has acted as a catalyst on the Spanish parties, dragging them all, in varying degrees, towards more Galicianist positions, perhaps through a realization of the politically legitimizing value of Galicianism whose standard would be dangerous if left solely to exclusively Galician parties. For instance, considering any autonomy statute insufficient, the UPG has, for several years, favoured the establishment of an independent constitution for the 'Galician nation'. Article 5 of the recent draft of the 'Preliminary Statute for Galicia', drawn up in Galician by the 'Assembly of Galician Parliamentarians' and following the guidelines of the 1978 Spanish Constitution, states: (1) the language of Galicia is Galician; (2) the Galician and Castillian languages are official in Galicia and all people have the duty to understand them and the right to use them; (3) the use of both is guaranteed and the use of Galician in all aspects of informative, cultural and public life will be encouraged; (4) all functionaries appointed to Galicia must learn Galician (*El Ideal Gallego* 1979).

The current terms of the future Autonomy Statute already drawn up and approved by this Assembly, in which the UCD has a comfortable majority, are presently being discussed with the opposition parties (the Galician branch of the Spanish Socialists and the Galician moderates of the PG), who refuse to accept a formula that does not exhaust the possibilities allowed by the Constitution itself, and who threaten otherwise to recommend abstention in the projected referendum which it must pass in order for it to become valid. The institutions foreseen for Galicia include a democratically elected legislative Assembly (with legislative powers); a government (Xunta) endowed with executive functions in designated areas of competency with a president appointed by the Assembly and responsible to it; and an independent judicial supreme court, subordinated to the state-wide one. Galicia is a procedurally privileged community (together with Catalonia and the Basque country) in the 1978 Spanish Constitution, as far as facilities for access to self-government are concerned, within an 'autonomist' Spanish state (arts 151, 2; 152 and 'disposición transitoria segunda').

The politicization of the socioeconomic cleavage

The desire for linguistic centralism is one of several factors of a single phenomenon, that is political centralism which is, in its turn, an instrument of economic centralism, the true instrument of domination.

Hans Schneider

Can we speak of development in Galicia without considering the agrarian structure which is suffering a depopulation crisis, destruction of the rural family and cutting of ties between the peasant and his lands due to a lack of planning and ordering of the primary sector...? Development without destroying Galicia is the big task.

X.B. Pena Trapero

Although the politicization of the cultural-linguistic cleavage was not evident until 1916, the socioeconomic cleavage crystallized much sooner. At first glance the two types of pro-Galician movements seem contradictory: the first originating from a situation understood as a chronic abandonment by the central government, and a more recent one adding to the previous motivation a deep sense of disgust for the way in which the central government intervenes in Galician affairs. The contradiction is more apparent than real, as the common denominator resides in the desire to augment the still meagre capacity for political decision-making in Galicia.

Although some historians maintain that Galician consciousness was born with the Enlightenment and not with the Romantic period, most authors agree in pointing out the crucial importance of the war against Napoleon in shaping the Galician conscience into awareness of the uniqueness of Galicia. Thus a Galician historian, after discussing the high involvement of the population — including priests — who rallied to the guerrilla war against the French ('all Galicia was a guerrilla'), goes on to say:

In 1808 the people did not trust government bodies. The people grouped, perhaps unconsciously, driven by a type of conservation instinct imposed by its social biology. These groups were typically local units: local and provincial Juntas headed by the Supreme Juntas in each kingdom. After centuries Galicia found herself again: the kingdom of Galicia organized the defence against the French, means were obtained, ambassadors sent to England, an army formed and taxes levied. She behaved as a kingdom. This means that the ancient vocation had not died. [Barreiro Fernández 1976: 131-2; also 1977]

In 1843 in the Federal Assembly of Lugo, the Galician Central Junta debated and voted on a proposal whether to 'declare this old kingdom independent or not'. The most surprising element of this separatist proposal was that it was defeated by only one vote. Although there are different historical interpretations over the Assembly, it at least emphasizes the importance of the degree of opposition by Galician liberals to the policies made in Madrid by their fellow party members. This tension became radicalized, and in 1864 a military coup took place against the central government. Antolín Faraldo, the ideological leader of the uprising, wrote in the manifesto:

Galicia, until now living an ignominious existence, having been converted into a true colony of the court, is going to throw off this humiliation and dejection...Galicia will achieve the influence she deserves, climbing to the high position she is called to occupy by the ancient 'suevos' kingdom. [cited in Piñeiro 1974b: 525]

This manifesto has been traditionally considered the first political cry in favour of Galicia. The 'revolution' was militarily suffocated, and its leaders — the martyrs of Carral — shot.

Today Galicia is a region of small 'minifundista' landowners, but it has not always been so. The present structure is the result of a transformation occurring over the last hundred years. The Galician 'casa', the unit of agrarian exploitation, originates from the medieval pattern. In exchange for rent, the great landowners — nobles and high clergy — allowed the peasants to work the land. This contract, called a 'foro', constituted an important source of income for the landowners, since it was they and not the peasants who produced the surpluses. Towards the end of the eighteenth century the practice of 'subforos' became extensive, creating a new intermediary who would also live at the expense of the peasant, already overburdened with taxes and rents and dragged down by a miserable existence (García Fernández 1975: 94-100).

The struggle for the redemption of the foros and ownership of the land for those who worked it constituted the factor of politicization that crystallized around agrarian movements at the beginning of the twentieth century. In 1907 Solidaridad Gallega was founded, and Agrarian Assemblies followed in 1908, 1910 and 1911. The Liga de Acción Gallega continued the battle after its foundation in 1912, giving the struggle a new impulse and authentic popular mobilization.

On comparing the ability of the Liga for mobilization with that which the Irmandades could have had, Alonso Montero (1973: 90-1) rhetorically asks: 'Did the peasant identify with the language of the socially innocuous and politically ambitious Irmandades?' and he answers:

> In 1912 the leaders of Acción Gallega offered a more concrete and inciting message while the [more intellectual] leaders of the Irmandades proposed a more distant programme about the Galician language which never really took hold. The truth is that the peasants were moved more by priest and agrarian leader Basilio Alvarez's Castillian than by Antón Villar Ponte's passion in Galician: 'Basilio Alvarez in spite of the language; Villar Ponte not even with the language'.

The conclusion is obvious: the right to ownership of the land and reduction of taxes worried the peasants at the beginning of the century much more than any linguistic claims. The socioeconomic question achieved primary interest in politicizing the masses. In the National Assembly of Lugo's 1918 Manifesto, in decalogue form, we find, apart from the need for autonomy and co-official status of the language as well as the teaching of Galician, the need for the land to become the tax-free property of those who work it, as a 'summary of all agrarian aspirations' (quoted by Risco 1976: 147).

After the fall of the Spanish monarchy and proclamation of the Republic in 1931, Galicianism, until then only a movement, became a political party. The centrist and inter-class Partido Galeguista took over the linguistic claims of the Irmandades and the socioeconomic ones of the agrarian movements, by proposing to eliminate feudal servitude and instilling co-operativism as a 'third road mid-way between capitalism and communism', in an attempt to solve agrarian problems. They were also in favour of free trade, which would permit uncontrolled importation of corn for Galician cattle and tin for the canning industry, but found themselves confronted by the Castillian cereal protectionist policies of Madrid. For this party, rural and fishing Galicia was the 'true' Galicia and they would dedicate themselves politically to it. An autonomy statute was the subject of a plebiscite in June 1936 with the following results: census, 1,343,135; votes in favour, 993,351; votes against 6,161; and void, 1,451. This could be considered the great success of prewar Galician politics, since the statute detailed, within the Republican Constitution's limits, the principal claims. The preeminent position accorded to Galicia by the 1978 Spanish Constitution stems from the positive results achieved in this historic plebiscite.

Previously, we referred to the existence of an important polarization in Galician society between countryside and city. In this sense language has an highly discriminatory value, since it is strongly associated with education, income and habitat. This cleavage then is not a cross-cutting one, but rather is formed by a number of mutually reinforcing ones. Therefore, we could expect that such mutually reinforced cleavages would be politically expressed in agrarian parties. Why has such a thing not happened? According to Lipset and Rokkan (1967), 'Distinctly agrarian parties have only emerged where strong cultural oppositions have deepened and embittered the strictly economic conflicts'. This means that cultural opposition alone is not enough to create an agrarian party, since 'the conflict between landed and urban interest was centred in the commodity market...': we are already familiar with the traditionally scarce economic transactions, confined to the indispensable minimum, between countryside and city in Galicia.

There was still at least a theoretical possibility that this tenuous economic cleavage would become stronger, since 'economic cleavages became more and more pronounced as the primary-producing communities entered into the national money and market economy' (Lipset and Rokkan 1967: 20). But this was not the case in Galicia, where only the industrial economy, and to some degree the ranching sector, achieved this 'large-scale, profit-oriented' state. Agriculture, with its semi-autarkic practices, remained isolated from the modernization process. This low saliency of the economic cleavage between city and countryside could perhaps explain its non-politicization into stable agrarian parties, in spite of the reinforcement of other cleavages. It had been crystallized only sporadically in the form of agrarian movements because of the foros and the presence of charismatic leaders.

Economically, the 1960s was the decade of planning and development. This statewide action would be carried out through the 'Planes de Desarollo Económico y Social'. In the first two plans, two poles of development were created for Galicia in La Coruña and Vigo. In the Third Plan, and by decree (1973), Galicia was considered the 'Great Area for Industrial Expansion'. The planning action had, among others, the following characteristics:

1. Insufficient advantage was taken of Galician natural resources, always responding to the specific needs of Spanish development rather than Galician (MPWU 1978: 15).

2. Those industries located in the 'poles of development' have the character of enclaves, cut off from their economic surroundings and involved in exporting from Galicia (MPWU 1978: 78).

3. Galician financial resources are exported through banks that invest in more developed regions (Barral Andrade 1975: MPWU 1978: 61).

4. This capital has been invested in other regions, thereby creating labour markets which are followed by Galician workers. Between 1960 and 1970, not only did the emigration drain not decrease; rather, it increased (Richardson 1975: 92).

5. Still more serious is the fact that the relatively small investments made in Galicia reflect an absence of 'specifically regional' criteria (MPWU 1978: 57).

6. The theoretically isolated 'poles', intending to avoid territorial economic imbalances, produced the opposite effect in Galicia with the corresponding concentration of activity and population in the littoral areas (Ferrol-Vigo axis), generating depopulation and impoverishment in the interior provinces (Richardson 1975: 161-2).

7. Equipment policies did not take into consideration the forms of population settlement (MPWU 1978: 106).

8. Another inconvenience, although having historical precedents, caused negative consequences to planning: the inadequacy of the Spanish administrative structure for Galician reality (MPWU 1978: 94).

In summary, there is a divorce between Galician needs and the development plans proposed. In Galicia there is a dual economy, something apparently unconsidered by the technocratically minded (undemocratic) planners, more interested in obtaining immediate and stunning successes, and in this way helping to legitimize politically the more and more contested Franco regime, than in a true transformation; more worried about 'efficiency' (a technocratic key word) than in equality. They believed that development equalled industrialization, forgetting the primary sector, which is the largest in the Galician economy. This is what caused the failure of the Planes and the need to redefine state action in Galicia as that which 'does not succeed in generating a break with the traditional model' (MPWU 1978: 16-17). The reasons pointed out in a government-requested report, analysing the inoperability of the planning 'a tavolino' from Madrid, are clear:

> All economic policy measures which do not take into account contents especially designed for Galician reality will lose strength and, at best, will have weak propelling effects, being unable to integrate an economic system that is precisely characterized by being relatively autarkic and resistant to change.

This report's conclusion is explicit:

> Definitively, the problem lies in the absence of a regional development policy which harmoniously integrates components as diverse as planning of infrastructures, price policy, measures of industrial development, procedures to modernize agrarian economy, and the lack of relevancy that such actions and measures have for Galicia. [MPWU 1978: 13]

The generation of postwar nationalists shaped in the University of Santiago, although indebted to the founders of the Irmandades and to the men of the Partido Galeguista, differ substantially from them in the substitution of the 'petit bourgeois' criteria of 'race' by the Marxist one of 'class'. With the mix of Marxism and Galicianism, this political banner, in former times seen by many as bourgeois, would be backed primarily, although not exclusively, by the left.

For the two Galician parties born around the middle of the 1960s, the communist UPG and the socialist PSG, Galicia is a nation colonized by the centralist Spanish state and is composed of fundamentally two classes: the oppressors and the oppressed. The former tag is pinned to the bourgeoisie, intermediaries and allies of 'Spanish capitalism', and the latter to peasants, fishermen and workers. This triangular dependency relationship would be labelled 'interior colonialism'.[13] The struggle for 'Galician national liberation' resembled that of Third World countries. Following these examples, the UPG created, in collaboration with the Basque ETA, an armed front which the police managed to break up in 1975.

These parties, with tactical and strategical differences, had basically similar objectives: 'the fight for national liberation for Galicia', and for the social liberation of its workers. With this basis and the failure of the centralist and technocratic economic plans, their leaders could not but present a radically different economic model, in accordance with their ideals and general conception of Galicia. Economically, the common denominator of the two programmes could be summarized thus:

1. the need to end the colonial situation where the state capitalists take advantage of the Galician economic surplus, cheap labour and raw materials or products with low Galician-added value;
2. Galician economic self-planning, based on its own resources (productive investment of Galician savings in the development of Galicia, etc.) within a comprehensive programme that would permit the structuring of an integrated economy;
3. transformation of raw material through to the final productive cycles to take place within Galicia;
4. nationalization of mining and energy resources;
5. large-scale production and commercialization through business co-operatives;
6. democratic workers' participation in the control of companies and planning organisms;
7. creation of regional territorial and administrative organization, reflecting the geographic, economic and social reality of the country.

We have contrasted here two conceptions of development, one enacted, the other programmed, but both originating from very different conceptions: Galicia as a region within a nation (Spain) whose interests are primary; or Galicia as a nation within a state (Spain) and therefore with rights, among which economic development should be in accordance with the area's needs and not with those of the rest of Spain. The problem, as is evident, is essentially political, and as such cannot but have repercussions in daily practice.

For example, hydroelectric development is extensive in Galicia, which produces 20.5 percent of all Spanish hydroelectricity while consuming less than the average. It could seem paradoxical that, with the existence of small areas still without electricity, Galicia continues to export. Neither this nor the fact that there has been no 'inducement of complementary or alternate means of irrigation', is surprising, since development was carried out according to the needs of extra-Galician areas. The hydroelectric development in reality 'has provoked conflicts by expelling agrarian activity from land flooded by its dams'. Land expropriation — sometimes of fertile valleys — has frequently occasioned violent conflict between expropriated peasants, supported and stimulated by Galician leftist parties, and the police. Similarly, the announcement of the construction of a nuclear power plant — completely unnecessary for a Galicia which exports energy — or expropriation of land for the construction of a toll freeway connecting the coastal industrial centres, judged superfluous by the nationalist left and capable of accentuating the already great differences between interior and littoral areas, have been the source of many protest marches, demonstrations, boycotts and violent confrontations with the Civil Guard. Other manifestations of the conflict between the concept of 'concentrated development for greater economic integration into the national [Spanish] economy' and the alternative of autonomous development are found in the modernization of the deep sea fishing fleet (Galicia provides more than one-third of the supply of fish in Spain), which was completely renovated in 1968, while the small craft fleet, which constitutes 'the productive base of Galician coastal municipalities belonging to the rural economic system', never received the anticipated financial aid (MPWU 1978: 44, 48, 114).

Parties and party systems

The emergence of a party system in Galicia is also related to structural cleavages, although not all cleavages automatically generate political parties. We can distinguish two general cleavages that have politically crystallized: the ideological (left-right), and the territorial (cf. Table 5). If we consider the electoral results as a test we would have to agree that the ideological cleavage has a higher salience than the territorial. However, the comparison of results from the legislative elections (1977 and 1979) as well as the municipal elections (1979) demonstrates a large and sustained increase of votes to Galician parties.

If a series of questions or problems could be considered a cleavage when they constitute a type of ideological pattern in the electorate's mind, then, in Galicia there are two political cleavages — ideological and territorial — whose cross-cutting gives rise to the system (or even systems) of parties.[14] Comparing both 'systems', we see that in the Galician system there are no right-wing parties. Until 1978, however, there were two centre parties (as weak as the structural fragility of the Galician bourgeoisie), the demo-christian Partido Popular Galego (PPG) and the social-democrat Partido Galego Socialdemocrata (PGSD). After the dissolution of the latter, some of its members founded the Partido Galeguista (PG), heir of the Republican homonym.

TABLE 5
Distribution of Spanish and Galician parties on the
left – right continuum

	Spanish parties		Galician parties
		1977	
Right	AN – 18 J		
	AP		
	UCD		PDG
Centre	RSE		PPG – PGSD
	ASD		
	PSOE	FDI	PSG
Left	PSPG	AET	BNPG
	PCG	FUT	
		1979	
Right	UN (former AN – 18 J)		
	CD (former AP)		
Centre	UCD		
	IR		
			UG
	PCa	PSOE	BNPG
	PCG	OCT	
Left	ORT	MCG	
	PTG	OC(BR)	
	LCR	PSOE (H)	

* Falange Española de las JONS (auténtica) — not classifiable.

Galicianism is better represented on the left by a Partido Socialist Galego (PSG) inspired by Marxism; by a 'Eurocommunist' party called Partido Obreiro Galego (POG), founded following the 1977 elections from a split in the UPG and attracting various ex-militants from other parties; and, finally, by the Marxist-Leninist Union do Pobo Galego (UPG), which suffered, even before these elections, a split to the left, creating the tiny Partido Galego do Proletariado (PGP), which was against participation in the elections. Except for UPG (1964), PSG (1965), and PGSD (1974), all the other Galician parties were founded following Franco's death.

In 1975 Galician politics seemed to be on the road towards a unification of the forces opposed to the regime. The greatest impediment was territorial polarization. There were constituted platforms which saw demo-christians working in harmony with communists (the Xunta de Galicia 1975-76), but it was impossible to find a meeting point between the UPG 'popular nationalist' communists and the PCG branch of the PCE to collaborate on a project of transition to democracy. The former refused Galician legitimacy to the PCG (whom they accused of opportunism for having added the 'G' to their name), while the latter censured the UPG as sectarians.

The alliances, then, were made on two opposing fronts: Galician parties versus Spanish parties, comprised respectively in the Consello das Forzas Políticas Galegas (CFPG) of January 1976, and the Táboa Democrática de Galicia born from the Xunta Democratica in July 1976, and which would reproduce in Galicia the opposition alliances already formed in Spain. The most notable strategic difference between both unitary platforms was that, while the Spanish parties and some moderate Galicianist personalities in the Táboa felt the need to claim an autonomy statute for Galicia — using the 1936 plebescite as a starting point — the radical nationalists, and particularly the UPG, demanded 'the right of self-determination for the Galician nation'.

With the first elections to be held in 41 years convoked for June 1977, the impossibility of finding class alliances between 'Spanish' communists or socialists with their 'Galician' party homonyms soon became evident. Only two small centre Galician parties (PPG and PGSD) managed to achieve an alliance. There was, however, an agreement among centre and left Galician parties (with the exception of UPG) and leftist Spanish parties to support joint lists for the Senate, the Candidatura Democrática Galega (CDG).

TABLE 6
Distribution of Galicianist vote: legislative elections

	1977		1979	
	Votes	%	Votes	%
BNPG	22,771	2.02	63,709	5.96
PSG/PPG/PGSD: UG	50,211	4.46	58,391	5.47
TOTAL	72,982	6.49	122,100	11.44

TABLE 7
Distribution of ideological vote: legislative elections

| | 1977 | | 1979 | |
	Votes	%	Votes	%
Right	150,355	13.39	156,833	14.70
Centre	647,149	57.65	522,203	48.94
Left	325,359	28.98	387,818	36.35

In the 1979 legislative elections, in view of the meagre results of the previous elections and under public pressure of one sector of the electorate, an alliance was arranged but, once more, along the territorial axis and not the ideological. The coalition, Unidade Galega (UG), was formed by three Galician parties: the recently created centrist PG, the PSG and the POG. Criticized by some sectors for its 'ideological incoherence' because of the inclusion of liberals, socialists and communists, but well received by others, the coalition accepted, within the limits of the 1978 Spanish Constitution, the defence of a broad autonomy statute guaranteeing the 'national rights of Galicia' achieving the maximum of self-government.

The results of the 1977 and 1979 legislative elections (Tables 6 and 7) should be examined in the light of two further features: the high number of parties in Galicia (17 and 19 respectively), considering the low level of politicization of the people; and the fact that Galicia had the highest abstention rate of all European regions, a rate that rose from 39 percent in 1977 to 50.3 percent in 1979. The following features of the vote distribution are observable:

1. the *sweeping results of the democratic right (UCD):* in 1977, 54.0 percent of the total vote, in 1979 lower with 48.7 percent;

2. the *stability of the authoritarian right (AP/CD)* in Galicia, in absolute figures, while in the rest of Spain it lost votes;

3. the growth, in comparative terms, of the marginal *extreme right*, going from 2,116 to 8,246 votes, though still a modest cipher which does not even reach 1 percent;

4. the *important rise of Galician forces* (UG and BNPG), which went from 6.5 percent in 1977 to 11.4 percent in 1979. The most surprising result was that of the radical Bloque Nacional Popular Galego (BNPG), a fusion of the UPG and the 'Asamblea Nacional

Popular Galega', who (given their youthful militancy) were favoured by the reduction of the voting age to 18, and which went from 2 to 6 percent;

    5. the *general increase of the left's* electoral weight, from 29.0 to 36.3 percent. In addition, extreme left 'Spanish' parties rose from 12,524 to 19,592 votes, owing above all, to the presence of the MCG — a strongly galicianized party — in the 1979 elections which received 5,184 votes.

TABLE 8
Votes and seats for the centre and left parties

	1977		1979	
	% votes	No. seats	% votes	No. seats
Centre	57.65	20	48.94	17
Left	28.98	3	36.35	6

Table 8 shows the lack of correspondence between votes considered 'centre' (in fact, democratic right and practically all UCD) and those considered 'left' and their respective number of seats in the Cortes because of the distorting effect of the d'Hondt electoral system, which strongly penalized the left for its excessive fragmentation into ten different lists. The same can be said of the Galicianist (UG and BNPG) parties, which, if they had entered the 1979 elections together, and assuming that they would obtain the sum of their actual respective votes, could have obtained two extra seats, in Coruña and in Pontevedra, from the CD, displacing it to a fourth place in those two provinces. Presently, the CD is the third party behind UCD and PSOE in both of these provinces and in the whole of Galicia.

    The parliamentarian reflection of these results is that in the Cortes the UCD went from 20 to 17 seats; AP/CD maintained their 4, 1 in each province; and the PSOE went from 3 to 6, in consonance with the general increase of the left. In the Senate the UCD maintained its 12 senators (3 per province), while the 4 remaining seats were divided between the PSOE, which improved from 1 in 1977 to 3 in 1979, and the AP/CD, which kept its previous senator from the most conservative province, Lugo. The progressive 'Candidatura Democratica Galega', which only entered the 1977 elections, had obtained 2 senators: these were taken by the PSOE in 1979. In short, progressive/socialist, rightist and governmental positions were all maintained.

In the municipal elections of 1979 the patterns initiated in the legislative elections of the same year were accentuated. There was an increase in the electoral weight of the left and Galician parties, above all in the cities. Improvements were also registered by the PSOE and PCG in the industrial belts around Ferrol and Vigo, and by the BNPG in some agricultural zones, where its union, Inter-sindical Nacional Galega (ING), has a certain implantation. The UCD, however, dominated the small municipalities as well as some cities, occasionally in alliance with the CD. In the more urbanized and industrialized municipalities, the left dominates. In La Coruña the coalition Unidade Galega won 5 council members, behind UCD (8) and PSOE (6), to find itself the arbiter between the centre-right and left as well as between radical Galician nationalists and the other parties: paradoxically, this city, the capital of the region and the most 'Castillianized' of all Galician cities, had a Galician left-wing nationalist mayor.

Over all, three general themes emerged from these elections: abstention, the success of the UCD, and the leftist vote. A high rate of abstention was registered in rural areas in the legislative elections and in urban areas in the municipal ones, where occasionally more than 50 percent of the registered voters did not exercise their right. This can be attributed to several factors: deficiencies in registra-tion, the high percentage of emigrants who found administrative difficulties in voting, rural population dispersion with the conse-quent difficulties of physical access to the polls, an aged popula-tion. Above all, there seemed to be a general political disinterest, which, as Tocqueville observed, brings with it a centralized state administration and, in this case, an evident lack of democratic habits. The skepticism and weariness of many Galicians were heightened by the seeming lack of relevance of the five voting occa-sions in less than three years (two referenda and three elections)[15] which caused no change in their lives. These are some of the explanatory hypotheses suggested for the high abstention rate in Galicia, which is the highest not only in Spain, but also in the whole of Western Europe.

The leftist vote, not surprisingly, corresponds to areas of greater participation, population density and industrialization, generally coinciding with coastal areas. The 'proletarian belts' of Ferrol and Vigo registered support for the left which sometimes was as high as

50 percent in the municipal elections, though only in small and clearly located areas. The limited industrialization of the rest of the country and the presence of peasant-workers with an owner mentality did not facilitate voters to support workers' parties.

The *personal characteristic* dominant in the Galician political culture is a consequence of the unprotected situation the peasants find themselves in before the 'authorities' and administration. It is the psychological basis of the clientele structure, in its turn a surviving mental pattern inherited from the old feudal relation of vassalage in which protection and favours were exchanged for votes to the local boss or the candidacy he designated. The functioning of this structure is based on two pillars: the privileged relations of the boss with the central power, or with its provincial delegate, and his local personal influence. Its efficient functioning is also more clearly influenced by two further conditions: the obviously greater power of the 'boss' over the 'client', which in turn depends on his ability (political and economic power, social influence) to help the protegé resolve his problems; and the lack of implantation and consolidation of intermediary structures between citizen and state, which could 'compete' with the boss in a bridge between peasant and administration. Clientelism, according to Gellner and Waterbury (1977: 3-4), signals the 'cleavage' between the 'legal country' and the 'real country', and is 'favoured by the existence of an insufficiently centralized state, a defective (political) market and a deficient bureaucracy' (cf. also Schmidt 1977). In rural Galicia the divorce between traditional institutions (parish, 'comarca') and liberal ones (municipality, province) is portrayed by the distortion produced in the superimposition of a liberal electoral system in a society that has conserved feudal type personal patterns.

AP/CD and UCD are the two parties among whose ranks can be found personalities from the Franco regime and who at the time still occupied positions in the only capillary organisms capable of reaching the farthest provincial corners — the Francoist agrarian, fishing, peasant and cattle-breeders' corporate unions — which, along with their personal influence, they utilized to the utmost in favour of both parties but especially the UCD, which came to hold the reins of power in Madrid.[16] Newspapers commented on the massive defection of large and small local bosses from the AP to the UCD on the eve of the 1977 general election, which greatly explains its resounding electoral success. At the same time, however, the prestige of the UCD leader, Suárez, who had successfully initiated the political reform, and the moderation that this option represented, were attractive to the Galician peasant with his decidedly owner mentality, which made him fearful of any possible attempt by the left at socializing his pittance.

NOTES

1. Political, administrative and linguistic uniformity were seen by the centre as indispensable modernizing factors for the free circulation of merchandise and labour, fundamental pillars of the liberal system.

2. Although the Royal Order of Aranjuez in 1768 dictated that 'in all of the kingdom actions and teaching shall be in Castillian', popularization of elementary state education was a product of the twentieth century (Rodriguez 1976: 227).

3. 'According to sources of the Ministry of Education and Science itself ('Educational Planning in Galicia', Madrid, 1970) only about 18 percent of the children between seven and thirteen achieve an academic level corresponding to their chronological age; the other 32 percent show a slack of up to six years' (MPWU 1978: 107). The consideration of Castillian as the cultivated version of Galician can be understood given the fact that, according to the Foessa Report, 76 percent of Galicians learned to speak their language, but only 4 percent to read it and 3 percent to write it. Catalonia, by contrast, has the following percentages: respectively, 15, 13 and 13. Lack of teaching in Galician explains the mental dissociation of Galician from culture (Gutiérrez 1978: 82; Anaya Santos 1970).

4. 'This resigned, submissive, self-defeating psychology is our tremendous social defect, it is our gravest illness as a people, because it is what prevents any collective will for change, for transformation, for protests against injustice, for struggle in favor of a transformation of this traumatic situation.' This author, profoundly knowledgeable of the Galician reality, continues by pointing out this 'terrible collective inferiority complex' as the origin of the Galician people's misfortunes, epitomized by 'this incapacity to react collectively to evil' (Piñeiro 1974a: 249).

5. An important link in this battle for the language is seen in the efforts of linguists definitely to establish its use. The unfinished process of standardization of Galician has been tremendously long if we consider that, already between 1862 and 1868, there were two dictionaries and four grammatical treatises. There are several positions in relation to the problem which, simplifying greatly, is reduced to a struggle between those in favour of the development of its own endoglossic standard (isolationist) and those in favour of an exoglossic standard or gradual incorporation of Portuguese orthographic norms (lusistas). Between these two extremes we can find independent tendencies: 'recuperate Galician within itself with modest concessions to lusism while not fearing it'; or reintegrationist tendencies, that want to 'restore the purely western character of this romance language, freeing it as much as possible from centralist contamination but without identifying it with Portuguese'. These two intermediate tendencies do agree, however, that the main danger is Castillian and not Portuguese. Aside from the incomplete norms elaborated by the Royal Galician Academy in 1971, there are 'Métodos de Galego 1, 2 and 3' prepared by the 'Instituto de Lingua Galega' and the recently published 'Bases pra unificación das normas linguísticas do Galego', which is a summary of the seminars held at the

University of Santiago between 1976 and 1977. All in all, it has been recognized that some publishers, particularly Galaxia, have done more in the last decades to standardize the language than all of the published norms (cf. Gutiérrez 1978: 39-47; Moralejo 1977).

6. 'In a peasant economy roughly half or more of all agricultural production is consumed by the peasant households themselves, rather than being marketed' (Thorner 1976: 202-8).

7. I tend to believe this was a printing error as all other reliable sources cite around 230,000. Cf. Richardson, 1975: 93 and *Situación actual y perspectivas del desarrollo de Galicia*, 1975, Vol. 3: 117.

8. About this Villar Ponte (1861-1936), founder and leader of the 'Irmandades' (Brotherhoods of the Language), wrote: 'Our language is the golden road to our salvation and our progress. Without the language we would die as a people and would never mean anything to the universal culture' (cited in Baliñas 1977: 19).

9. Risco (1976: 146-7). Carlos Durán wrote: 'It must be repeated one hundred thousand times that everything will be lost if the language is lost. If we lose Galician, we will also lose the reason for existence of Galicians' (cited in Montero 1973: 199).

10. The Civil War meant dispersion: exile, prison or execution, the latter the fate of Alexandre Boveda, one of the leaders of the Partido Galeguista.

11. In 1954 the order of the General Director of the Press strictly prohibiting the use of Galician was denounced by UNESCO (cf. Montero 1973: 182).

12. For the new attitudes of the Galician clergy, see Os Irmandiños, 'A lingua e a Igrexa' in *O porvir da Lingua Galega*. 1968. Also Chao Rego, et al. 1972, X.M.R. Pampín. 1976, '¿Unha Igrexa Galega?' (1977). For the position of the pro-Galician hierarchy cf. M. Anxo Arauxo (Bishop of Mondoñedo-Ferrol), 1975.

13. The term and the concept that it reflects are from Lafont (1967). Beiras (1972: 180-98) saw the relevancy of its application to the Galician case.

14. Definitive scholarly studies on the origins and development of contemporary Galician parties remain to be written: cf. Prada Fernández (1977: 193-205), a useful description of the pre-electoral period. Of a journalistic and informative character with a certain left-wing Galician slant, cf. Rivas and Taibo (1977); Gaciño (1978: 701-15). For the Galician elections, cf. Coma (1978: 707-36); and Díaz López (1978: 81-101). For a diachronic comparative analysis cf. also Díaz López (1980).

15. These were the political reform referendum (15 December 1976); the general election (15 June 1977); the constitutional referendum (6 December 1978); the general election (1 March 1979); and the municipal election (3 April 1979).

16. Among the various ways of exerting pressure on the politically unsophisticated peasants is the often repeated case of the civil servant in charge of paying monthly subsidies to them and who, when elections approach, vaguely comments that it might be the last time they see the subsidy if the socialists or communists win the elections.

REFERENCES

Alonso Montero, X.A. (1973). *Informe dramático sobre la lengua gallega*. Madrid: Akal, Arealonga.

Alonso Montero, X.A. (ed.) (1974). *Encuesta mundial sobre la lengua y la cultura gallegas y otras areas conflictivas: Cataluna, Puerto Rico*. Madrid: Akal, Arealonga.

Anaya Santos, G. (1970). *La depresión cultural gallega.* Vigo: Galaxia.

Arauxo, M.A. (1975). *A Fé cristiá ante a cuestion de Lingua Galega.* Santiago: Sept.

Baliñas, C. (ed.) (1977). *Pensamento Galego 1.* Santiago: Sept.

Barral Andrade, R. (1975). *O Aforro e a inversión na Galicia.* Vigo: Sept.

Barreiro Fernández, X.R. (1976). 'Historica Política', pp. 95-117 in G. Fabra Barreiro et al. (eds), *Los Gallegos.* Madrid: Ediciones Istmo.

Barreiro Fernández, X.R. (1977). *El levantamiento de 1846 y el nacimiento del galleguismo.* Santiago: Pico Sacro.

Beiras, X.M. (1972). *O atraso económico de Galicia.* Vigo: Galaxia.

Bozzo, A.A. (1976). *Los partidos politicos y la Autonomía en Galicia 1931-1936.* Madrid: Akal, Arealonga.

Chacón, R. (1978). 'La Problemática Linguística', in J.A. Durán (ed.), *Galicia. Realidad económica y conflicto social.* La Coruña: Banco de Bilbao (edition not distributed).

Chao Regio, X., et al. (1972). *As Relacions Eirexa-mundo en Galicia.* Vigo: Sept.

Coma, M. (1978). 'Las Elecciones', pp. 717-36 in J.A. Durán (ed.), *Galicia. Realidad económica y conflicto social.* La Coruña: Banco de Bilbao (edition not distributed).

Cores Trasmonte, B. (1976). *El Estatuto de Galicia (Actas y documentos).* La Coruña: Librigal.

CRPO (1978). Caja Rural Provincial de Orense, *La Agricultura Gallega 1977.* Orense: Estudio Económico.

Díaz López, C.E. (1978). 'Algunas hipótesis explicativas de los resultados electorales en Galicia', pp. 81-101 in Equip de Sociología Electoral (ed.), *Estudis Electorals.* Barcelona: Universidad Autónoma de Barcelona.

Díaz López, C.E. (1980). 'El sistema de Partidos en Galicia entre la dos primeras elecciones generales'. Paper presented to the Congress of the Spanish Political Science Association, Barcelona.

Gaciño, J.A. (1978). 'Partidos y grupos políticos', pp. 705-15 in J.A. Durán (ed.), *Galicia. Realidad económica y conflicto social.* La Coruña: Banco de Bilbao (edition not distributed).

García Fernández, J. (1975). *Organización del Espacio y Economía rural en la España atlántica.* Madrid: Siglo XXI de España Editores, S.A.

Gellner, E. and Waterbury, J. (eds) (1977). *Patron and Clients in Mediterranean Societies.* London: Duckworth.

Gutiérrez, E. (1978). *A Lingoa é noso escudo.* Pontevedra: Escola Aberta.

INE (Instituto Nacional de Estadística) (1976). *Síntesis Estadística de Galicia,* Madrid: Presidencia del Gobierno.

Lafont, R. (1967). *La Révolution régionaliste.* Paris: Gallimard.

Lipset, S.M. and Rokkan, S. (eds) (1967). *Party Systems and Voter Alignments.* New York: Free Press.

Moralejo, J.J. (1977). *A Lingua Galega Hoxe.* Vigo: Galaxia.

MPWU (Ministry of Public Works and Urbanism) (1978) *Informe Preliminar: situacion actual y diagnóstico: Plan Director Territorial de Coordinacion de Galicia.* Madrid: Direccion General de Ordenacion y Accion Territorial.

Os Irmandiños (1968). *O Porvir da Lingua Galega.* Lugo: Círculo de las Artes.

Pampín, X.M.R. (1976). *Nova concencia na Igrexa Galega.* Vigo: Sept.

Piñeiro, R. (1974a). *Olladas no Futuro.* Vigo: Galaxia.

Piñeiro, R. (1974b). 'El regionalismo gallego', *Razón y Fé* no. 917; 519-29.

Prada Fernández, J.L. (1977). 'El Sistema de Partidos Politicos en Galicia: Una aproximacion descriptiva', in P. de Vega (ed.), *Teoría y práctica de los partidos políticos*. Madrid: Edicusa.

Richardson, H.W. (1975). *Politica y planificación del desarollo regional en España*. Madrid: Alianza Universidad.

Risco, V. (1976). *O problema politico de Galiza*. Vigo: Sept (first published in 1930).

Rivas, M. and Taibo, X. (1977). *Os partidos politicos na Galiza*. La Coruña: Edicions do Rueiro.

Rodríguez, Fr. (1976). 'La lengua', in G. Fabra Barreiro et al., *Los Gallegos*. Madrid: Edic. Istmo.

Rokkan, S. (1970). *Citizens, Elections, Parties*. Oslo: Universitetsforlaget.

Schmidt, S.W. (ed.) (1977). *Friends, Followers and Factions: A Reader in Political Clientelism*. Berkely/Los Angeles: University of California Press.

Thorner, D. (1976). 'Peasant Economy as a Category in Economic History', in T. Shanin (ed.), pp. 202-18 in *Peasants and Peasant Societies*. Harmondsworth: Penguin.

'Unha Igrexa galega?' (1977). *Encrucillada*, 3 (May-June).

Vilas Nogueira, X. (1977). *O estatuto galego*, La Coruña: Edicions do Rueiro.

Villar Ponte, A. (1971). *Pensamento e Sementeira*. Buenos Aires: Ediciones Galicia, Centro Gallego de Buenos Aires.

Zubillaga, C.A. (1974). *El Problema Nacional de Galicia: Génesis y estructura del Estatuto autonómico de 1936*. Montevideo: Edic. do Patronato da Cultura Galega.

11

Conclusion: Perspectives on Conditions of Regional Protest and Accommodation

Derek W. Urwin
University of Warwick

It is merely banal to say that ethno-national movements are only one kind of regional political expression within a state. Nevertheless, it may be useful to begin with a reminder of the obvious. As the contributors to this volume indicate, there can be many kinds of regionalist demands which may be translated into political movements. Their impact and import, moreover, can be variable across the several strata of the national and regional populations. In some instances, their salience may be muted at the mass level, no matter how much this has excited the elites and disrupted the political system. In Belgium, for example, some commentators have argued that the mass interest in linguistic issues after their increased prominence in the 1960s has remained relatively low, no matter how agitated the politicians, and that the latter have had to take this more placid mass position into account in their deliberations and manoeuvrings (cf. Zolberg 1977: 104). In Galicia too, the masses seem scarcely to have been disturbed by the activities of the linguistic nationalists: the region has a tradition of aligning itself with the dominant political forces in Madrid, a tendency which has survived the transition from the Franco era to a democratic regime.

In this context it is highly appropriate to repeat Greenwood's point (1977: 101) that 'ethnicity is highly malleable and responsive to the circumstances in which groups find themselves'. Malleability in the widest possible regional perspective, refers simply to the contextual framework. We stressed in the Introduction that while it is undoubtedly important to incorporate a regionalist perspective in any centre-periphery analysis — to remember, for example, that while Belfast may be highly peripheral in the eyes of both London and Dublin, to many of those standing in Belfast it is the two state capitals that are peripheral — it is essential also to remember that the centre, because it is the centre, is crucial in influencing the generation, orientation and consequences of regionalist agitation and mobilization. At a fairly simple level, the so-called regionalist renaissance in France during the past two decades provides an illustration of political malleability. Regionalist demands had long been associated with conservative programmes as one weapon with which to assail the centralizing Jacobin state. A parting of the ways came in the 1960s as the left and the regionalists began to discover each other as potential allies in opposition to the Gaullist and post-Gaullist regimes: many regionalists swung from conservative association to radicalism.

What the chapters in this book illustrate is the varying balance across countries between centre and periphery, and the contribution of each to that balance. They also point to the fact that the ethnic party — or, if you like, the regionalist party in the broadest sense of the term — constitutes only one form of expression in what amounts to a complex mosaic, and per se it may be a relatively minor one (cf. Urwin, forthcoming). Ethnic consciousness, or a concern with ethnic identity, may be universal; nationalist movements are not. We may identify ethno-national groups and accept that they are equal theoretically. To assume that they are all equal empirically is misleading. While it may be true that regionalist demands, of whatever kind, are potentially traumatic for a state, under certain circumstances their political impact can be negligible. The lingering affection displayed in referenda in West Germany for the resurrection of defunct political territorial units such as Oldenburg and tiny Schaumburg-Lippe has not affected political trends in the country as a whole or in the areas concerned, while one conclusion that can be drawn from the study of Alsatian developments since 1918 is that regionalist aspirations and agitation scarcely disturbed the monolithic French state.

Regionalist political movements — and parties perhaps even more so — form on the surface a heterogeneous category. Other kinds of political organizations, such as Christian Democrat or Socialist parties, are more visible and significant in their respective political systems, and seem to have much more in common across state frontiers. Regionalist organizations are more disparate in the specificity of their demands. There is little in the way of a common economic policy or a universalistic view of the structure of society; not all are linguistically based; and all tend to be small, in many cases even miniscule, in size. But one feature is general. What separates these parties and movements out from the European mass is the nature of their claim upon the state. They identify with, and make claims upon the centre on behalf of, territories and groups which are not coincident with state boundaries and national populations. Indeed, identification with some piece of territory is about the only thing that they all have in common. Because of the nature of their claims, they tend to be broadly aggregative in intention and bifurcated in behaviour between 'party' and 'pressure group'. Their significance for a political system lies in their potential disaggregative impact upon the latter, since their objectives almost inevitably imply drastic structural change.

While peripheralization is an ongoing process, the kinds of peripheries that can be identified in Western Europe have retained the same characteristics over long periods of time. It was the city and the alphabet that heavily influenced the fate of Western Europe. The geography of trade routes made for differences in the resources for state-building, while the invention of printing aided the emergence of vernacular standards of communication which in turn prepared the ground for the later stages of nation-building at the mass level. In effect, there was in the past a progressive freezing process in the territorial and group structure of Western Europe, and current peripheral political protest must be considered not only as a reaction to contemporary economic problems and culture and political trends but also against the backdrop of the geoethnic, geopolitical and geoeconomic heritage of the continent (cf. Rokkan et al. forthcoming). It is within this context that we should consider the politicization and activation of group differences along the two interrelated dimensions of membership space and geographical space. Two bases of political mobilization would seem to be essential: territory and 'group' identity.

Territory itself is almost a sine qua non of regional politics. Possession, according to the proverb, is nine points of the law. Any group that is not territorially concentrated will be a much more difficult mobilization target for any movement with such ambitions: where its target population is geographically dispersed, its efforts can more easily be frustrated by the institutional arrangements of the state. For example, quite apart from the constitutional safeguards granted to the Swedish Finns, it could be suggested that to some extent the need to seek a compromise between the two linguistic groups and the later 'moderation' of the Svenska Folkpartiet are due to the relatively high degree of geographical dispersion of the Swedish-speaking minority, especially the physical presence of large numbers, including those putatively most qualified to provide a territorial leadership, in the Finnish centre. Similar considerations are true of the electoral arena. For example, in the United Kingdom with its plurality electoral system, only Irish Nationalists have ever looked — and that to an insignificant extent — beyond their own bailiwick to attempt to tap the potential 'ethnic' support within England. A similar reluctance seems also to apply in more liberal proportional systems. The Belgian linguistic parties, for instance, confine themselves largely to their 'own' regions, and the Svenska Folkpartiet to those constituencies with substantial numbers of Swedish Finns.

The salience and consciousness of group membership is variable over time, but always some kind of ethnicity lies at its base. All contemporary regional movements lay claim to some form of ethnic identity. In multi-ethnic situations minority populations hold, or are expected to hold, multiple identities: Scottish and British, Flemish and Belgian. Multiple identities may be benign or antagonistic, that is, they may serve to reduce the probability of conflict or they can nurture a sense of ethnic distinctiveness and awareness. Regionalist mobilization is much easier where hostility prevails: otherwise, the problem for regionalist protagonists is to break the benignancy of the relationship. Mobilization is also easier where there is linguistic distinctiveness, the case of Scotland notwithstanding. Language is the key resource for many who stress ethno-nationalism. A distinctive means of communication is a powerful political resource for a potential movement, and one which easily leads to political conflict. For language is a collective good that must be shared by all, including at the extreme those in the centre. That at least is the ultimate implication of linguistic parity and bilingualism. Moreover, as Zolberg (1977: 140) has acutely observed, a state can pretend to be blind, but not deaf mute.

Religion has historically been a powerful motor force in generating distinctiveness and mobilization in Europe, frequently providing territorial minorities with a major, if not sole, means of indigenous elite recruitment, as well as being a major structural force in the formation of party systems. While religious issues have largely disappeared from the Western European agenda, either by their complete removal from the public sector or by the formalization of equity within the public sector, the displacement of religion as a partisan structuring force has not followed on at the same rate. Religiously based parties persist, and as a part of this survival religious unity or distinctiveness can be a potential resource for territorial mobilization, as seems to be the case in the Bernese Jura, or in Northern Ireland where it is worn almost as a tribal badge.

However, resources by themselves constitute only a potential. How catalysts metamorphose this potential involves at least two separate questions: the formation of regional movements, and their expansion. The first is perhaps easily answered. Such movements tend to be creatures of the national revolution (Rokkan, 1970) which refined and stabilized the state structure, or — as for example with the Slesvig Parti in Denmark or the Südtiroler Volkspartei in Italy — a consequence of the collapse of the central European empires in 1918. The two major contemporary exceptions, lacking any meaningful predecessors, are the Front Démocratique des Francophones and Rassemblement Wallon in Belgium.

The second question is more complex. At a minimum, there have been two waves of mass mobilization. The first arose from the democratic and industrial revolutions, more specifically from the contagion of democratic principles. A greatly extended suffrage permitted the organization in the electoral arena of parties for the protection of minority and territorial interests. But democratic extension also implied greater standardization and centralization, while industrialization created a new and wider cross-local labour market and allowed more rapid communication and greater territorial interaction and exposure. The most obvious field where identities clashed was in language, with education perhaps as the most bitter arena: to a considerable extent the long retreat of minority languages within states can be traced to the introduction of and degree of effective central management of universal education.

The second wave, arising more from the existence and awareness of economic discrepancies, is more modern. At a territorial level, regional economic resources may be considered vis-à-vis those of the centre, and five broad scenarios might be postulated: backward, declining, similar, improving, superior. Little attention was paid to economics in the first cultural wave of mobilization. After 1945 economic attention focused more on backward areas (true geoeconomic peripheries) — the Italian Mezzogiorno was outstanding. A heightened tempo of structural change persuaded governments to switch attention to the decline of older industrial concentrations, such as Wallonia, Scotland and Wales. While it is true that such industrial regions can pack more political muscle than rural peripheries, without any kind of distinctive cultural identity-base their ability to threaten the regime may be posed more in electoral (in changing the national party system or orientation of the state) than in territorial terms. A more dangerous territorial challenge may well come from superior or improving economic regions with a real or potential economic weight that more demonstratively can support independence and/or counter-balance the political resources of the centre. Flanders, Catalonia, the Basque country and, after the discovery of North Sea oil, Scotland would all fall into these two categories.

When considering groups in this context, we can perhaps conceive of regional balance and group accommodation occurring when total interdependence characterizes the whole society or when there is a clearly established and accepted hierarchy of functions (Rogowski, 1973). Conflict is more readily generated by economic change, up or down, within dominant or subordinate groups. And where these are stratified internally, then the more intense conflict emerges among those gaining or losing. Demographic changes may also be a catalyst, but these may be both cause and consequence of conflict, and so inappropriate as explanations of the waxing and waning of the latter. It may well be that the accumulation of cultural and economic inferiority provides fertile soil for regionalist movements: more significant, perhaps, is the question of inconsistency between economic strength/potential and cultural status. In a broader sense the two waves have something in common — a confrontation with change: processes of state- and nation-building such as linguistic and educational standardization; economic expansion or decline; and a heightened awareness of the outside world through improved communications, geographic mobility, and a more penetrative mass media. However, although economic change in the broadest sense may be necessary, it is not sufficient. There must be something upon which regional mobilization can be built, and that something is a form of cultural, usually ethnic, identity within the region. Together, cultural base and economic catalyst seem to offer a satisfactory approach to regional political mobilization.

There are several movements which less ambiguously are part of the first wave. The Svenska Folkpartiet is perhaps the best example, with a sense of solidarity that was strengthened by the attempts at finnicization. A second category would consist of movements in regions which previously had not only been relatively isolated from the centre, but which also enjoyed sociopolitical privileges and/or a distinctive institutional infrastructure: the Basque provinces and Scotland are examples. The contrast between the two is instructive, for one experienced early and significant regionalist mobilization, the other did not: any comparison would have to build on cultural distinctiveness to incorporate styles of central control, such as the centralizing efforts of Madrid versus the more indirect collaboration preferred by London, and upon economic factors such as patterns of migration. For the mobilization of other movements, economics occupies a more central role. The Volksunie is unique in coming from a region which holds a majority of the state's population. Its advance was accelerated when the psychological perception of cultural inferiority could be coupled with demonstrated economic improvement. The continued economic strength of the Basque provinces, and actual and potential central discrimination, have helped to sustain the possibility of regional mobilization. By contrast, the Rassemblement Wallon is an example of a party stimulated by economic decline and a challenge from other, more prosperous regions. In the Jura too, the Rassemblement jurassien and separation corresponded to negative sociostructural and economic change. But whatever the economic movement involved, all are also grounded in some form of cultural identity.

Common links in mobilization do not necessarily mean uniformity in their subsequent political fortunes. Deploying a party's vote as a continuous variable, several measures are available for assessing electoral change in terms of trends and fluctuations. An analysis of those regionalist movements which have contested national elections on at least three occasions since 1945 pointed towards extreme steadiness at the national level, and only limited volatility at the regional level (Urwin, forthcoming). It served also to demonstrate the limited size of the movements and that there was no conformity in the direction of the trends or nature of the fluctuations. Some of these points are also borne out by survey data. While the latter do tend to suggest that regionalist parties lean towards a definite (and usually linguistic) base and to be broadly aggregative across socioeconomic groups (though skewed towards middle-class groups, a tendency which is more prominent among the leaders and activists), the range of variations in support is considerable. A simple reference to ethnic identity cannot be sufficient to explain what they have in common. If that were the case, then all perhaps should be branded as failures, for none has achieved anything like its potential strength. This is particularly true of those which have not put themselves to the electoral test. Most of the 'new' regionalist parties have been electoral failures. The most impressive performances have been those of the Front Démocratique des Francophones and the Scottish National Party; even so, these two peaked at only half the voting power of long-established electoral parties such as the Svenska Folkpartiet and the Südtiroler Volkspartei. But while such parties have not won the allegiance of their total potential support and have not succeeded in making population and territory synonymous, sympathy for at least some of their claims may be more widespread, albeit not translated into votes.

Electoral competition, however, is only one means of formulating regional political expression and/or discontent. Depending upon circumstances and inclinations, others may be more appropriate. The most dramatic form is undoubtedly extra-legal activity. In Spain and Ulster ETA and IRA terrorism overshadows the activities of the parties. However, though depending upon what the group wishes, the results may be rather limited and the costs high, a lesson learnt in the Alto Adige during the terrorist phase around 1960 (Alcock 1970: 358-60). Alternatively, regionalism may operate within the political system. One option is to establish a cross-party umbrella organization that seeks to mobilize support by endorsing individual candidates, of whatever party, for public office — the strategy of the Rassemblement jurassien. A further option, more usual of miniscule groups or those in decline, is to seek an alliance with a state-wide party. A further possibility is negative mobilization. Where regional representation is so small that it will be swamped in a national legislature, a regionalist movement may urge its supporters to follow the path of abstention and non-participation.

These alternatives are perhaps an admission of weakness or uncertainty. They are linked to barriers which can hinder regionalist mobilization. One important obstacle is the combination of institutional structures and previous mobilization. The regionally skewed support of state-wide parties, for instance, may be more meaningful for regional demands. The power of Labour in Wales may be at least as meaningful for Welsh demands as it is for British integration, and more so than the presence and activities of Plaid Cymru. In Spain politicians from the small-town and rural region of Galicia have always been prominent in Madrid, a practice that was continued after Franco with the national and provincial electoral success of the Unión de Centro Democrático. In Finland the Svenska Folkpartiet holds only some two-thirds of the Swedish-speaking deputies: it cannot claim to be the exclusive spokesman of the minority group. A second barrier is internal group or territorial variation, which may turn potential into negative resources. Language, for instance, can be a two-edged sword where there is incongruence between it and identity or territory. Plaid Cymru's linguistic identification hinders its penetration of non-Welsh speaking areas. The Jura has been consistently divided over its future since the Congress of Vienna, with French Catholic juxtaposed against French Protestant and German areas.

Opinions within a region can range from upholding the status quo, through demands for limited legislation in specific fields and decentralization, to autonomy and independence. Most regionalist movements seem to concentrate, at least as a first step, upon some form of decentralization. In many ways this option, the one which might attract a broader spectrum of support, is the most susceptible to attack. The others are easier to concretize, and hence to mobilize — but perhaps only within more restricted parameters. But whatever the demand, it is something which existing nation-wide parties must take into account. The latter have the option of intensifying territorial pressures by deliberately appealing to regionalist sympathies. In Galicia, for example, most parties formally describe themselves as Galician. Alternatively, parties can ignore the territorial question by saying that it is not an issue, the British stance up to 1970 and perhaps also of France. By contrast, parties may consciously aim at cross-regional mobilization and internal integration, a difficult task once territorial issues become salient. In Belgium, for instance, the old parties did not succeed in applying the traditional accommodational practices to the linguistic question: one effect was deep internal dissension within each. Finally, parties may simply tolerate or accept the differences, though whether this is genuine tolerance or fear of the consequences may be difficult to determine.

Party responses to territorial demands are linked to regime responses. At the extremes governments may accept secession or resist totally. The latter has been more nearly approached by France, while the former would be a highly unnatural act. More typically, governments will respond by seeking to accommodate the demands or to defuse the situation within existing boundaries. While the number of specific options might shade into infinity, two broad alternatives are available: to seek either group or territorial accommodation. Federalism, as in West Germany or Switzerland, is the classic pattern of territorial accommodation. Group accommodation is a form of 'verzuiling', seeking consociational solutions where power and public goods are shared or duplicated among the conflicting groups. The prototype is the Proporzpaket of 1969 and 1972 in the Italian Alto Adige. For years the Bernese government attempted to resolve the Jura dispute by suggesting consociational devices, before reluctantly accepting the territorial option of a new canton available under the Swiss institutional structure. By contrast, only a form of group accommodation was possible in Finland. Belgium, perhaps, has a mix of the two approaches, more territorial in Flanders and Wallonia, a kind of 'proporz' in Brussels. The United Kingdom structure lends itself more to a territorial solution (as abortively pursued in Scotland and Wales in the late 1970s): however, the power-sharing experiment in Northern Ireland was more a group accommodation that failed, perhaps because the protagonists viewed the conflict more in territorial terms. By contrast, the devolution of some powers to the Basque country entailed the application of the territorial principle in a situation where, because of in-migration and the linguistic divide, some form of group accommodation might also be necessary — though the latter would certainly be rejected by Basque nationalists. In any case, the options available to any regime are circumscribed by the group configuration, the institutional style of centralizing politics, and the costs involved. Without an existing decentralized structure, as in Switzerland and West Germany, a territorial solution entails a radical break in the style of government, and may well be possible only in the case of miniscule, remote and relatively placid communities. In the last resort, however, everything boils down to cost. For any state, the prime objectives are to preserve the integrity of the territory and to ensure within these boundaries the legitimacy of its existence through obtaining popular support for and acquiescence to its political authority; and regimes are willing to pay a high price to maintain legitimacy and control within the existing structure of territorial integrity.

It is easy to over-dramatize the strength of regionalist movements. None is large, and very few are more than medium-sized within their own territory. All have strong links to historical questions which today are mainly irrelevant or almost impossible to disentangle. Alternatively, they could be part of an intensified identity crisis, an 'ontzuiling' process which some observers have seen emerging over the past two decades. But if there has been a reawakening of ethnic consciousness and an upsurge in regional protest and volatility (which is by no means universal), it should be viewed in the widest possible perspectives. And whatever regionalist movements are, there are no grounds for assuming that they will disappear if their wishes are granted. Though in decline, the Svenska Folkpartiet has not, nor has the Rassemblement jurassien disbanded completely. One wonders whether the IRA or ETA would accept voluntary liquidation if independence were ceded.

Where reason to persist remains, a movement's freedom of action is circumscribed by its context. Outright opposition to the regime might lead to illegality; the retention of regional purity may run the risk of alienation from the bulk of the population; while regional mobilization might produce counter-mobilization. Collaboration may have similar consequences for, as the Svenska Folkpartiet has discovered in its long occupation of office and the Belgian linguistic parties more recently, collaboration — or even merely participation in the regime's institutions — means involvement with wider issues that do not fit easily with their own regional concerns. And where there is no hard issue, collaboration might lead to assimilation. 'Going national' is a choice fraught with dangers, and probably impossible in multi-ethnic situations. The only major postwar example is the Norwegian Kristelig Folkeparti: however, the fate of the West German Deutsche Parti — decline and virtual death — is perhaps more common.

What counts in the long run is the resolution of territorial complaints. The presence of even weak regional political activity may force the centre to take cognisance of the fact. It is almost two decades since Pizzorno (1964: 276) wrote, 'since the Industrial Revolution, Europe has known more struggles for independence than struggles for union', a point which remains valid today. Even so, although ethnicity may have re-emerged as a more salient basis of mobilization, no ethnic group is totally monolithic. Other loyalties and cleavages characteristic of modern societies have not been submerged. The net result is a mosaic, where explanations that seem to be valid in some situations appear not to be so in others. The regional resurgence of the late twentieth century, where it has surfaced, may be less a recurrence of ethnonationalism in the traditional manner, and more an aspect of a more general problem affecting most European states. While the great political debate of the twentieth century has been focused on the issue of redistribution at the individual level, there has arisen a new concern with a territorial redistribution of resources, and an increasing unease over centralization and governmental management of structural socioeconomic change within a welfare state framework. Where scarce resources are also distributed unevenly cross-territorially, those regions which are worse off have tended to press for redistribution, while the richer have resisted or have ignored the problem. Where there exist territorially concentrated groups with distinctive cultural traits, then they can serve as a more obvious base of mobilization: where not, there may be a different form of disruption through, for example, the transference of support to new nation-wide and/or protest movements which are not particularly interested in regionalist qualities or dilemmas. Of course, old battles are still being fought, but the success of regionalist movements depends more upon contemporary problems and reactions to these by both the elites and the masses in both centre and periphery. And while such movements may modify the dimensions of conflict in a political system, it is dubious whether they can repeat the great movements of the past for national independence. Nevertheless, wherever territorially-based complaints remain unresolved, so does the possibility of political disruption.

REFERENCES

Alcock, A.E. (1970). *The History of the South Tyrol Question*. London: Michael Joseph.

Greenwood, D.J. (1977). 'Continuity in Change: Spanish Basque Ethnicity as a Historical Process', in M.J. Esman (ed.), *Ethnic Conflict in the Western World*. Ithaca: Cornell University Press.

Pizzorno, A. (1964). 'The Individualistic Mobilization of Europe', in S.R. Graubard (ed.), *A New Europe?* Boston: Houghton, Mifflin.

Rogowski, R. (1973). *Rational Legitimacy*. Princeton: Princeton University Press.

Rokkan, S. (1970). *Citizens, Elections, Parties*. New York: McKay.

Rokkan, S., Urwin, D.W. and Aarebrot, F.H. (forthcoming). *Economy, Territory, Identity: The Politics of the European Peripheries*. London: Sage Publications.

Urwin, D.W. (forthcoming). 'Harbinger, Fossil, or Fleabite?: "Regionalism" and the Western European Party Mosaic', in H. Daalder and P. Mair (eds), *Working Papers on Western European Party Systems*. London: Sage Publications.

Zolberg, A. (1977). 'Splitting the Difference: Federalization without Federalism in Belgium', in M.J. Esman (ed.), *Ethnic Conflict in the Western World*. Ithaca: Cornell University Press.

Notes on Contributors

Frank Aarebrot graduated in 1976 from Bergen University and now works there as a research assistant in the Institute of Comparative Politics.

Risto Alapuro is a lecturer in sociology at the University of Helsinki and is the author of several articles on political mobilization in Finland.

David Campbell graduated from Carleton University, Ottawa, and has completed post-graduate study at Duke University and the University of Geneva. Currently he works in the private sector in Montreal and is researching the development of Quebec nationalism.

César Díaz López gained his PhD from the University of Madrid and his MA from the University of Washington. At present he is Assistant Professor of Political Science at the University of Madrid.

André P. Frognier is Professor of Political Science at the Catholic University of Louvain, Belgium. He has published several studies of mass politics and elites in Belgium.

Solange Gras is a lecturer in history at the University of Strasbourg II.

Marianne Heiberg is a research assistant in anthropology at the London School of Economics.

Michel Quevit is Associate Professor in the Centre for the Study of Social and Political Change at the Catholic University of Louvain, Belgium.

Stein Rokkan was Professor of Political Sociology at the University of Bergen until his death in 1979. He had also served as President of the International Political Science Association, President of the International Social Science Council, and was the first Chairman of the European Consortium for Political Research. His many publications on European politics include *Party Systems and Voter Alignments* (with S.M. Lipset); *Quantitative Ecological Analysis* (with M. Dogan); *Mass Politics* (with E. Allardt); *Citizens, Elections, Parties; Building States and Nations* (with S.N. Eisenstadt).

Marie Stenbock is a research associate in the Centre for the Study of Social and Political Change at the Catholic University of Louvain, Belgium.

Derek Urwin is Professor of Politics at the University of Warwick. Until 1980 he was Associate Professor of Comparative Politics at the University of Bergen. He has been editor of *Scandinavian Political Studies*, and is currently co-editor of the *European Journal of Political Research*. His several publications on European politics include *Scottish Political Behaviour* (with I. Budge); *Western Europe Since 1945; From Ploughshare to Ballotbox.*

ECPR Publications Committee

Jean Blondel *University of Essex, UK*
Hans Daalder *University of Leiden, Holland* (Chairman)
Serge Hurtig *National Foundation of Political Sciences, Paris*
Kenneth Newton *University of Dundee*

SAGE Modern Politics Series
sponsored by the European Consortium
for Political Research/ECPR

1
Interorganizational Policy Making
Limits to Coordination and Central Control
edited by Kenneth Hanf and Fritz W Scharpf (1978)

2
Democracy, Consensus and Social Contract
edited by Pierre Birnbaum, Jack Lively,
and Geraint Parry (1978)

3
Power, Capabilities, Interdependence
Problems in the Study of International Influence
edited by Kjell Goldmann and Gunnar Sjöstedt (1979)

4
Models of Political Economy
edited by Paul Whiteley (1980)

5
Policy Analysis and Policy Innovation
Patterns, Problems and Potentials
edited by Peter R Baehr and Björn Wittrock (1981)

6
The Withering Away of the State?
Party and State under Communism
edited by Leslie Holmes (1981)

7
Patterns of Corporatist Policy-Making
edited by Gerhard Lehmbruch and Philippe Schmitter (1982)

other books sponsored by the ECPR

Challenge to Governance
Studies in Overloaded Polities
edited by Richard Rose (1980)

Balancing the Books
Financial Problems of Local Government
in West Europe
by Kenneth Newton (1980)

The Local Fiscal Crisis in Western Europe
Myths and Realities
edited by L J Sharpe (1981)

The Impact of Parties
Politics and Policies in Democratic Capitalist States
edited by Francis G Castles (1982)